# Bionics for the Evil Genius

## 25 Build-It-Yourself Projects

## Evil Genius Series

*123 Robotics Experiments for the Evil Genius*

*Electronic Gadgets for the Evil Genius: 28 Build-It-Yourself Projects*

*Electronic Circuits for the Evil Genius: 57 Lessons with Projects*

*123 PIC® Microcontroller Experiments for the Evil Genius*

*Mechatronics for the Evil Genius: 25 Build-It-Yourself Projects*

*50 Awesome Automotive Projects for the Evil Genius*

*Solar Energy Projects for the Evil Genius: 16 Build-It-Yourself Thermoelectric and Mechanical Projects*

*Bionics for the Evil Genius: 25 Build-It-Yourself Projects*

*MORE Electronic Gadgets for the Evil Genius: 28 MORE Build-It-Yourself Projects*

# Bionics for the Evil Genius

## 25 Build-It-Yourself Projects

NEWTON C. BRAGA

**McGraw-Hill**
New York Chicago San Francisco Lisbon
London Madrid Mexico City Milan New Delhi
San Juan Seoul Singapore Sydney Toronto

The McGraw·Hill Companies

**Cataloging-in-Publication Data is on file with the Library of Congress.**



1 2 3 4 5 6 7 8 9 0 QPD/QPD 0 1 0 9 8 7 6 5

ISBN 0-07-145925-1

*The sponsoring editor for this book was Judy Bass and the production supervisor was Richard Ruzycka. It was set in Times Ten by MacAllister Publishing Services, LLC. The art director for the cover was Anthony Landi.*

*Printed and bound by Quebecor*

This book is printed on recycled, acid-free paper containing a minimum of 50 percent recycled de-inked fiber.

# Contents

# About the Author

Mr. Braga was born in Sao Paulo, Brazil, in 1946. His activities in electronics began when he was only 13, at which time he began to write articles for Brazilian magazines. At 18, he had his own column in the Brazilian edition of *Popular Electronics*, where he introduced the concept of "electronics for youngsters."

In 1976, he became the technical director of the most important electronics magazine in South America. *Revista Saber Eletronica* (published at that time in Brazil, Argentina, Colombia, and Mexico). He also became technical director of other magazines of *Editora Saber,* such as *Eletronica Total,* and became the technical consultant for the magazines *Mecatronica Facil, Mecatronica Atual*, and *PC & CIA*.

During this time, Mr. Braga published more than 100 books about electronics, mechatronics, computers, and electricity, as well as thousands of articles and electronics/mechatronics projects in magazines all over the world (United States, France, Spain, Japan, Portugal, Mexico, and Argentina among others). Many of his books have been recommended at schools and universities around the world and have been translated into other languages, with sales of more than 3 million copies worldwide

The author currently teaches mechatronics at Colegio Mater Amabilis, is a consultant for distance learning organizations, and is engaged in education projects in his home country of Brazil. These projects include the introduction of electronics, bionics, and mechatronics in middle schools as well as the professional training of workers and teachers who need enhanced knowledge in the field of electronics, mechatronics, bionics, and technology. Mr. Braga lives in Guarulhos (near Sao Paulo, Brazil) with his wife and 16-year-old son.

# Preface

This book has been written for anyone looking for projects linking electronics with biology. It doesn't pretend to be a complete resource for the bionics evil genius, but it certainly will offer a large assortment of useful information and ideas for projects not found anywhere else.

For more than 25 years, the author, as a collaborator with American, European, and Latin American electronics and mechatronics magazines, has published a large assortment of practical circuits. Many of those projects linked electronics to biology, meaning that that they can be classified within a science called *bionics*. Many of those projects and ideas are included in this volume, most of which the reader can build using common tools and components.

The projects range from experimental types through practical types to amusement types. Of course, other devices can be used to teach you something about the fantastic science of bionics. So the purpose of this book is not only to teach the evil genius many tricks and techniques used to build bionic devices, but also to provide ideas and complete projects that can be built easily using low-cost and easy-to-find parts.

The audience for *Bionics for the Evil Genius* includes beginner, intermediate, and advanced builders who want new ideas for projects, and educators who want to introduce the use of technology in their schools, helping students find a vocation. Of course, the most important reader is the evil genius, who can make incredible things using his or her imagination, skills, and some parts gathered from old equipment and appliances, "robbed" from a younger sibling's toy, or bought in a local electronic parts dealer.

If you think that it is impossible to build interesting things using simple materials and technology, you are wrong. Three types of technology are used to build electronic and mechatronic projects.

The simplest or "traditional" technology depends on the use of electric parts, such as motors, cells, and passive components. This technology is simple enough that even elementary school children can understand it. Some interesting projects can be done using this technology combined with your imagination and skills.

Intermediate technology uses something more advanced than the passive components of traditional physics. The projects in this category include components such as semiconductors (diodes, transistors, *silicon-controlled rectifiers* [SCRs], and *light-emitting diodes* [LEDs]) and some *integrated circuits* (ICs), but they are not as advanced as the ones using microprocessors, *very large scale integration* (VLSI) chips and *digital signal processors* (DSPs), which fall in the third category.

The great advantage of intermediate technology is that it is accessible to all. Discrete components such as transistors, resistors, and diodes can be easily handled without the need for special tools. Using these components can help reveal vocations and talents, starting with a person's natural aptitudes. Since you will not need special tools to handle these components, and they are strong enough to withstand an inexperienced evil genius, it is very easy to build any project described here.

The important point for our readers is that they can build or create different devices that can only be found in the movies, on TV, or in science fiction magazines. Using low-cost parts and simple technologies, the reader can explore nature to create these bionics projects. He or she can build devices for fish and plants or interact with his or her own body or other living beings with the same approach you see in TV series on the Discovery Channel.

You'll be able to do all that, and we intend to give you some of the necessary tools, ideas, and techniques in this book. You'll only have to complete these items using your skills and the super-imagination only found in a real evil genius.

The book is divided into five sections. In the first two sections (Chapters 1 and 2), we will explain what bionics is and introduce some basic concepts of this science. We will convey what the reader will have to know in order to handle electronic devices, especially when used with living beings. The technology used in the projects is also explained.

We will also dedicate some space to educators who want to reveal the evil geniuses among their pupils by building projects and making experiments with them. The educators will see how easy it is to link many of these projects with their middle or high school science classes, using the projects as cross-themes for their science curriculum.

We begin the projects in Section 3. Twenty-five of them have been chosen from among the large collection of the author's files, and many of them have been specially created for the readers of this book.

The projects are complete, with all the necessary information for building a basic version that works alone as a complete device. A brief description of the projects will show what its primary outcome should be. Following this part, the reader will find information on the operation principles and how to mount the parts, as well as a complete parts list.

Following the building process, many additional circuits and ideas are provided for upgrading the project, building variations, or creating new projects based on the same principles. This is ideal for the reader who wants to explore the creation of further experiments and devices. The complete additional circuits given actually increase the real number of projects in this book to more than 100. In some cases, ideas for educators will be discussed that link the projects as cross-themes for the science classes, and additional information completes each project. This approach makes the book easy to use as a large reference for bionics projects, teaching the reader much about the use of technology.

We hope that as a reader and potential evil genius your face will light up and your eyes will be full of mischief from the ideas provided in this book.

Newton C. Braga

P.S. In the volume *Mechatronics for the Evil Genius* by the same author, a detailed section explains how to mount electronic devices (Sections 2.1 and 2.2). If the reader is not experienced with these techniques, we recommend taking a look at that publication.

# Acknowledgments

I would like to thank all the people who helped make this book possible:

Jeff Eckert, my book agent who helped me with all the bureaucratic procedures involved in the production of the book.

Carmen Regina Silvestre, teacher of biology at Colegio Mater Amabilis, a great friend who helped me with a lot of information in the bionics field, in which she is a specialist.

Carlos Eduardo Portela Godoy and Marcelo Portela Godoy, who gave me support when working with my pupils at Colegio Mater Amabilis in Guarulhos (Brazil), revealing among them many an evil genius, and who let me use the laboratory for many experiments described in this book.

Helio Fittipaldi, who allowed me to use many illustrations and photos from articles I had published in the magazines *Mecatronica Facil* and *Eletronica Total.*

Edson de Santis, my great friend who supplied me with many parts of the components I used in this book's projects.

My wife Neuza and my son Marcelo, who have both been supportive of my efforts.

Newton C. Braga

# Introduction



The living world faces a wide range of biological problems, yet within our world one can find a fantastic source of high-performance solutions to these problems. These solutions have been attained as a result of natural selection, which has been taking place over the last 3 billion years of life's evolution on Earth. The perfect operation and structural organization of Earth's biological systems are the result of specific principles and laws that guide and control the functions of all living matter.

The most important of all the principles involved in the evolutionary process, or acting upon living matter, is the *Optimal Design Principle* or ODP. It cannot be demonstrated by formulas or theories, but it is considered valid because its existence can be verified in the organization of many living beings, from the molecular level to the human being.

ODP is the methodological consequence of the basic concepts within Darwin's evolution theory: the struggle for survival and natural selection. According to the Darwinian evolution theory, living beings are changing their structure all the time in a way that mutants are created at each generation. The mutants replace the other individuals, and some of them are superior according to some specific criteria. The criteria include best performance, considering energetic costs for maintenance, propagation (reproduction), and operations. This means that the fantastic performance of most living beings in our time is the result of billions of years of evolution. Scientists are convinced of the optimization of all creatures' abilities in a world where a fight for survival is constant.

So why not use the characteristics found in nature and observed in living beings to help us with our technologies? This is exactly the aim of bionics and another science considered to be a branch of bionics, known as biomechatronics. Bionics can be defined as a science that utilizes the results of biological evolution.

Of course, not always, but very often, a natural result can be the basis for the best solution to many problems faced by a technology designer (and even in other fields such as architecture or mechanics). Starting from the previous concepts and from the definition of bionics, we can show certain methods that can be used by the bionics researcher, linking solutions given by nature to solutions created by our technology.

Technology can be linked to the solutions given by nature in three ways, as shown in Figure 1.1. In the first case, we have the direct application of the bionics definition. We can use the results of biological evolution to create artificial devices. Observing the solutions given by nature, we can imitate those natural solutions using components and devices created by our technology.

For example, the sonar abilities of bats and dolphins were the basis for the creation of devices designed to detect objects underwater and even in the air. Artificial legs and arms, as well as *shape memory alloys* (SMA), were developed after observing how our muscles work.

The second way to work with bionics is also shown in Figure 1.1. We can combine special functions created by technology directly with nature, such as connecting a living being with an artificial device.

For example, an electric fish, generating a signal that changes its frequency according to the pollution level, can be used as a sensor directly connected to an electronic circuit. It can also be used to monitor the purity of water resources as an extremely sensitive sensor. A plant, used as a sensor, can be connected to an electronic circuit to monitor the level of oxygen in the air, to monitor light, or to monitor for the possibility of an earthquake. Even a person can be connected directly to an electronic circuit to monitor his or her degree of stress under determined conditions.

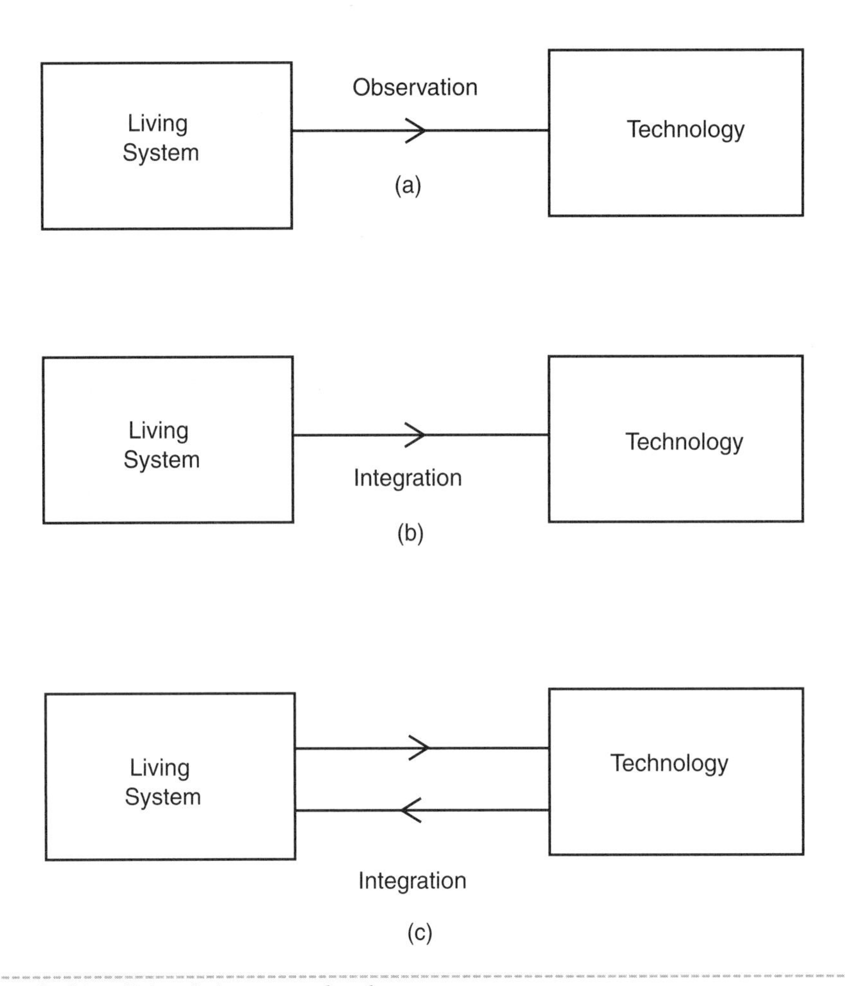


Figure 1-1 *Bionics, linking living beings to technology*

In these cases, a living structure is used as part of a more complex device, performing tasks that artificial parts can't do, all of which are considered bionics.

Finally, a living being can be used to send signals to a circuit, acting as a sensor and at the same time receiving signals to make him or her perform some task. The living being is part of a more complex device, acting at the same time as a sensor and an actuator.

When we describe biofeedback, we are referring to this type of bionic application. A signal is created by the person connected to the electrodes, and a circuit processes the information sent by the person. The circuit provides feedback, commanding some kind of actuator that interacts with the person. Examples of this would be a lamp that changes its flashing frequency or an oscillator that changes its tone, acting on the eye or ear of the person and commanding some kind of action. A stroboscopic lamp controlled by the resistance of a plant leaf is another example of this third application of bionics.

The practical projects described in this book can be classified in one of the three categories described earlier. The reader will find projects that can stand alone as a complete device or circuit, or that can be plugged into other applications or devices. Additionally, many other circuits derived from the basic projects will be given to the reader who wants to go beyond the frontiers of the primary.

This book is a rich resource of ideas and practical applications for the bionic evil genius, and it has as its limits only the imagination and the ability to work with living beings and electronic parts.

# Biology and Electronics



Electronics and biology have many points in common, as electronic circuits and living beings work thanks to electric currents flowing through them. Living beings work with weak currents of only a certain number of millivolts that pass through cell membranes, excite nerves and muscles, and perform other tasks vital to their lives. Electronic circuits work with a wide range of currents, starting from microamperes in microprocessors or low-signal transistors, and reaching many amperes in high-power semiconductors as *silicon-controlled rectifiers* (SCRs), power *metal-oxide-semiconductor field effect transistors* (MOSFETs), *insulated gate bipolar transistors* (IGBTs), and Triacs.

The point of all this is that it is possible to interface living beings with electronic circuits in a way that they can work together in harmony. Today we observe around the world that researchers have started to combine biology and electronics. The results have led to new sciences such as bionics, molecular electronics, artificial intelligence, biophysics, biomechatronics, and many others. The synergy of biology and electronics can be used to deliver many real products, and this book provides some practical examples.

Although we believe this fusing of biology and electronics has great potential, we have learned to be cautious. A convergence exists between electronic circuits and biology, because both sides can work at the same scale. Bionics is just a result of the combination of electronics and biology. As we saw in the introduction, nature and technology can be combined in three ways.

For the reader, it is important to consider that in this book we will provide electronic solutions for the bionic applications that are discussed. Since we are not biologists, the science aspect will not be approached as deeply as the reader may like, and readers are encouraged to research more information according to their interests. Resources such as the Internet, books, magazines, and many other sources can be used to discover necessary additional information.

We also must consider the importance of safety when working with electronic circuits and living beings, especially if the living being is a human. You might not expect that your experiments with bionic devices would require some special protection, such as Asimov's "Laws of Robotics" in the case of robots or any device interacting with people. Although most of the projects in this book use common parts, the relative inexperience of the builder will increase the likelihood of an accident. Your simple bionic project could cause harm to yourself or another. This means, in simple terms, that you are not free from accidents, such as receiving an electric shock when touching a high-voltage circuit or cutting your hand when trying to implant an electrode in a plant or another creature.

This book is not directed to professionals, and since the bionics used in the projects don't have a professional aim, they are very simple and sometimes experimental, but to avoid accidents it is important to follow some basic simple rules of safety. You are working with living beings, including humans and, of course, yourself. If you lose control of your experiments, they can do harm.

You must also consider that, even when you are not working with dangerous devices, the plants, fish, and other living beings used in these projects are very fragile creatures. Even a current of few microamperes flowing through their bodies can injure or even kill them. So we believe that the inclusion of a short chapter about safety is very important.

## Safety Rules

Accidents caused by bionic projects, such as the ones described in this book, tend to have three sources: electronic, biological, and chemical.

### Electronic

Low-power circuits, such as the ones used in our projects, are not particularly dangerous because they can be placed inside boxes or operate from low-power supplies. The following are some safety measures to be considered:

- Protect all areas where high voltages are present. Avoid the use of metallic parts or metallic enclosures with high-voltage circuits.
- Do not power the circuits from transformerless power supplies, mainly the ones that will be used with humans (biofeedback, etc.).
- Include fuses or current-limiting circuits in all sensitive circuits to avoid short-circuit problems.
- Do not connect high-voltage circuits or circuits powered from the AC power line to any person.

### Biological

Circuits used with living beings, such as humans, are dangerous because they cause changes in the control we have on our body. Biofeedback, for instance, can cause a state of hypnotism or temporary confusion if not correctly used. Some recommendations are given here to avoid accidents:

- Do not cause any injury to the living being used in the experiments. Be sure that all the conditions it needs to be comfortable are present.
- Observe if the experiment causes some discomfort to the living being or human. Stop the experiment if this occurs.
- Experiments involving humans must be made with the supervision of an adult.

### Chemical

Chemical substances used in some experiments will create additional hazards. The substances produced by the specimens under experimentation used to produce special effects can cause accidents if used improperly. To prevent accidents with chemical substances, observe the following rules:

- Provide a secure exhaust if any gas is produced during the experiments.
- Be sure that any product used in the experiments will not be injurious to humans or even the specimen if exposure or contact occurs with the eyes and skin.
- Avoid using chemical substances in closed or unventilated rooms.

Following these recommendations, biology and electronics can be used together with the best results. The projects given in the next section are only a small part of the wide universe of bionics today and in the future.

## How to Mount

The projects described in this book are the result of combining electronic technologies with biology. How to handle the living beings used in the projects and perform the experiments will require some care, and in order to mount the electronic parts of the projects, the reader must be able to use some tools and know the basics of electronics.

Starting from the fact that the reader knows something about electronics, let's indicate the basic techniques we intend to apply in the projects described in this book as well as the tools the reader will need to mount them. It is also important to take a look at the components and devices recommended.

The cost of a device built from parts acquired in the specialized market is many times higher than the same device bought ready for use, but the construction with carefully selected parts is still important if one considers the fun, the opportunity to learn, and the satisfaction that readers will have by "doing it themselves." As has been described in other books, today we have many technologies that are used to construct electronic projects and among them are ones that can be used in the bionics field.

The manufacturers of consumer appliances use tools, special methods, and even machines for attaching small parts to the appliances, including an approach called *surface mount technology* (SMT), which employs very small parts (known as *surface mount devices* or SMD). Without the use of these special tools and machines, it is difficult to handle these parts and certainly it is very difficult to mount any project with them. This is different from the basic technology that uses old parts, such as tubes, and electric components, such as lamps, switches, fuses, and other parts.

For most readers, the ideal is to start with the use of an intermediate technology. In the case of bionics the large components are too large to be handled with special tools and therefore are ideal for mounting by hand. The larger components do the same tasks that the small SMDs do. They only need more space in a board. The smaller devices, on the other hand, require special tools and familiarity with those tools. Beginners may not have these tools or skills in handling very small parts.

So, the first thing you have to do before choosing the projects to build alone, with your friends, or pupils, or for your work in a science fair is to learn something about the tools you have to use, the components you are going to assemble, and the main process: soldering.

## The Tools

A bionic project requires two particular areas: a workbench for mounting the electronic circuits and a space for working with living beings or even humans.

## How to Mount

Small electronic components need some support to be kept in place and wired in a circuit. Several techniques can be used to keep the components in working condition. The simplest way to place a component in a circuit is to use a terminal strip, as shown in Figure 2.1.

The components are soldered to the terminals, and the interconnecting wires are soldered to the corresponding terminals. The way the components are placed and wired determines what the circuit will do.

This is not the best way to mount a project these days, but it has the advantage of being simple. It also doesn't need special tools or resources. Many projects described in this book, mainly the ones with experimental purposes, use this technique. They are intended for beginners who are not familiar with more advanced techniques, such as the ones that use *printed circuit boards* (PCBs).

Another way to wire a circuit is to use a terminal strip with screws, as shown in Figure 2.2. This method of building circuits has the advantage that the components don't need to be soldered, but the disadvantage is that you must take special care to place the connections firmly. Any bad contact can affect the circuit's performance or even cause no
performance at all. Another disadvantage is that this mounting method can only be used with simple circuits.

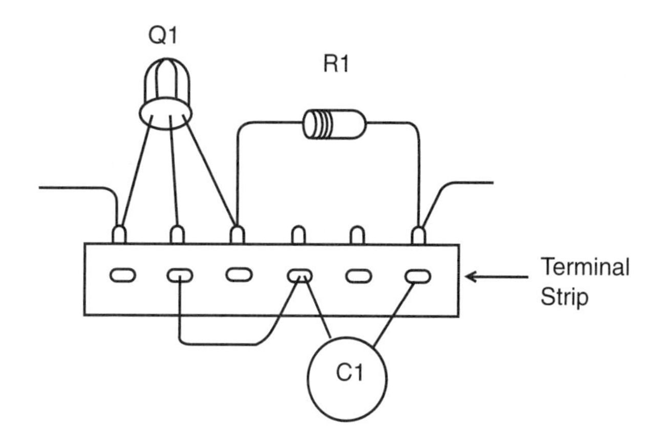


Figure 2.1 *A terminal strip used as a chassis for a simple project*

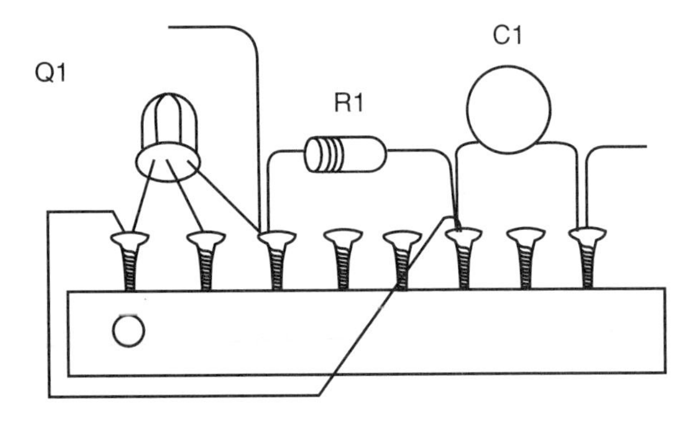


Figure 2.2 *Wiring a circuit using a terminal strip with screws*

Finally, many experimental projects can be mounted on solderless boards, such as the one shown in Figure 2.3. The components' terminals are placed into holes where metallic terminal strips enable them to be interconnected according to a predetermined pattern. The advantage of this technique is that you don't need to solder the components, and they can be reused in other projects. It is also very easy to replace components, making experiments achieve the best values for the desired performance.

All the projects described in this book can be mounted on solderless boards, because they are ideal for experimental purposes. The projects can also be easily altered and mounted in many versions.

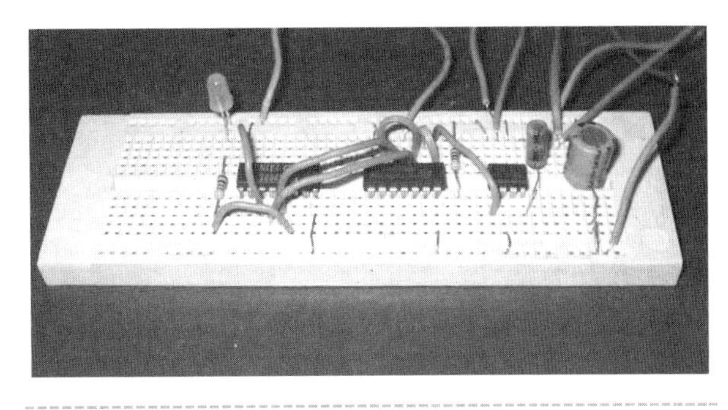

Figure 2.3 *Solderless board*

## The Printed Circuit Board (PCB)

The small components used in electronic equipment can't stand alone without any physical support. They need some kind of support to keep them firmly in their operating positions and at the same time provide the electric connection with the rest of the circuit.

Observing any electronic equipment, the reader will find that the small parts or electronic components are mounted in a special board of fiber or another insulating material. This support or chassis for the components is called a *printed circuit board* or PCB. The board, as shown in Figure 2.4, is made of an insulating material where copper strips are printed on one or both sides.

The copper strips are the wires conducting the currents from one component to another, and the pattern of the strips is determined by the function of the circuit. All the strips are planned before the manufacturing process to provide the necessary connections between the components and the desired function. This means that a PCB produced to receive components that form a radio can't be used to mount TV circuits or any other equipment.

The small components are placed on the board with their terminals passing through holes in the PCB. On the other side of the holes, the terminals are soldered to the copper lines, as shown in Figure 2.5.

A mounting process found in consumer electronics and professional equipment manufacturing uses very small components. This process uses automatic machines (for example, SMDs) controlled by computers to place the components on a PCB according

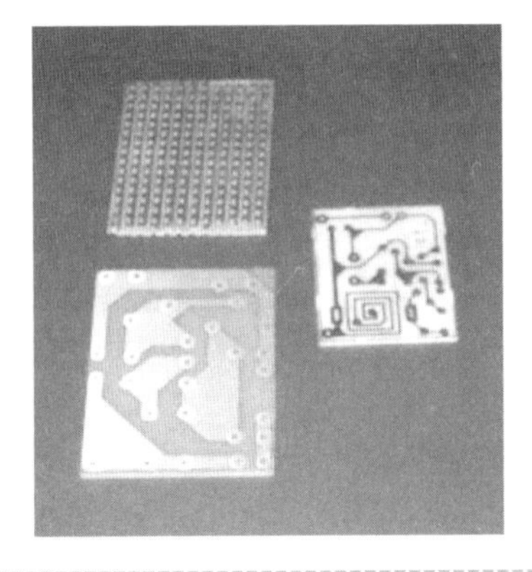

Figure 2.4 *Common PCB*

Figure 2.5 *Soldering connects the components' terminals to the PCB.*

to a program that identifies what the circuit is being created to do. They are programmed to be fixed by automatic machines, as in SMDs. They are placed and soldered to the board at the same side of the strips, as shown in Figure 2.6.

Installing common components on a PCB is a delicate operation that the reader who wants to work with electronic mounting circuits must know. Special techniques are required for this work because the components are small and fragile.

The components are generally soldered to terminal strips, and the copper lines of a PCB. The solder used in an electronic assembly is an alloy formed by 60 percent tin and 40 percent lead with some rosin. It is common to call this kind of solder a transistor solder, a radio-TV solder, or a 60-40 solder.

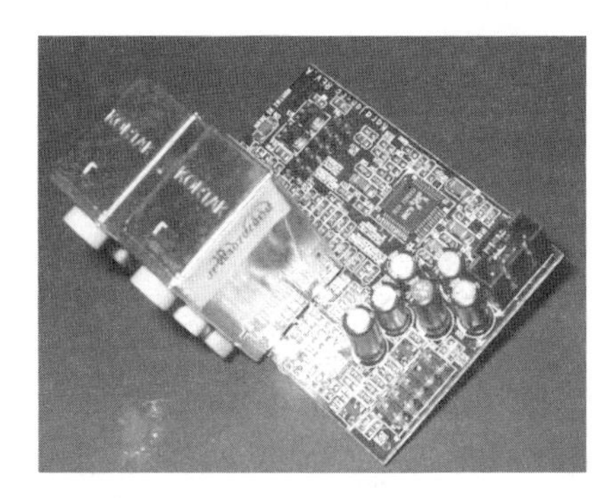

Figure 2.6 *SMD components are soldered directly to the copper lines.*

When heated to about $273^{\circ}$ Celsius ($523^{\circ}$ Fahrenheit), the solder melts and forms a little ball at the terminal of the components that fixes it to the board while at the same time providing electric contact with the copper strips or other components. When working on mounting or replacing components, the reader will need some solder and a soldering iron. The solder can be bought in small quantities, as shown in Figure 2.7, and a soldering iron is shown in Figure 2.8.

A 25- to 40-watt soldering iron with a shiny tip is recommended when working with the small, electronic components found in the circuits described in this book. Of course, the builder could also use a heavy-duty soldering iron to remove or place larger components like those found in some electric and electronic projects.

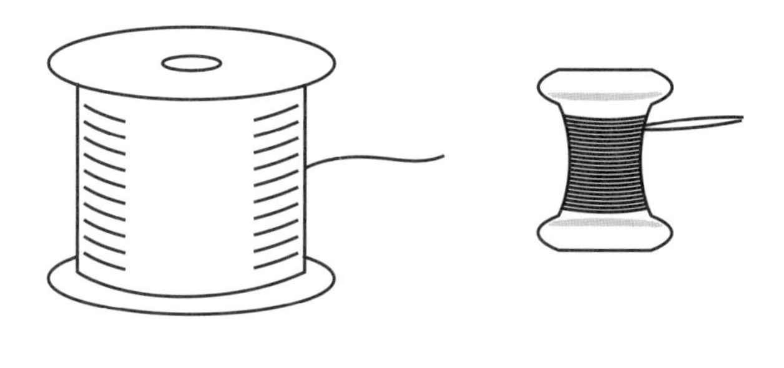

Figure 2.7 *Common solder*

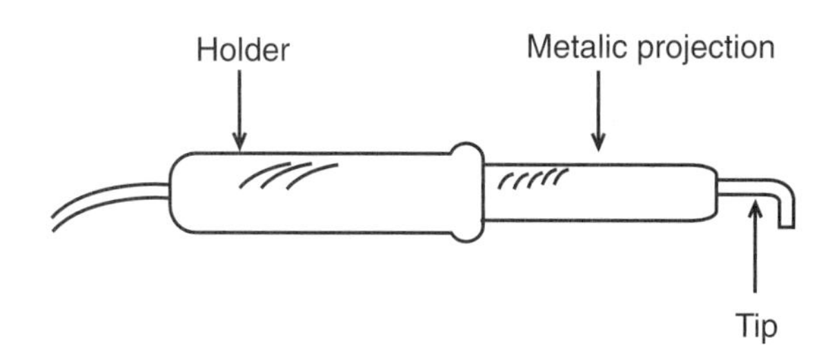


Figure 2.8 *A common soldering iron for the builder who wants to build the projects described in this book*

Soldering is a simple operation, and anyone experienced with electronic mounting will be familiar with this kind of work. Electronic devices are very delicate and care must be taken not to damage them. Many electronic devices are easily damaged by excess heat or an incorrect soldering procedure. The basic procedures for soldering electronic components (removing or installing on a PCB) are as follows:

1. Plug the soldering iron in and let it heat for at least 5 minutes. This will be long enough to bring the tip to the correct temperature for a good soldering operation.
2. Touch the soldering iron to the component for a very short time, allowing the component (or the connection) to heat up, and then touch the solder to the connection, not the iron, as shown in Figure 2.9. You will notice that when the solder melts it penetrates every part of the solder joint.
3. Remove the iron and do not move the joint until it has had time to cool. It is easy to see when the joint has cooled. A peculiar haze will pass over the metal, after which the joint is cool enough and strong enough to withstand movement.

Figure 2.10 shows a perfect solder joint and some solder joints with problems. One of the principal causes of problems in electronic equipment is the "cool solder." The solder seems to contact the component, but no electric contact is made because the joint was not heated enough to penetrate the metal, creating an isolated layer of moisture or oxide to form between them.

## Other Tools

The soldering iron isn't the only special tool needed to work with the electronic circuits used in bionic projects. When working with electronic parts, many of the tools used in electric installations or automotive electricity are suitable. Probably the reader has many of these tools at home, but many electronic components are very small and delicate and need special tools and care.

The use of improper tools when working with these components is often the cause of damage. If the reader intends to work with electronic circuits, we suggest having at least some of these tools:

- Cutting pliers, or diagonals (often called dykes), from 4 to 6 inches long
- Chain-nose or needle-nosed pliers with very narrow tips from 4 to 5 inches long
- Two or more screwdrivers between 2 and 8 inches long

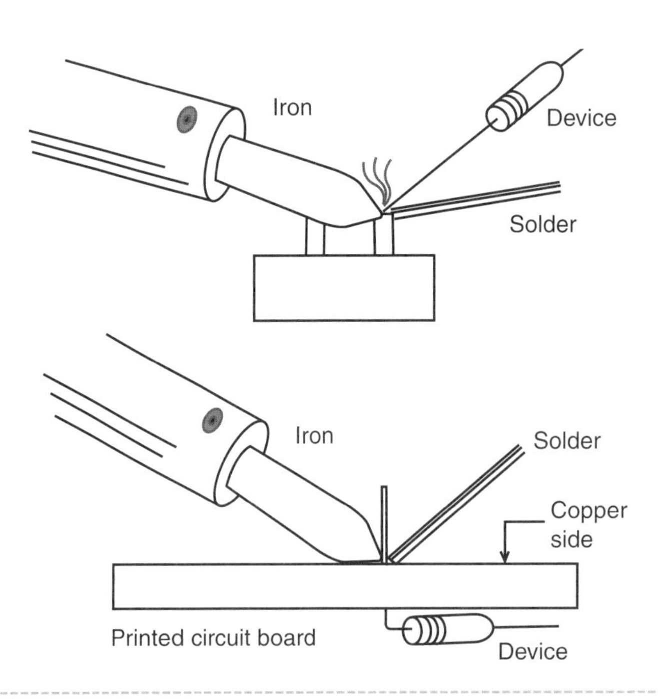


Figure 2.9 *Soldering a component to a terminal strip*

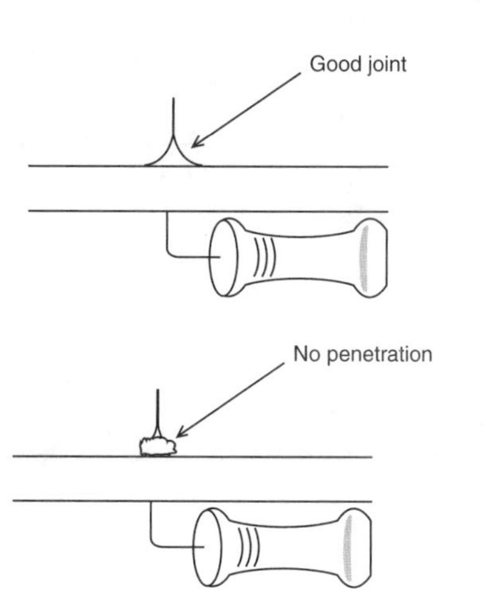


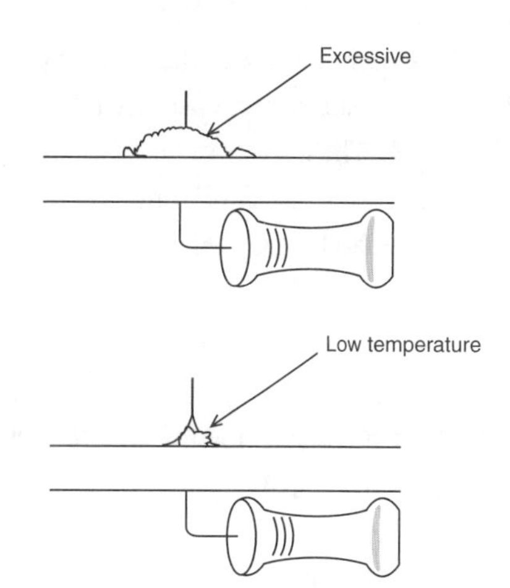


Figure 2.10 *A good soldered joint*

- Crimping tools, a stripper and cutter, for 10 to 22 wire gauges
- Precision tool set (10 to 16 tools) with small screwdrivers of hex, common, and Philips types
- Soldering and desoldering accessories, such as a desoldering bulb and a soldering iron holder/cleaner
- Extra hands to hold the work, such as a mini-vise with a vacuum base or a project holder
- Mini-hand drill

Many other tools can be found in electronic and tool catalogs.

## Diagrams and Symbols

Schematic diagrams or simply *schematics* are used to represent how the many parts of a piece of equipment are interconnected. The builder of bionics projects must be able to show the components and the way they are interconnected via a schematic diagram. In a schematic diagram, the components are not represented by their real shape or format; they are represented by symbols.

How to interpret a diagram is fundamental to the reader who is unfamiliar with electronics but wants to build something using this technology. Learning the meanings of the symbols used to represent each component and what they do in a circuit is an important step toward becoming familiar with schematics.

Let's start with an example of how to read a schematic diagram. Figure 2.11 shows a diagram of simple electronic equipment, an audio oscillator used in animal conditioning.

This schematic diagram represents all components by their symbols, and in many cases, the identification, values, and other important information are given. At the side of each component's symbol is the identification number. This is important because it can help the builder find the component on the PCB, the terminal strip, or inside the equipment.

By a general convention, all the resistors are identified by the letter R, followed by the number of the device in the project. This means that if many

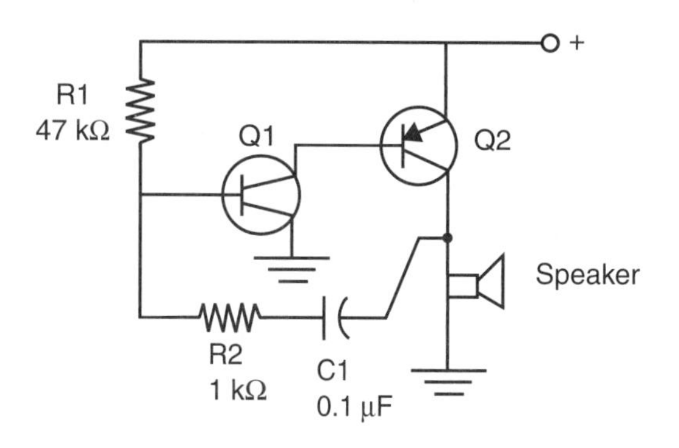


Figure 2.11 *Representing a circuit by symbols*

resistors exist in a device, they will be identified by R1, R2, R3, and so on. Capacitors are usually noted by the use of the letter C. The capacitors of a circuit are numbered starting from C1, C2, C3, and so on. Transistors can be identified by Q, T, or TR. They can be represented as Q1, T1, or TR1.

In many cases, a second number can identify the "block" or "stage" in which the component is placed. The resistors of the first stage can begin at R101 and the resistors in the second stage start at R201. Near the identification and the symbol, we can also find the value or type of component.

Resistors have the value of resistance to the side, such as R1, 1,000 Ω (ohms), or 1k. If it is a transistor, you might find BC548, meaning that, when used, the transistor must be replaced with a BC548 or, when mounting, the transistor placed there must be a BC548. The usual identification number for transistors begins with a 2N, but many manufacturers have started using groups of alphabetic characters signifying their names, such as TIP (Texas Instruments) and MPS or MM (Motorola). A European code uses BC or BD as the identification of devices, and a Japanese configuration uses 2SB, 2SC, or 2SD to indicate transistors.

In this book, the transistors for general-purpose, low-power applications will generally be the BC548 and BC558. Medium-power transistors will be TIP31, TIP32, TIP115, BD135, or BD136, and high-power transistors will be TIP41, TIP42, or 2N3055.

Depending on the circuit, other important information can be found in the schematic diagram, such as the voltage at different points in the circuit. In the example presented, it is indicated that, between A and the ground (normally taken by reference or 0 volts), the measurement 6 volts is found when a multimeter is used to take the voltage (see Figure 2.12). Also found on a schematic are procedures for installation, diagnostic problems, equivalence, and so on.

To make things easier for the reader, besides the schematic diagram we will also provide figures detailing the building steps and offer a parts list. In the parts list, we will describe the main characteristics of the components used in the projects, including the color codes for resistors, voltages, and current rates for other components such as capacitors, transformers, and diodes.

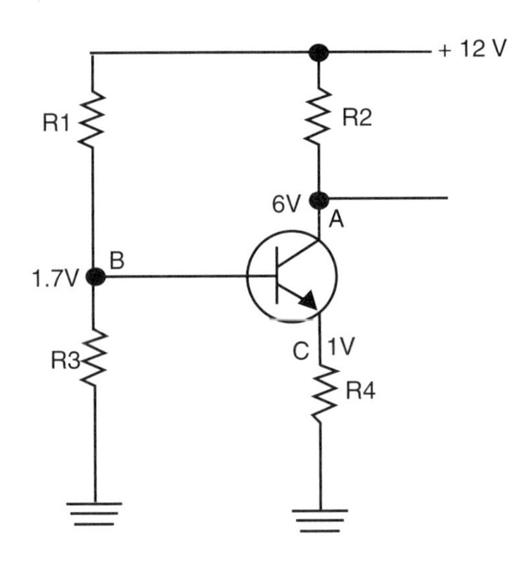


Figure 2.12 *A diagram showing the voltages of some points*

## Working with Living Beings

Many projects described in this book will involve direct contact with living beings. Although in many cases they could be insects or plants, whom you certainly won't cause any harm, when working with humans you will have to double your efforts to be careful.

Although the projects described in this book don't have any potential to cause great damage or pain to anyone, you must be careful when using them. In case of any doubt, consult an adult or make the experiment under the supervision of an adult.

It is also important to have the appropriate place to work with living beings. A biology lab offers special workbenches for this task, and in your home find a place where you can work with plants, insects, or even fish.

Remember that life is fragile. If you don't take the necessary care, your specimens will die before you have time to make the experiments. Keep in mind that they are living creatures and we must respect their lives, not cause any injury to them, or let them die from neglect.

# 25 Practical Projects



The purpose of this part is to describe projects, ideas, and experiments involving bionics based on simple circuits. The circuits and experiments described here can be altered to change their performance or to be used with other intentions.

The reader will find 25 basic projects, each standing alone as a complete device, and a large assortment of new circuits and ideas to be explored. Many circuits are new versions of the basic projects using different configurations to reach the same end, but others are completely new devices, exploring only the basic principles. This means that although the book claims to contain only 25 projects, the reader will find much more—in fact, over a hundred of them.

## Project 1—Experiments with an Electric Fish

The Amazon Basin is a mysterious place where many strange creatures have been known to live, even today, although we're not talking about prehistoric monsters or man-eating plants.

Although the author lives in Brazil, he is from the southern part of the country, very far from the Amazon (more than 2,000 miles!), and he can suggest some interesting experiments and projects with small

Figure 3.1.1 *The Itui Cavalo or Black Ghost*

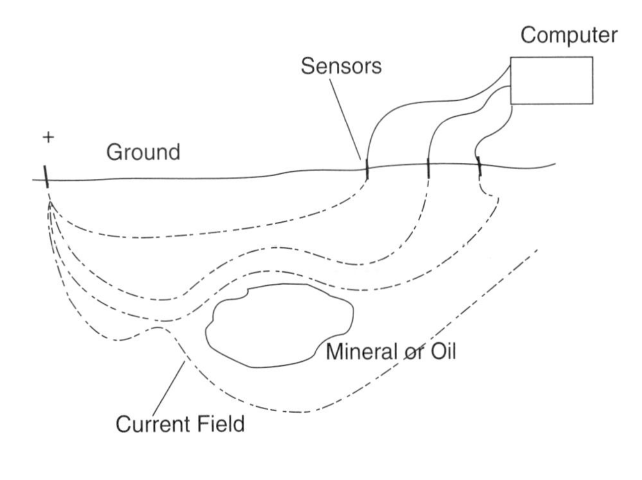


Figure 3.1.2 *A current field in the water or earth can be used to detect minerals, caves, and large objects.*

(and docile) creatures from the Amazon easily found even in the United States. In some pet shops (yes, pet shops!), it is possible to find some of these strange creatures that live in the Amazon and use them in bionics projects like the ones described here.

One of those curious creatures is the electric fish Black Ghost (or *Itui Cavalo* in Portuguese), known by many who keep small aquarium fish (see Figure 3.1.1). The Black Ghost (Genus Apteronotus, and families Electrophoridas, Gymnotoidae, Rhamphichthidae, Hypopomidae, and Sternopygidae) and some catfish have the ability to produce electric, low-frequency fields around their bodies that are used to detect other beings and objects.

The same principle of operation is used by equipment that is used to detect mineral deposits and even large objects in the sea. The idea is that a "current field" is deformed by the presence of objects of different conductance, as shown by Figure 3.1.2. Electrodes can map the currents and determine the location of the objects that produce the deformation of the field.

The fish uses this principle to detect objects and other creatures in the dark water where he lives. The Itui is nearly blind and must depend on his electric sense to find food and escape from enemies. Figure 3.1.3 shows how the Itui produces an electric current field in the water both to detect objects and communicate with other fish of the same species.

The most interesting points to consider are the characteristics of the fish's electric field and the sensors developed by the fish's nature. The Itui produces a very stable, low-frequency field with frequencies that depend on the size of the specimen and the chemical conditions of the water around him or her.

For our experiments, we recommend the use of the Itui Cavalo or Black Ghost (Apteronotus Albifrons, the Gymnotoidae Family), which produces very stable, low-frequency fields in a range between 400 and 4,000 Hz. Figure 3.1.4 shows a typical wave shape of the current field generated by the electric fish.

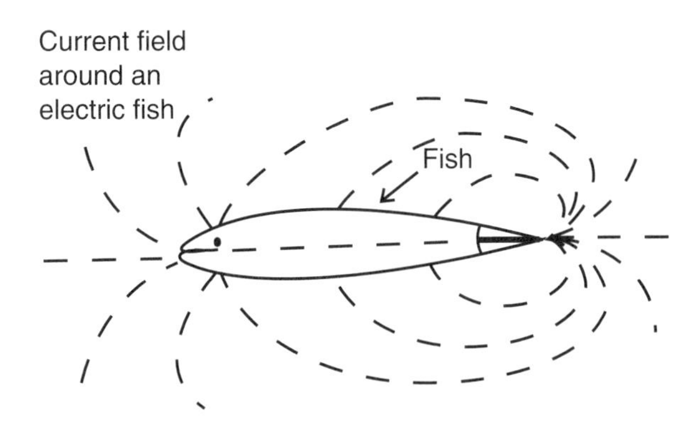


Figure 3.1.3 The current field around the Dark Ghost

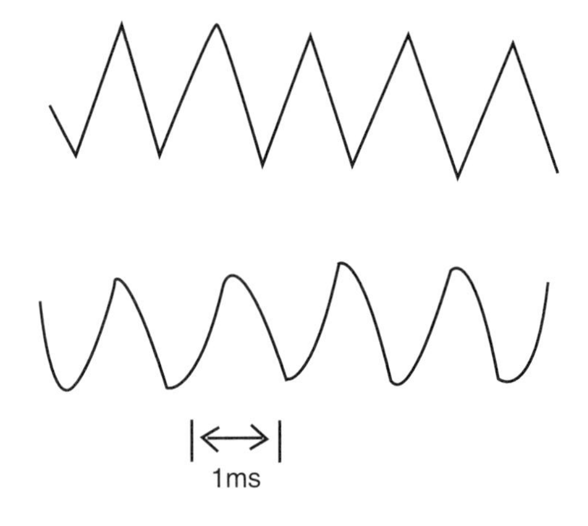


Figure 3.1.4 *The wave shape of the signal generated by an electric fish*

Placing metal electrodes near the fish, it is possible to pick up the signal produced by the fish and use it to make interesting bionic experiments, as we will describe in the this section. Of course, the reader can go ahead and create his or her own experiments with the electric signals produced by this strange creature.

Since the signal generated by the fish is very stable in frequency, the reader can consider it as a living "quartz crystal" or "tuning fork," which can be used to drive computers, microprocessors, clocks, and even a musical instrument, as suggested in the next section.

## The Bionic Experiments

The signals generated by the Black Ghost reach an amplitude ranging from 3 to 4 volts, which is enough to drive many external electronic circuits.

The simplest experiment, described here, is made by picking up the signals and using them to drive a small audio amplifier. When amplified, the low-frequency signals can be heard as a pure audio tone in a very interesting experiment involving animal behavior. In a science fair or even in the lab, it is nice to see how impressed people are when the pure tuning-fork sound produced by the fish is heard. You can also perform other projects such as the following:

- The stable signal can be used to drive a digital or mechanic clock.
- A signaling system can be driven by the fish, indicating to his or her owner that the fish is alive. A *light-emitting diode* (LED) will blink, controlled by the signal produced by the electric fish.
- It was shown that the frequency of the signal produced by the fish changes slightly when the pH or the $CO_2$ concentration of the water changes. This means that, monitoring the frequency of a Black Ghost, you can develop a bionic sensor of pollution for water sources. The signal can be picked up and applied to a computer to monitor the water of a source using a living sensor.
- The frequency of the signals also changes when the fish is disturbed by external influences. You can create an earthquake or tsunami detector based on the fish's reaction —an interesting form of research to be made these days.
- Dividing the signals by fractions of 8 (1/8, 2/8, 3/8, etc.), you can build a musical instrument operating from the signals generated by the fish.
- Knowing the frequency of your fish, you can build an oscillator to produce the same signal and make experiments to see how it can be disturbed.

## The Electronic Circuit

Our basic project consists of a simple, low-power audio amplifier used to pick up the signals generated by the fish. The signals are amplified and applied to a loudspeaker, or reproduced by a loudspeaker as shown in Figure 3.1.5.

Although you can use any audio amplifier for this purpose, the experimenter can build his or her own circuit and make experiments with the signals, placing new blocks in the same solderless board.

The circuit is based on the *integrated circuit* (IC) LM386, which can be powered from four AA cells, producing a good audio signal to a small loudspeaker. The gain of the circuit is determined by the capacitor placed between terminals 1 and 8. In our case, because the signal picked up by the electrodes is

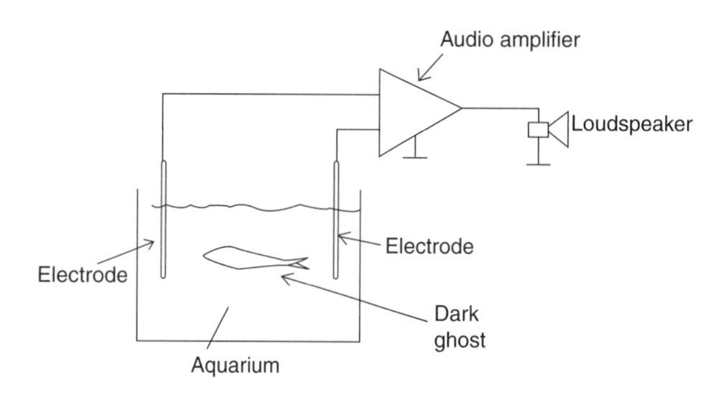


Figure 3.1.5 *Picking up the signals generated by the Dark Ghost*

Figure 3.1.6 *Electrode placement in the aquarium*

weak, we recommend working with high gain. So, a 10 $\mu$F signal is used, resulting in a voltage gain of 200.

The electrodes are pieces of bare wire immersed in the aquarium with the fish. It is recommended that the wires be 20 to 30 cm long and that they are separated by 20 to 50 cm, as shown in Figure 3.1.6.

It is not recommended that you power this circuit from the AC power line, even using a good power supply. The 60 Hz hum can disturb the operation of the circuit, appearing in the loudspeaker.

## How to Build

Figure 3.1.7 shows the schematic diagram of the audio amplifier used in the basic experiment. The components can be placed on a solderless board, as shown in Figure 3.1.8.

This mounting process is particularly recommended to beginners and for experiments made in a lab, because the materials used to etch a PCB are not

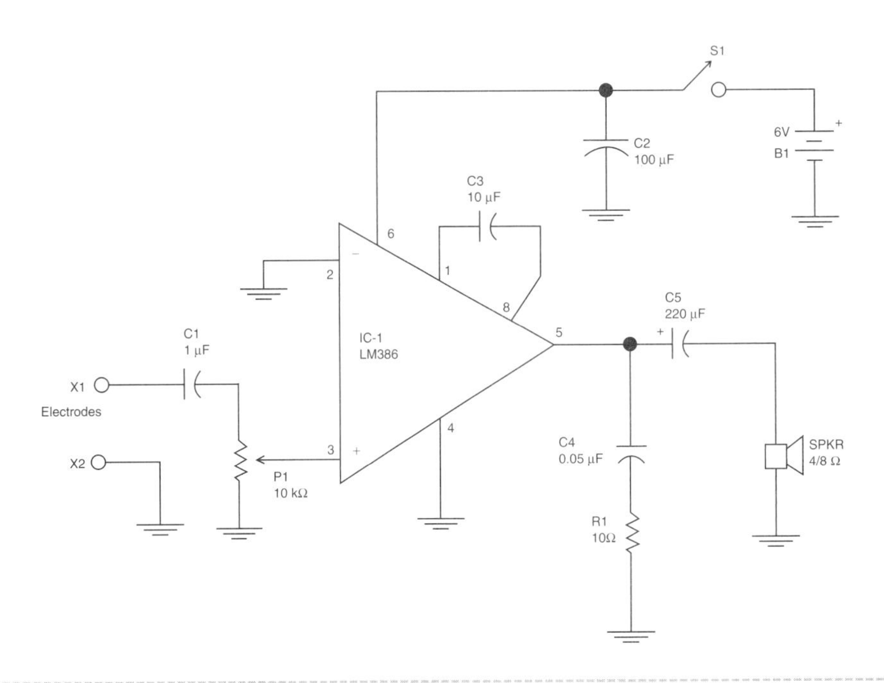


Figure 3.1.7 *Complete diagram for the audio amplifier*

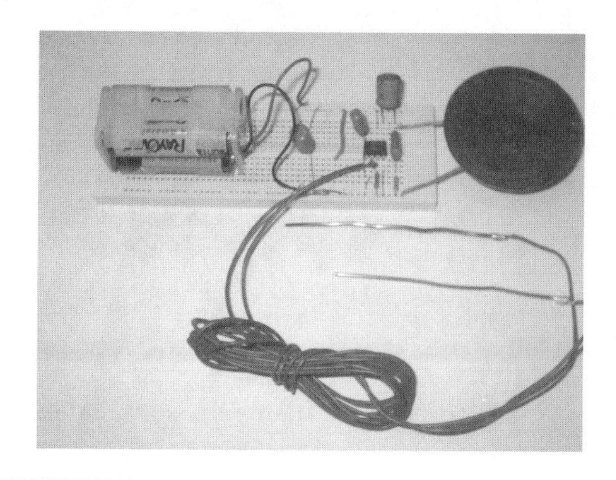

Figure 3.1.8 *The circuit is assembled on a solderless board.*

necessary. Of course, if the reader has the resources to etch a PCB, he or she is free to choose this mounting version for his or her amplifier. The PCB version is shown in Figure 3.1.9. The pattern for this board is shown in Figure 3.1.10.

The positions of the polarized components, such as the IC and the electrolytic capacitors, should be observed. The loudspeaker can fit inside a small plastic or wooden box, and an appropriate holder should be used for the cells. In this mounting, as the circuit works with low-power audio signals, it is very important to keep all the connections and the components' leads as short as possible to avoid instability or picking up undesirable hums.

## Testing and Using

Turn the unit on by using S1 or by placing the cells in the holder and plugging in an audio source, such as a ceramic microphone. If you intend to use an electret microphone, a 10,000-ohm resistor should be added, as shown in Figure 3.1.11.

To pick up the signals generated by the Black Ghost, you must place the electrodes into the aquarium and connect them to the input of the audio amplifier using common wires (no shielded wires are needed if the distance to be covered is short). If your aquarium has a fluorescent light and/or a heater, it is recommended that you unplug them from the power line to not introduce noise to the circuit.

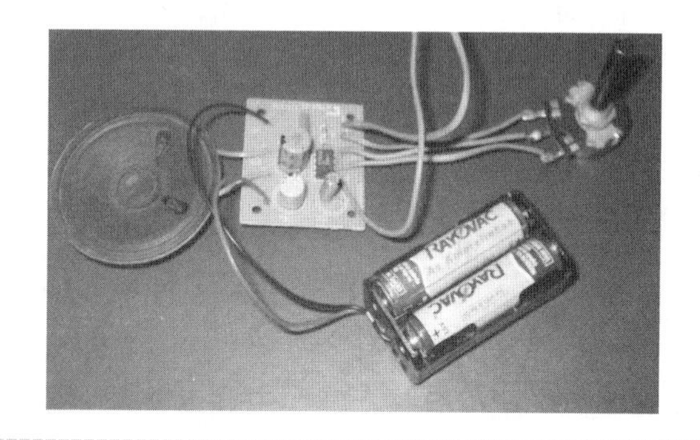


Figure 3.1.9 *The same circuit mounted on a PCB*

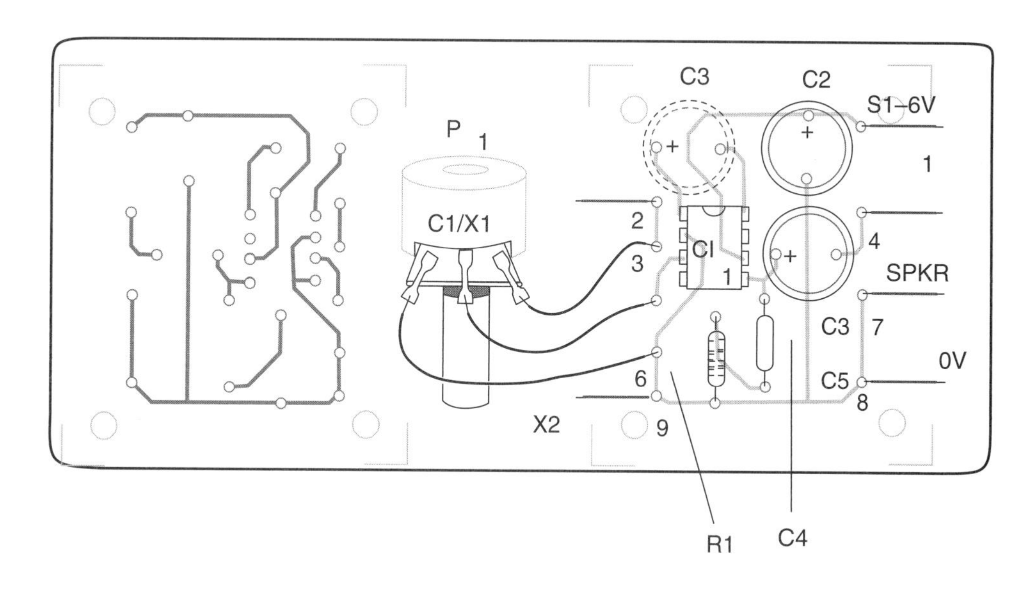


Figure 3.1.10 *The PCB for the audio amplifier using the LM386*

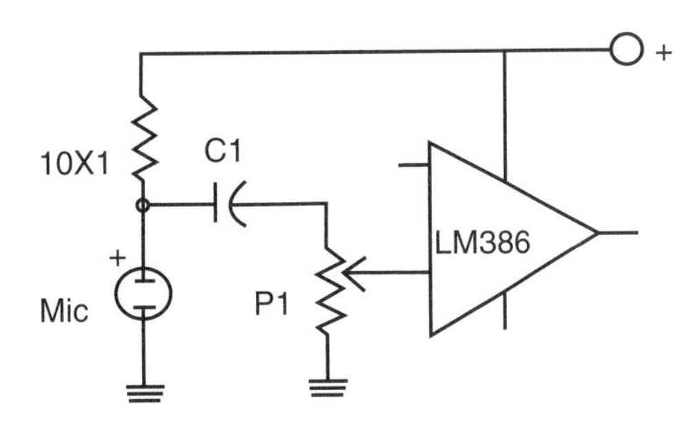


Figure 3.1.11 *Testing with an electret microphone*

Since the water is almost conductive, the aquarium acts as a low-impedance signal source. This means that if you haven't any other powerful interference source near the aquarium, such as electric devices, no hum will appear in the loudspeaker.

Opening the volume control, you will hear a pure tone, like an audio oscillator or tuning fork, which will change in volume when the fish moves to another place inside the aquarium. This tone is the signal produced by the fish.

Unplug the electrodes to be sure that the picked-up signals are really produced by the fish. If the noise (hum) is reproduced by the speaker, see if any appliance connected to the AC power line that is placed near the aquarium is the cause.

## Parts List: The Audio Amplifier

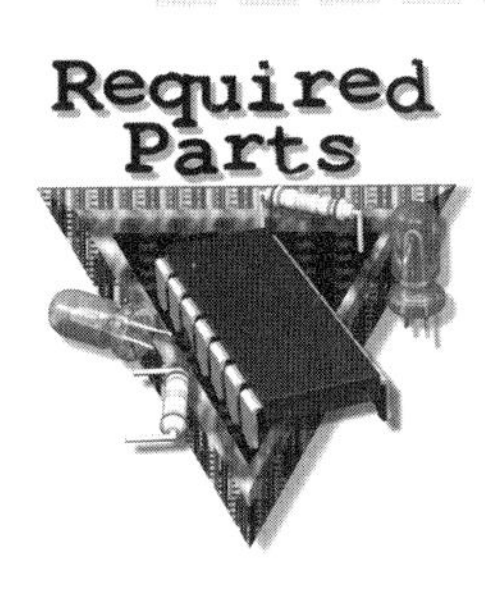


- IC-1: LM386, audio amplifier, integrated circuit
- C1: 100 μF x 12 V electrolytic capacitor
- C2: 10 μF x 12 V electrolytic capacitor
- C3: 220 μF x 12 V electrolytic capacitor
- C4: 0.05 or 0.047 μF ceramic or polyester capacitor
- P1: 10 kΩ log or lin potentiometer
- R1: -10 Ω x 1/8 W resistor, brown, black, black
- S1: Single pole, single throw (SPST) on/off switch
- B1: 6 V, 4 AA cells and holder
- SPKR: 4 or 8 Ω x 5 to 10 cm small loudspeaker
- X1, X2: Metal electrodes (see text)
- Other: Solderless board or PCB, wires, solder, plastic box, knob for P1, etc.

## Additional Circuits and Ideas

Starting from the basic experiments, where we only gave the reader the elements for picking up and hearing the signals produced by the fish, we can suggest many other experiments using the signals. In fact, bionics means the integration of living beings with electronic devices or using technology that adopts solutions provided by nature after millions of years of evolution. In the next section, we will give the reader some ideas for new projects.

### Observing the Signals in an Oscilloscope

If you have an oscilloscope, you can directly observe the signals generated by the Black Ghost. It only requires plugging the electrodes to the input of the instrument and adjusting the amplitude and frequency to the appropriate value of the signals. Figure 3.1.12 shows how to make the connections.

The signals can change in shape and frequency according to the specimen. Figure 3.1.13 shows some patterns observed in these fish. If your oscilloscope has a frequency meter, you can determine the exact frequency of the signal produced by your fish.

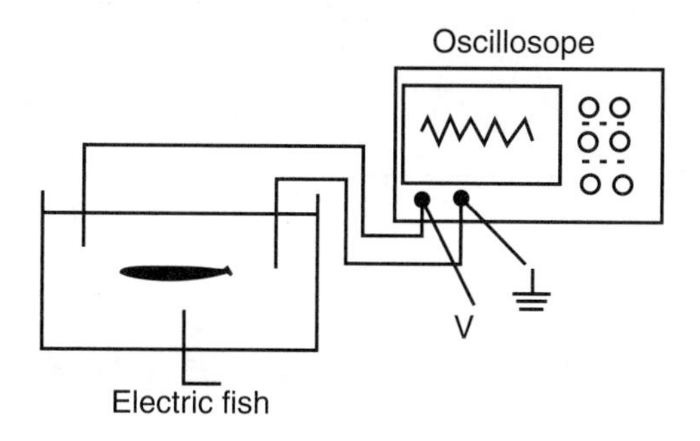


Figure 3.1.12 *The waveforms observed in the oscilloscope screen*

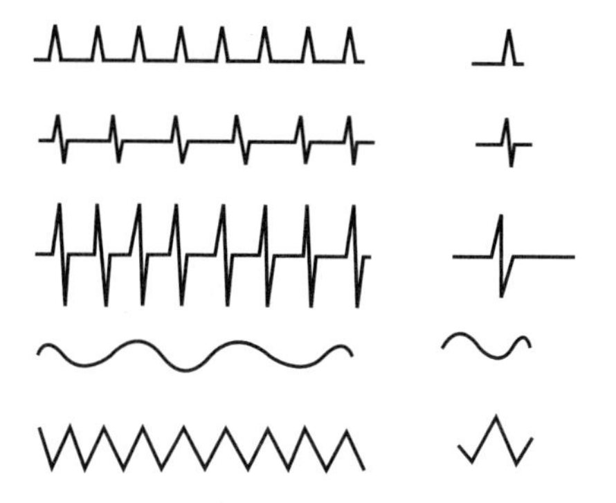

Figure 3.1.13 *Waveforms observed in the oscilloscope*

## Using the Computer

If you have a program in your computer for observing audio signals, you can use it to monitor your electric fish. It is enough to run the program and plug the electrodes into the microphone input, as shown in Figure 3.1.14.

It is important to keep the wires as short as possible and not have any electric device powered from the AC power line near the aquarium, including the fluorescent lamp. In some cases, if the amplifier used to convert the signals from the mic input to the digital form is not sensitive enough, the reader will have to use a preamplifier or even the amplifier with the LM386, picking up the signals from its output, as shown in Figure 3.1.15.

The loudspeaker can be replaced by a 10-ohms × 1/2-watt resistor to act as a load to the signals if you don't want to hear them during the experiments.

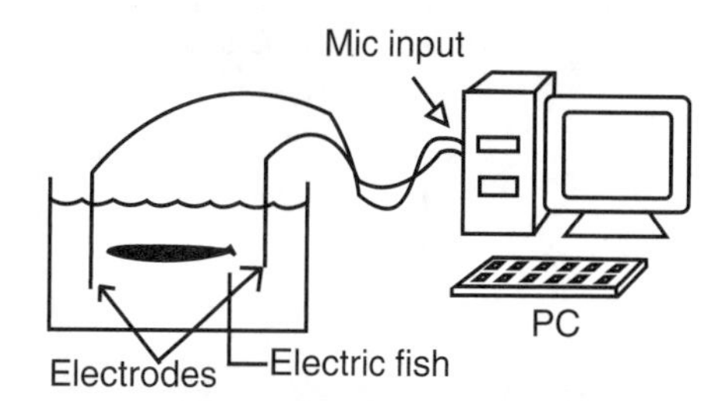


Figure 3.1.14 *Observing the signals in the PC*

## Using a Common Audio Amplifier

If you have an audio amplifier with a microphone input, you can use it to pick up the signals. All you have to do is plug the electrodes into the input using wires that are as short as possible. It is again recommended that no other device be plugged into the AC power line near or in the aquarium.

## Driving External Circuits: An LED Blinker

A very interesting experiment is one that can result in practical bionic devices using the Black Ghost. The basic idea is to determine the frequency of the signal generated by the fish and divide this frequency by the some integer number, obtaining a 1 Hz signal (1 pulse per second). This stable signal, generated by the fish, can drive electronic devices such as clocks,

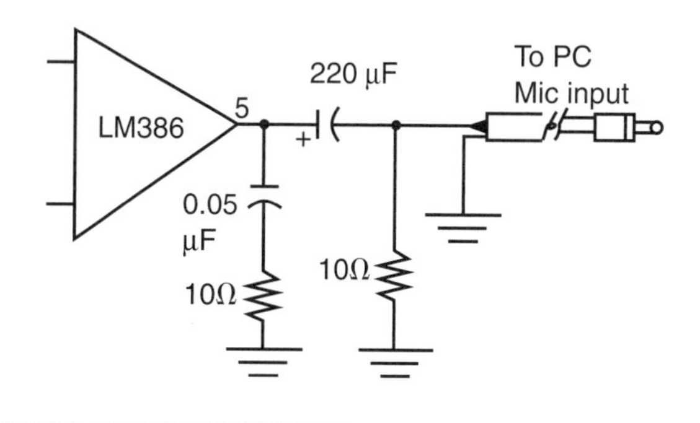


Figure 3.1.15 *Using the LM386 to drive a computer audio input*

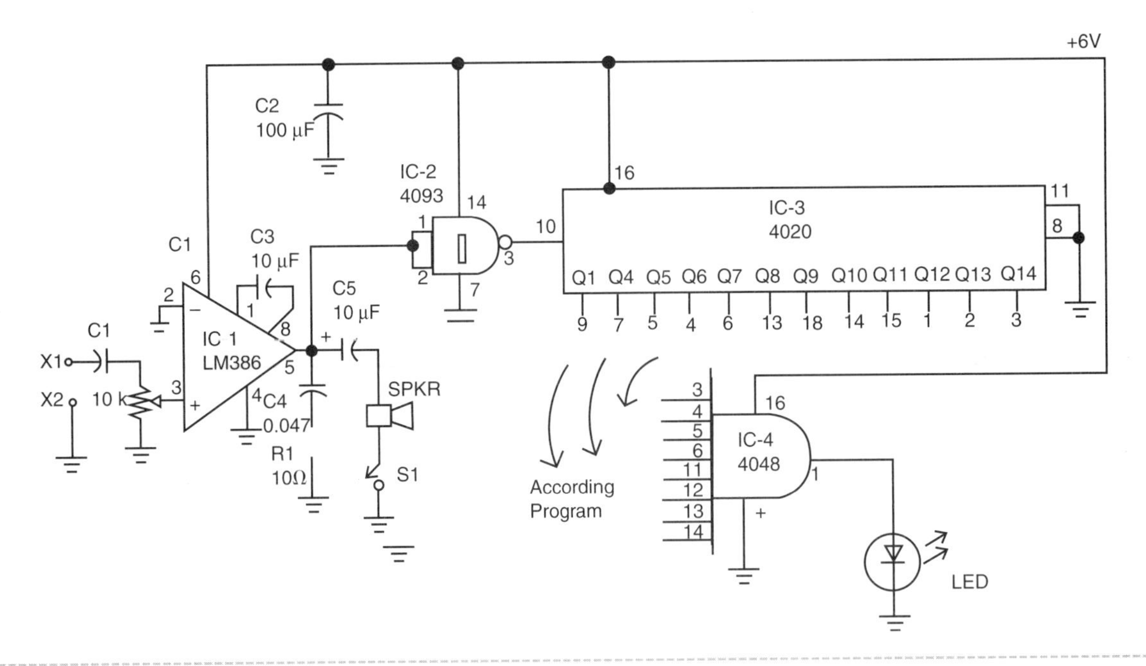


Figure 3.1.16 *Driving an LED and a beeper with a 1 Hz signal*

chronometers, and digital instruments. The simplest application is shown in Figure 3.1.16.

The exact frequency of the signal generated by the fish can be measured with a frequency meter or oscilloscope. For the frequency meter, it may be necessary to use an amplifier, as shown by Figure 3.1.17.

The basic amplifier using the LM386 is the ideal circuit for driving a 4093 IC, which produces a square wave from the waveshape of the signal produced by the fish. To divide the frequency by any number, we used a 4020 *complementary metal oxide semiconductor* (CMOS) IC and a 4048 CMOS IC.

As Figure 3.1.18 shows, the 4020 consists of a 14-stage ripple carry binary counter having 12 outputs that are a division of power-of-2 numbers of the input frequency. This means that the circuit can be used to divide the input frequency for numbers up to 16,384 ($2^{14}$).

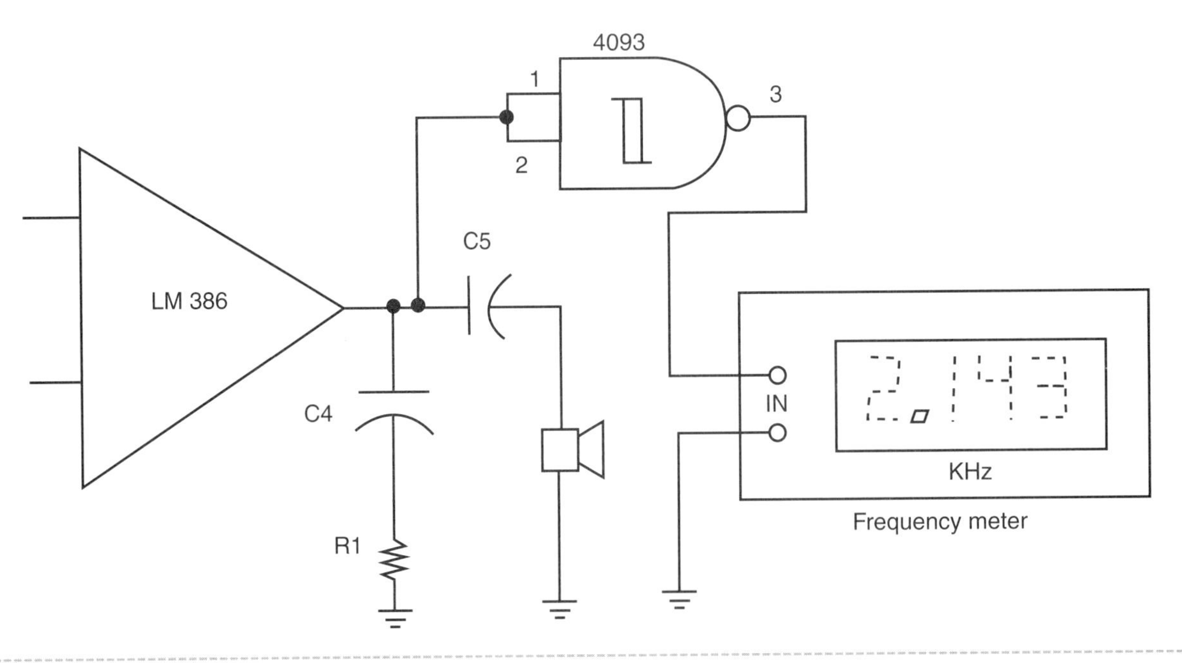


Figure 3.1.17 *Measuring the fish frequency with a frequency meter*

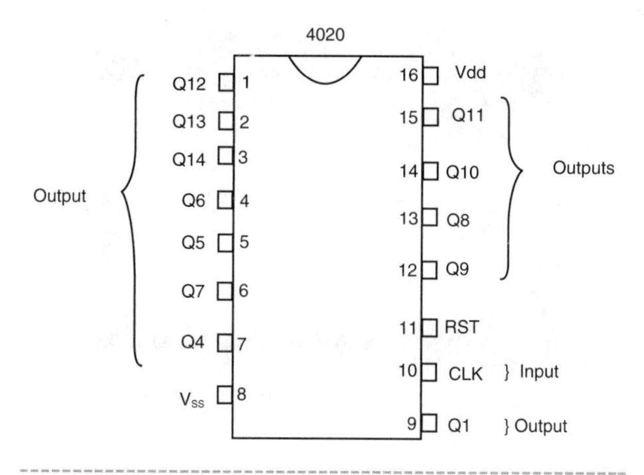


**Figure 3.1.18** *The 4020*

Since the frequency range of the signals produced by the Black Ghost is lower than the limit reached by the 4020, it is easy to find the necessary number to have a 1 Hz signal or even other frequencies such as 0.5 Hz, 0.2 Hz, or even 0.1 Hz (one pulse at each 2, 5, and 10 seconds).

Let's see how we can do that. Observing the 4020, we can see that the outputs are Q1, Q3, Q4, Q6, and so on. They correspond to the number represented by a power of 2, by which the frequency of the input is divided. For instance, Q4 is 16 ($2^4$), so the signal in this output is the input frequency divided by 16.

The circuit doesn't have outputs corresponding to 4 ($2^2$) and 32 ($2^5$), but the remaining outputs are enough to be combined and allow the designer to find the necessary number to establish 1 Hz with satisfactory accuracy.

Taking a numeric example, let's imagine that our fish generates a 1,556 Hz signal. Taking this number and making the decomposition in powers of 2, we will have the following table.

Because the 4020 does not have all the powers of 2 for outputs, the number obtained is the sum of the 1's of the last column, or

$$1{,}024 + 512 + 16 + 2 = 1{,}554 \text{ Hz}$$

This result has an accuracy better than 0.2 percent, which is enough for experimental purposes. But to get the 1 Hz signal, we must combine the corresponding outputs in a logic circuit, which is the 4048 programmed as an AND gate, as shown by Figure 3.1.19.

| *Output* | *Power of 2* | *Components of the decomposition for 1,556* |
|---|---|---|
| Q14 | 16,384 | 0 |
| Q12 | 8,192 | 0 |
| Q11 | 4,096 | 0 |
| Q10 | 2,048 | 0 |
| Q9 | 1,024 | 1 (lasts 524) |
| Q8 | 512 | 1 (lasts 12) |
| Q7 | 256 | 0 |
| Q6 | 128 | 0 |
| Q4 | 16 | 1 (lasts 8) |
| Q1 | 2 | 1 (lasts 2) |

The outputs Q1, Q4, Q8, and Q9 must be connected to the AND gate 4048. In this configuration, by applying the signal of the outputs determined by the decomposition table to each of the 1,546 pulses at the input of the 4020, we will have one pulse at the output of the 4048. An LED connected to this output will blink at a 1 Hz frequency (one pulse per second), as desired.

Of course, the number of outputs that will be connected to the input of the 4048 will depend on the value of the input frequency. If you are lucky enough, your fish can produce a frequency corresponding to a power of 2 (1,028 Hz, for instance), so you will not need the 4048 and can plug the LED directly to Q8, keeping the 1 k Ω resistor in series, of course.

The complete circuit that makes the LED blink at a frequency near 1 Hz is shown in Figure 3.1.20, using only one output of the 4020.

Make sure one gate of the 4093 is used to drive a piezoelectric transducer, producing beeps at twice the rate at which the LED blinks.

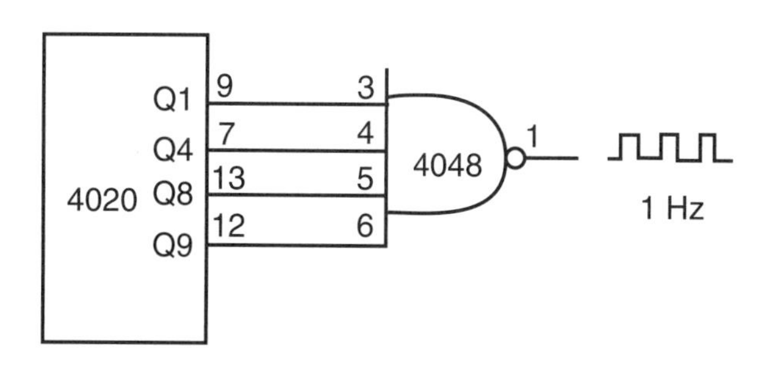


**Figure 3.1.19** *Using the 4048*

Since the loudspeaker is not in use, it is replaced by a 100-ohm resistor, acting as a load for the amplifier. Of course, if another output is used, the beeps can be produced at half the rate of the LED blinks. Figure 3.1.21 shows the circuit mounted on a solderless board.

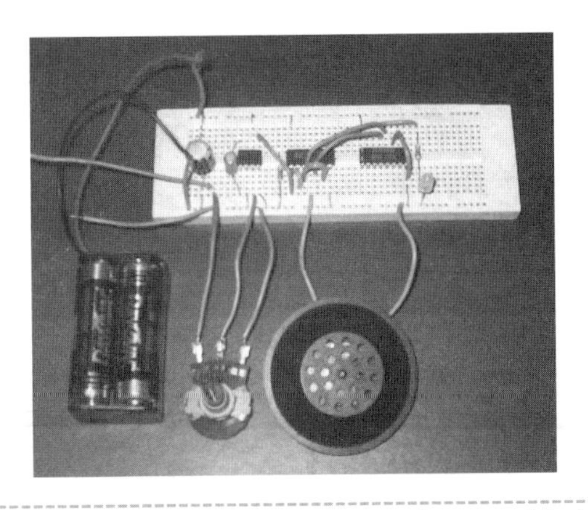

Figure 3.1.21 *The circuit mounted on a solderless board*

## Parts List

IC-1: LM386 integrated circuit, audio amplifier

IC-2: 4093 CMOS integrated circuit

IC-3: 4020 CMOS integrated circuit

LED: Common LED, any color

C1: 100 µF x 12 V electrolytic capacitor

C2: 10 µF x 12 V electrolytic capacitor

C3: 220 µF x 12 V electrolytic capacitor

C4: 0.05 or 0.047 µF ceramic or polyester capacitor

P1: 10 kΩ log or lin potentiometer

R1: 10 Ω x 1/8 W resistor, brown, black, black

R2: 100 Ω 1/8 W resistor, brown, black, brown

R3: 1 kΩ x 1/8 W resistor, brown, black, red

S1: SPST on/off switch

B1: Four 6 V AA cells and holder

X1, X2: Metal electrodes (see text)

Other: Solderless board or PCB, wires, solder, plastic box, knob for P1, etc.

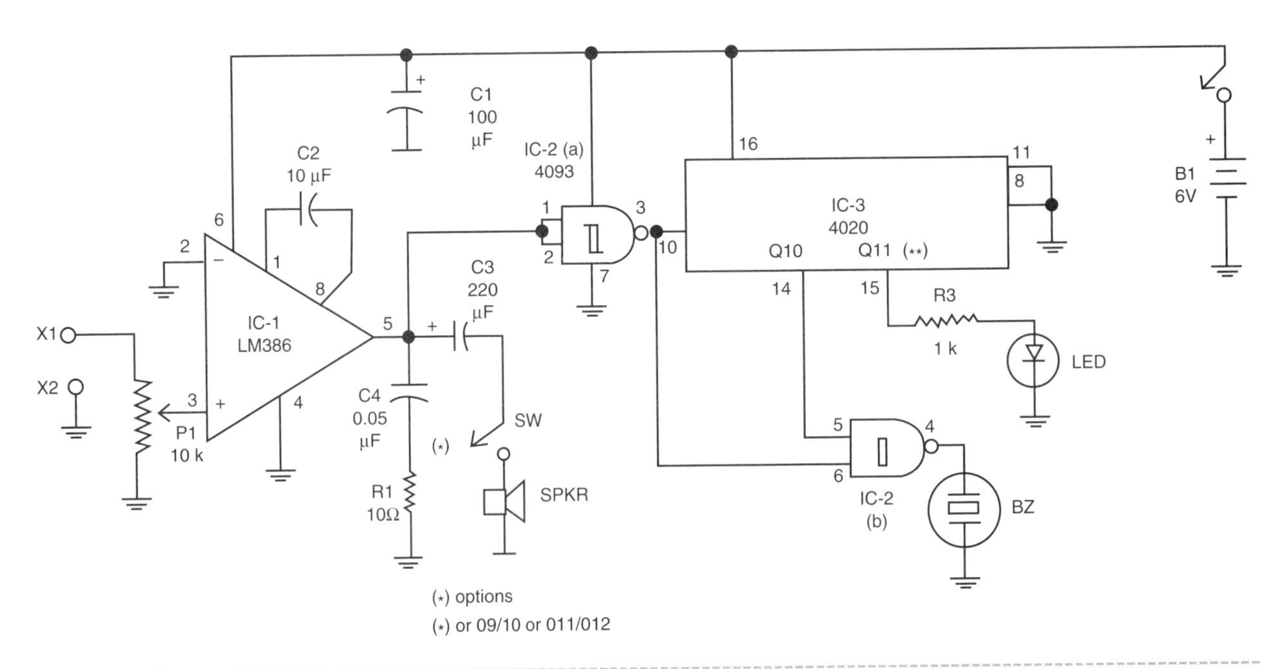


Figure 3.1.20 *This circuit makes the LED flash once per second and produce beeps at a double rate.*

## A Clock Driven by the Fish

A practical application for the signals generated by the fish is a clock. You can mount a digital clock or, if you prefer, use a cheap mechanical clock that is powered by an internal crystal that the fish's signal will replace. This internal crystal generates 1 Hz to power the small motor, and the clock's indicator advances 1 second at a time from each pulse produced by the circuit. All the reader has to do is disable the crystal and apply the 1 Hz signal produced by the fish to the motor.

Another possibility is to mount an experimental clock using a gearbox with a small electric motor. Figure 3.1.22 shows the circuit that will drive the motor from 1 Hz pulses produced by the fish and the circuit for an experimental clock.

All you have to do is determine how much the indicator advances at each pulse and make a scale for the clock. You can adjust the clock to a complete turn at each minute. For instance, you can add a gear to drive an hour indicator, as shown in Figure 3.1.23. The ratio between the gears must be 1:24 or 1:12, so the second indicator will show the hours.

The second alternative uses a digital clock. Figure 3.1.24 shows a typical circuit of a common digital counter using the 4511 and 4518 CMOS IC driven by the 1 Hz pulses generated by the fish.

See that this circuit counts up to 99. It must be programmed to divide by 60, with another two-digit block that must be programmed to divide by 60, giving the minutes, and a final two-digit block dividing by 12 to give the hours. Thus, you can receive the 1 Hz signal from the fish and have a 12-hour clock.

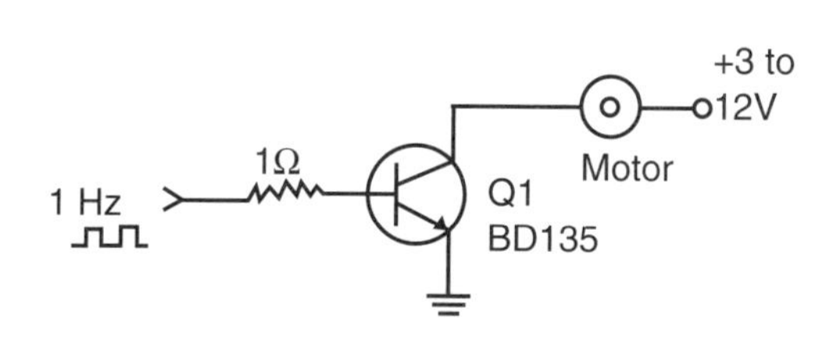


Figure 3.1.22 *Using the 1 Hz to drive a mechanical clock*

## The Fish as a Wireless Sensor

Today the most advanced devices in computing, communications, and even home appliances are wireless. These devices communicate data and "talk" with one another using high-frequency radio waves, as recommended by the 802.11 standards.

Even in a variety of industries, sensors that are plugged into machines with the aim of controlling processes are now becoming wireless. The physical aspects (temperature, pressure, force, and so on) are converted to digital signals and sent to a remote receiver where they are processed, as shown by Figure 3.1.25.

But what can nature and the electric fish teach us about communicating data without the need for cables or other physical media? The Black Ghost is a very sensitive creature. Small changes in the pH of the water, the temperature, and the presence of pollution alter the frequency of the signal generated by the fish, who can be used as a remote sensor for these changing factors.

Fish have been used to monitor water purity, for example, and this process has been patented. Figure 3.1.26 shows how this monitoring process can be done.

Going a little further, many creatures can detect small changes in the stress of tectonic plates. Their sense of danger can motivate them to leave a certain

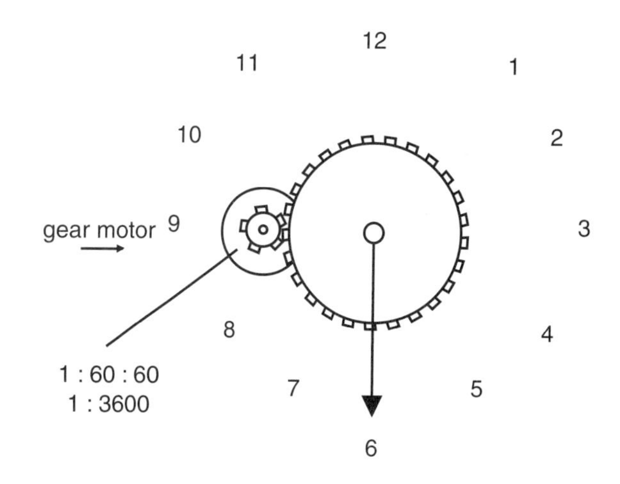


Figure 3.1.23 *The gearbox system to drive the indicators*

area before earthquakes or even tsunamis occur. It would be an interesting discovery if the Black Ghost could be used to detect such events. A simple circuit for detecting frequency changes in the fish's signal is shown in Figure 3.1.27.

The circuit is adjusted by P1 to tune the central frequency of the signal generated by the fish. When tuned, the transistor Q1 will not conduct, and a relay is then opened. P2 is used to adjust the sensitivity of the circuit and then the bandpass.

If the frequency changes, the output of the NE567 goes to the HI logic level and the transistor Q1 starts to conduct, closing the relay. An external circuit as an alarm can be activated at this moment.

## Other Fantastic Experiments and Ideas

Of course, the strange way the Itui Cavalo or Black Ghost "sees in the dark" using low-frequency current fields can be used in many new bionic devices. What can be done depends only on the imagination of the reader. Some ideas are given here.

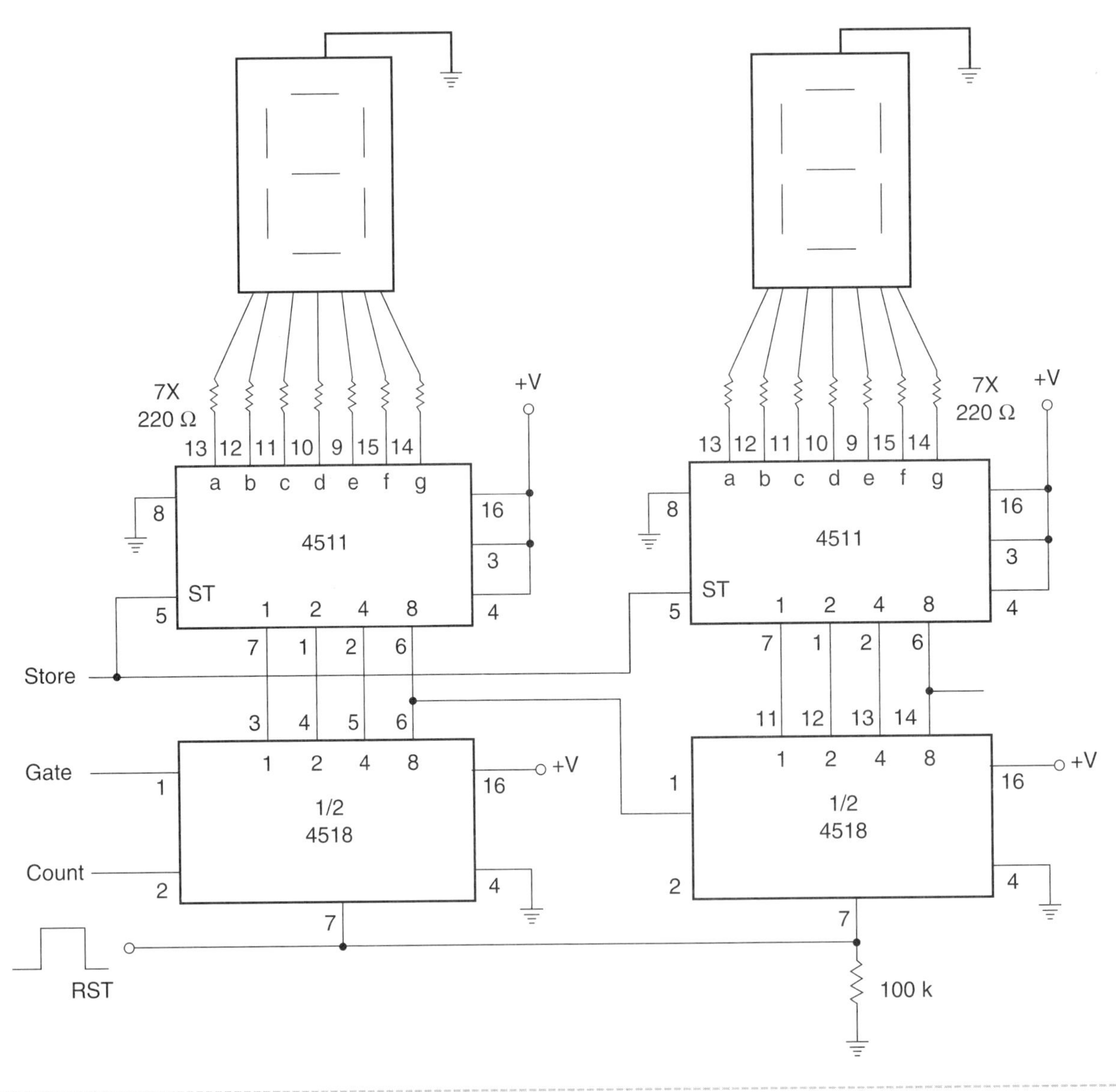


Figure 3.1.24 *CMOS digital counter to implement a digital clock*

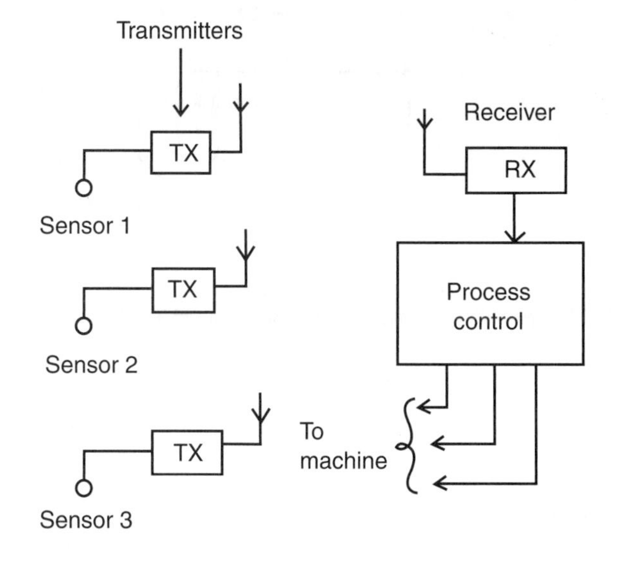


Figure 3.1.25 *The world is becoming wireless.*

## Domestic Alarm

As already discussed, the Black Ghost is a very sensitive creature. The presence of noise or light changes the fish's signal frequency. Placed in an aquarium with sensors plugged into an alarm, the fish can trigger it with the presence of intruders.

## Air Pollution Monitor

The presence of toxic or poisonous gases in an ambient area is easily transferred to the water within a large surface aquarium. If a Black Ghost is there, the frequency changes can be used to detect those substances.

As an example, in a semiconductor factory, a pet fish was kept in an aquarium where people worked with a very poisonous gas derived from fluorine. Since fish are much more sensitive than humans to the presence of such a gas, the workers knew that if the fish died, the gas had escaped and they would need to evacuate the area as quickly as possible.

The advantage of the electric fish is that no one would have to observe the fish to see if it is okay. The Black Ghost could trigger the alarm by itself!

## Radar for Blind

Based on the fish's system of "electric vision," the reader can create a system based on magnetic fields or sound fields to help blind people move around the objects in a home.

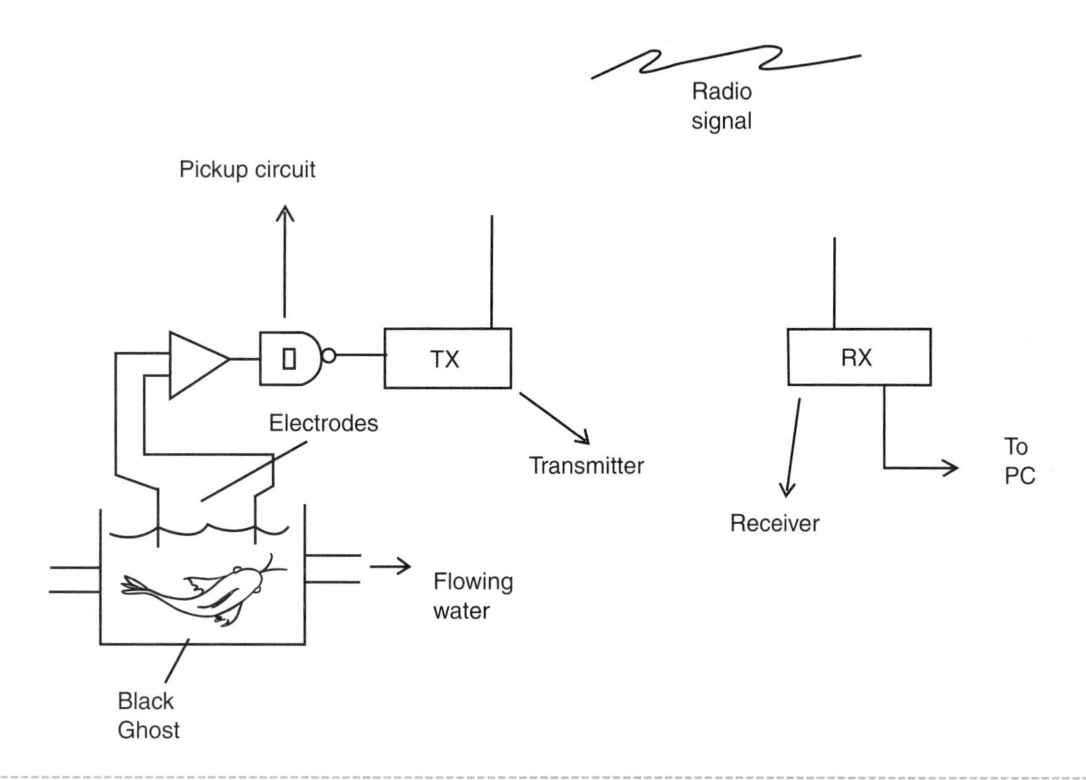


Figure 3.1.26 *Remote sensing using the fish as a wireless transducer*

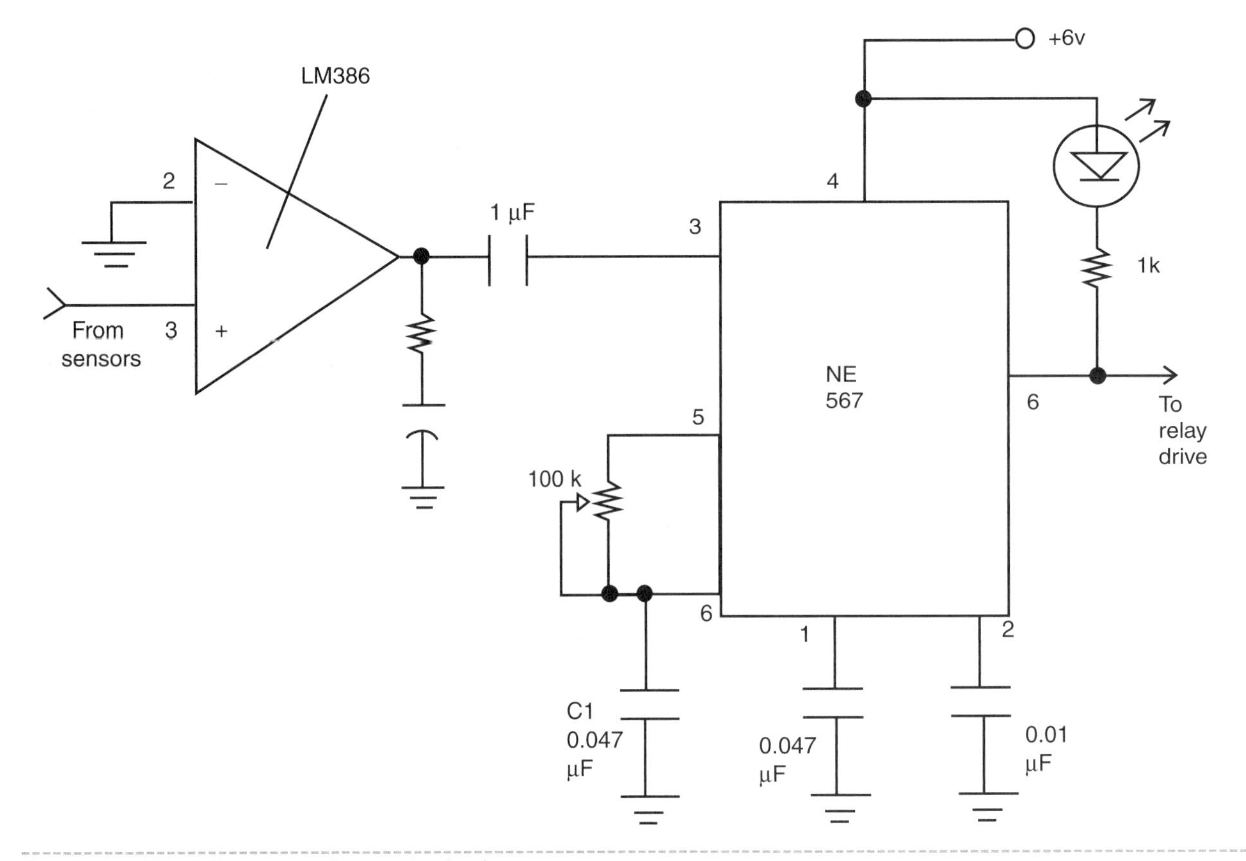


Figure 3.1.27 *Detecting changes in the frequency*

## External Influences Research

The extreme sensitivity of the Black Ghost as a sensor can be used to detect how external influences can affect its life. For instance, the use of light sources of different colors can be used to see how the fish alters its circadian rhythm.

Placing a coil in the aquarium and connecting it to low- or high-frequency circuits, the reader can study how magnetic fields alter the fish's behavior. A circuit for this task is shown in Figure 3.1.28.

## More About the Black Ghost

The current field of the Black Ghost is produced by a specialized organ called an *electric organ discharge* (EOD). The electric organ is made up of modified muscle or nerve cells placed in a series to gather the voltage produced by each muscle or nerve. Figure 3.1.29 shows how they are placed. For most electric fish, the EOD is located in the tail.

Strongly electric fish, such as the poraque (the electric eel), have an EOD that is powerful enough to stun their prey. The typical EOD amplitude for these animals can reach 300 volts.

A weak electric fish EOD produces less than 10 volts. The discharges are too weak to stun prey but are used to detect objects underwater in a process called *electrolocation*.

Another use of the signals is to communicate with other electric fish in a process called *electro-communication*.

In different species of electric fish, the EOD waveform takes two general forms. In some species, the signal is a continuous wave that is almost sinusoidal, as in the Dark Ghost and other members of the same family. In other species, the EOD waveform consists of brief pulses separated by long gaps, as shown in Figure 3.1.30. The fish with continuous wave operation are called *wave* species and the ones that produce pulses are known as *pulse* species.

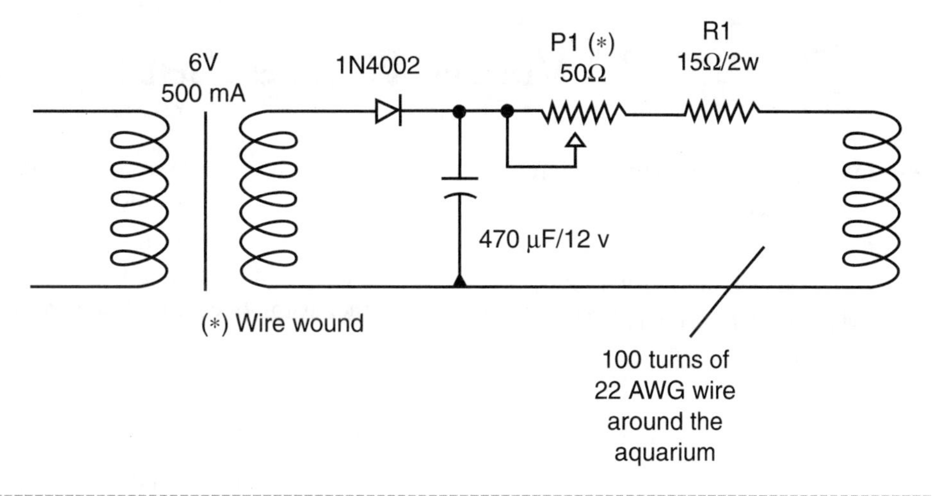


**Figure 3.1.28** *Studying the influence of magnetic fields on the fish's behavior*

## Keeping Your Black Ghost Alive

Tropical fish, such as the Black Ghost, are very delicate creatures. If the appropriate ambient environment is not provided, they can die in few minutes. In this final section, we will provide some idea of how to keep your fish alive for many years while powering your experiments and bionic devices.

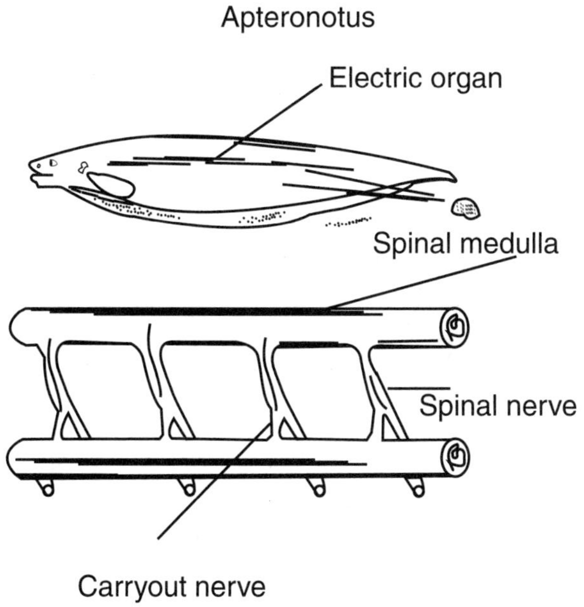

Apteronotus

Electric organ

Spinal medulla

Spinal nerve

Carryout nerve

**Figure 3.1.29** *The EOD of an electric fish*

## Feeding

The Black Ghost must be fed a varied diet, which includes live food, beefheart, shrimp pellets, snails, small fish, and worms. Tubifex and bloodworms are said to be a favorite.

They are big eaters, and if they are not properly nourished they may hunt other creatures in the aquarium. As always, live foods are beneficial, and Black Ghosts are very quick at detecting the smell of food.

## Water Chemistry

The water should be slightly soft with a pH of around 6.5 to 7.0 and a temperature range from 72 to 80 degrees F.

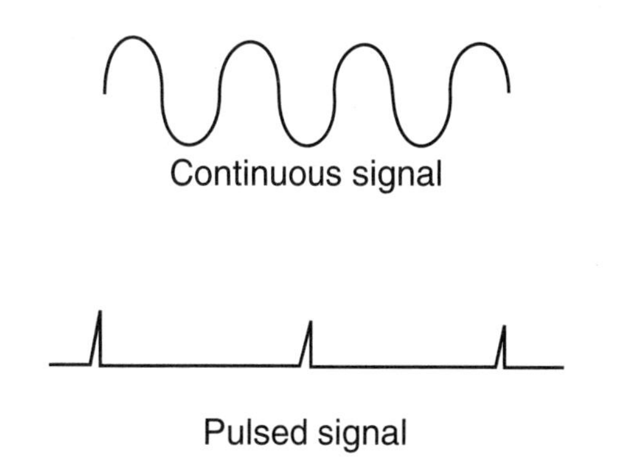


**Figure 3.1.30** *Waveforms for the signals generated by electric fish*

# Project 2—Visual Biofeedback

Many experiments in conditioning or animal behavior can be performed using electronic devices. The integration of animals and people with electronics is an important branch of bionics, and this integration can be made with devices using *biofeedback*. The objective of this project is to provide the reader with the resources for experimenting with biofeedback.

The aim of a biofeedback device is to use electronics to form a loop involving a living being and an electronic circuit in a manner that one can interact with the other, as shown in Figure 3.2.1. So, by controlling one's emotions or reactions, the living being can actuate the electronic circuit, which controls external devices or gives a response that can be observed by the researchers.

For this device, the term *visual* comes from the fact that the indications of the control of the specimen on the circuit comes in a visual form. A *light-emitting diode* (LED) pulse frequency depends on the current flowing across the loop formed by the living being and the electronic circuit. We also suggest that the reader see the next project (audio biofeedback) and take a brief look at the "Ideas to Explore" section suggested there.

This version is different from that audio project, as it gives a visual response because the electrodes are attached to the subject's fingers or hands. The subject presses on the electrodes to cause changes in the resistance sensed by the circuit. The two LEDs will flash in a rate determined by the operation frequencies of the two oscillators. One of the oscillators is adjusted to a fixed frequency, and the other's frequency depends on the pressure of one's fingers on the electrodes. The subject should control the pressure of his or her fingers or hands on the electrodes in oder to have a constant light pulse rate produced by the LEDs. The electrodes can also be placed on other parts such as the face, head, arms, or legs, according to the experiment the reader wants to conduct.

Many experiments can be performed using this circuit:

- Test your capacity to control the circuit using your mind. For example, changes in the mind can cause changes in the electric resistance of the skin in many parts of the body. This, in fact, is the operating principle of the electroencefalograph.
- Test the changes in the subject's ability to control the LED when the subject is under stress.
- Make collective experiments with many persons controlling the same circuit at the same time.
- Test if a living being can alter the flash rate under conditions determined in an experiment.

## How It Works

Most of the biofeedback circuits use changes in skin resistance to alter their operational characteristics. The changes can alter the frequency of a signal, change the intensity of a lamp, or trigger some kind of alarm. This circuit uses the skin resistance to control the frequency of a low-frequency oscillator driving the LEDs.

When the resistance is low, the time constant of the circuit is reduced and its frequency increases by

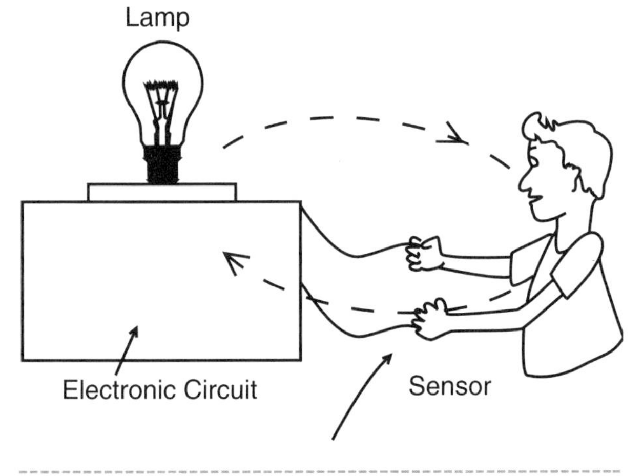


Figure 3.2.1 *The operation principle of biofeedback*

the same amount. This means that small changes in the resistance will be translated into small changes of the LEDs' flash rate.

To show this, we use a 4093 *complementary metal oxide semiconductor* (CMOS) *integrated circuit* (IC) formed by four NAND Schmitt triggers as the start point. Two of its gates are used as oscillators, and the other two are used as digital buffers driving two LEDs. One oscillator produces signals in a frequency given by C1 and adjusted by P1. The other oscillator generates a signal with a frequency determined by C2 and the resistance between the electrodes.

Observe that one LED lights up when the output of the buffer IC1-b (pins 5, 6, and 7) goes high and IC2-c (pins 8, 9, and 10) goes low, and the other LED lights up at the inverse condition. The following table outlines these conditions:

| Pin 4 | Pin 10 | LED1 | LED2 |
|---|---|---|---|
| Lo | Lo | Off | Off |
| Lo | Hi | On | Off |
| Hi | Lo | Off | On |
| Hi | Hi | Off | Off |

Figure 3.2.2 shows the schematic diagram of the visual biofeedback. The circuit is mounted on a *printed circuit board* (PCB), as shown in Figure 3.2.3.

C1 and C3 are electrolytic capacitors rated to 12 *working voltage DC* (WVDC) or more, and their position should be observed, as they are polarized components, like the LEDs and the power supply. The LEDs should be red and green.

The circuit can be powered from supplies ranging from 6 to 9 volts, and R2 depends on the supply voltage. For a 6-volt supply, use a 1,000 Ω resistor and for a 9-volt supply use a 1,500 Ω resistor.

The electrodes are small metal rods. The subject must press the rods and observe the effect on the LED frequency. Figure 3.2.4 shows how the electrodes are built. Other types of electrodes can be created by the reader according to the experiments to be performed.

Figure 3.2.5 shows how to use two electrodes inserted into a plant pot to observe the frequency of the circuit when there are differences in light, stress, noise, or other external influences.

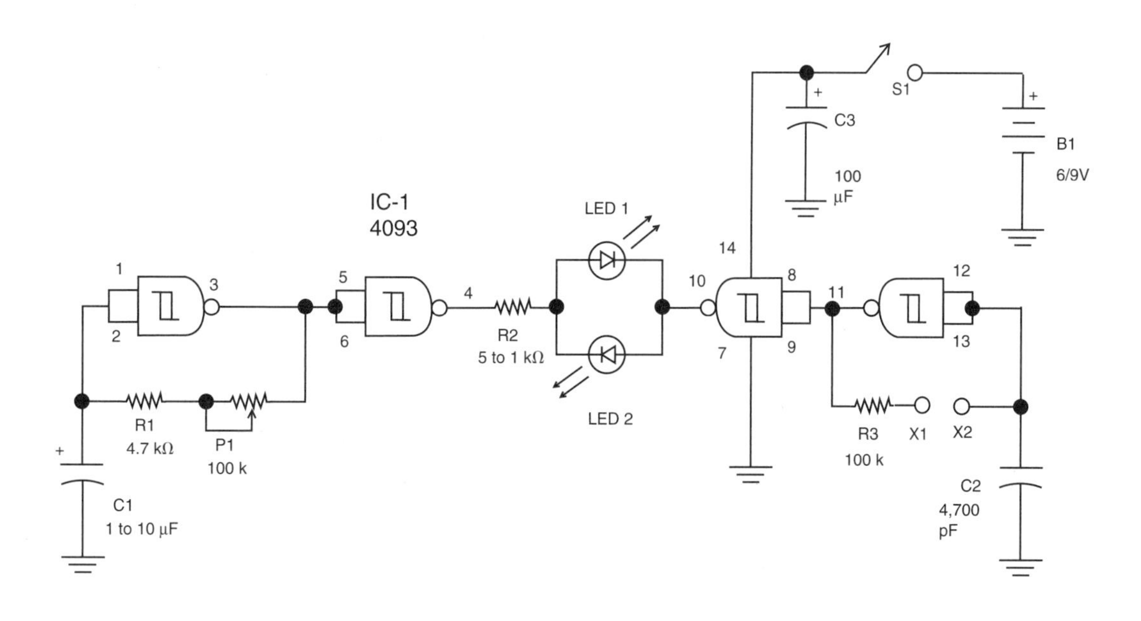


Figure 3.2.2 *Schematic diagram for the visual feedback*

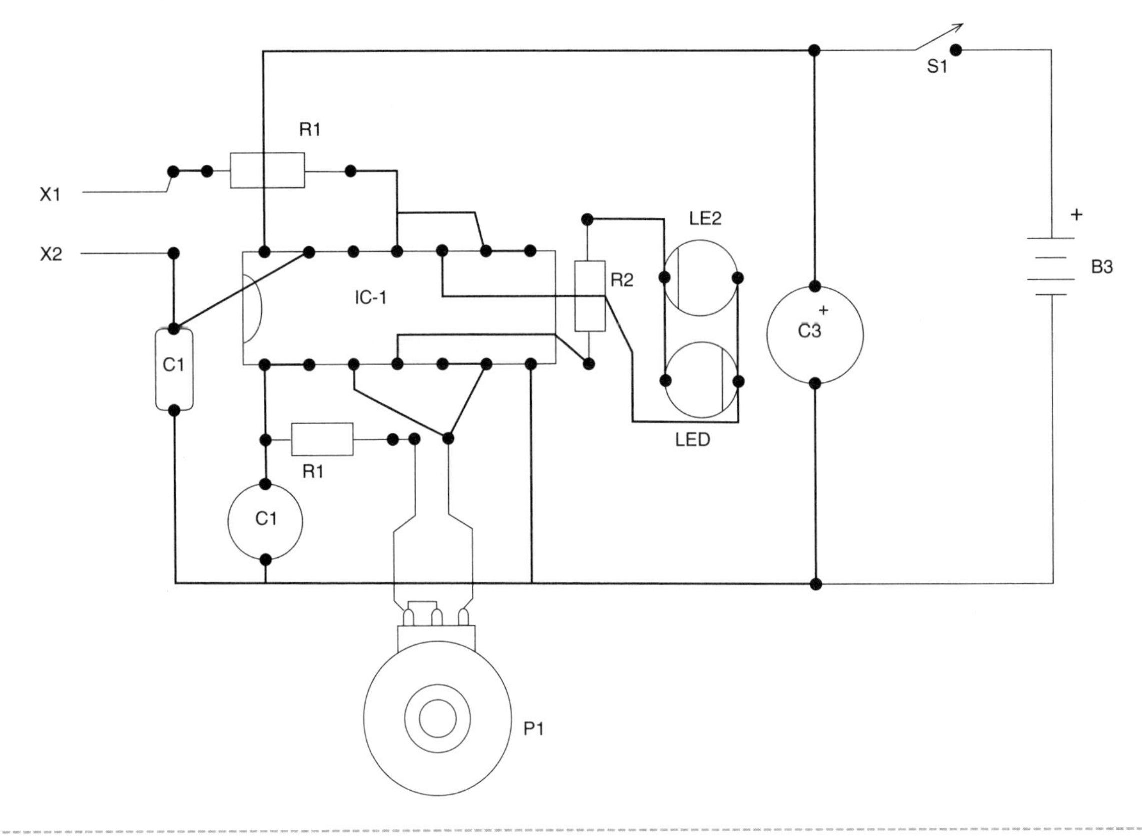


Figure 3.2.3 *PCB for the visual feedback*

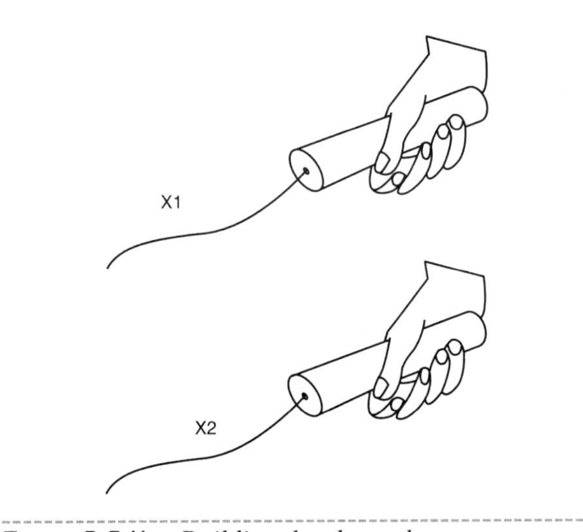


Figure 3.2.4 *Building the electrodes*

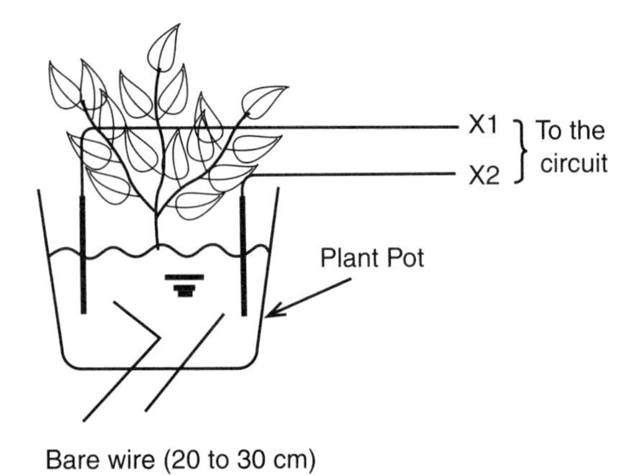


Figure 3.2.5 *Performing experiments with a plant*

## Testing and Using

Turn the power on using S1 and adjust P1 to a flash rate of about one or two flashes per second (1 or 2 Hz). This adjustment should be made with the electrodes in your hands but without one touching the other. After this, try to control the electrodes' flash rate, one in each hand, by changing the pressure on them.

## Parts List: Visual Biofeedback

- IC1: 4093 CMOS integrated circuit
- LED1, LED: Common red and green LEDs
- R1: 4.7 kΩ, 1/4-watt, 5% resistor, yellow, violet, red
- R2: 1 kΩ, 1/4-watt, 5% resistor, brown, black, red
- R3: 100 kΩ, 1/4-watt, 5% resistor, brown, black, yellow
- P1: 100,000 Ω potentiometer
- C1: 1 to 10 μF/12 WVDC electrolytic capacitor
- C2: 4,700 pF to 0.47 μF ceramic or metal film capacitor
- C3: 100 μF/12 WVDC electrolytic capacitor
- S1: Single pole, single throw (SPST) toggle or slide switch
- B1: Four 6 to 9 V AA cells or battery
- X1, X2: Electrodes (see text)
- Other: PCB, cell holder, plastic box, wires, solder, etc.

# Additional Circuits and Ideas

A flashing lamp can be controlled from the skin resistance monitored by sensors in many different ways. Besides the one described in the basic version, others are given here, using different techniques and components.

## High-Power Visual Biofeedback Using the 555

Figure 3.2.6 shows visual feedback that can drive a small 6- to 12-volt lamp, much more powerful than the LEDs of the basic version.

This circuit uses a 555 IC configured as an astable multivibrator. The frequency is controlled by the resistance between the positive rail of the power supply and pin 7.

P1 adjusts the basic frequency, and Q1 is used as a variable resistor that changes the resistance according to the amount of current flowing across the base.

Because the base current depends on the resistance between the electrodes (X1 and X2) and the resistance is given by the skin of the subject, it is clear that the frequency at which the lamp flashes depends on the amount of feedback of the person connected between the electrodes. The circuit is very sensitive due to the presence of the transistor, and even resistances of many megohms will cause the lamp to flash.

Of course, it is not just people that can be connected to the electrodes. Animals, plants, and even aquariums can be used.

To control the sensitivity, a 4.7 MΩ potentiometer can be wired between the base of Q1 and the 0-volt rail of the circuit. C1 determines the center frequency of the oscillator and can be changed according to the application. The reader is free to experiment with values between 0.1 and 10 μF.

The circuit can be powered from cells or a power supply with voltages between 6 and 12 volts. The voltage of the supply will determine the voltage of the lamp. Lamps with currents between 50 and 300 milliamps can be used. If lamps with more than 100 milliamps are used, the transistor Q2 must be mounted on a heatsink. For small lamps (up to 100 milliamps), AA cells can be used. For higher-power lamps, if cells are used, they must be C or D types.

If a power supply is used, it must include a transformer. Do not use transformerless power supplies, because they are not isolated from the AC power line and the electrodes could cause severe shock hazards.

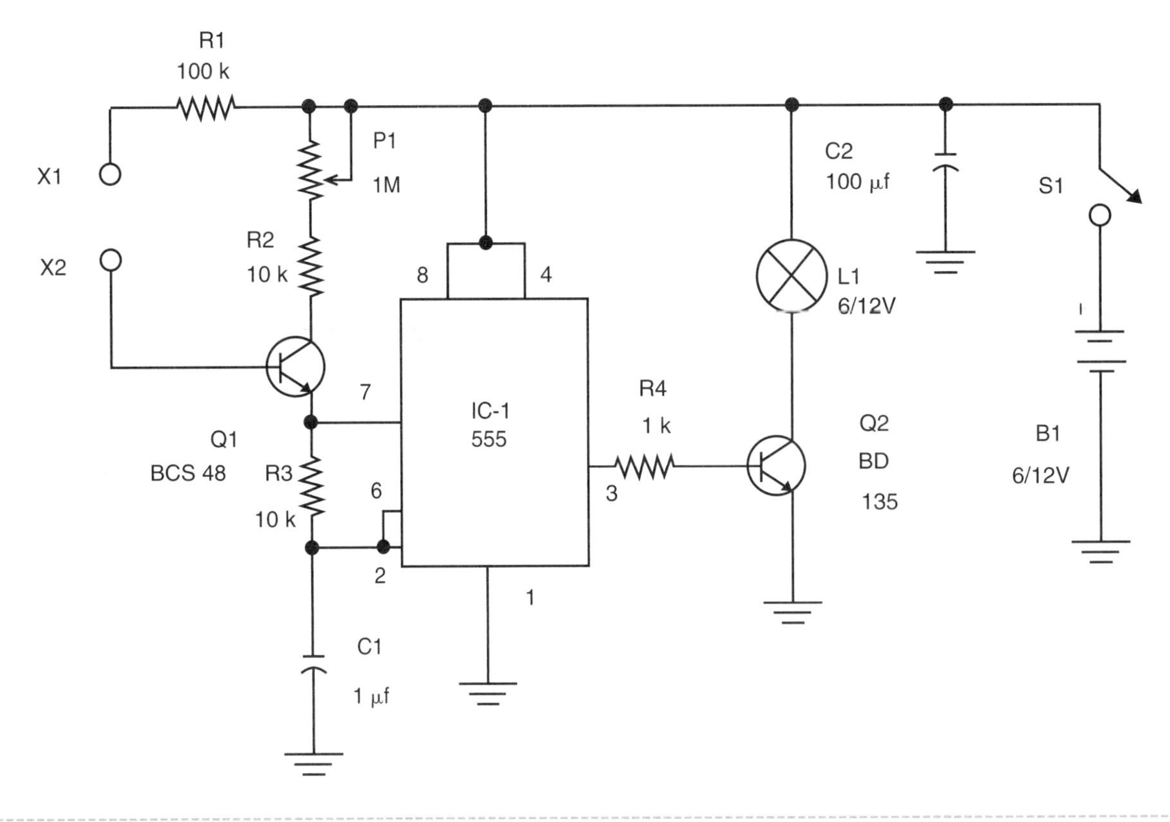


Figure 3.2.6 *Driving a lamp with a 555 IC for visual biofeedback*

The circuit can be mounted using a PCB or a solderless board. The electrodes from the basic project can be used, depending on the application.

A Darlington power transistor, such as the TIP120, can be used, replacing Q2. It is enough to increase R4 to 4,7 kΩ. This transistor can drive lamps up to 4 amps from a 12-volt supply.

## Testing and Using

Power on the circuit and observe the lamp. If the electrodes are separated, the lamp will not flash.

Placing your fingers on the two electrodes at the same time and acting on P1, the lamp will flash. You will see that the flash rate will change when you press on the electrodes.

## Parts List

IC1: 555 integrated circuit

Q1: BC548 or the equivalent NPN general-purpose silicon transistor

Q2: BD135 or equivalent NPN medium-power silicon transistor

X1, X2: Electrodes

L1: Incandescent lamp, 6 or 12 V, up to 300 mA

R1: 100 kΩ x 1/8 W resistor, brown, black, yellow

R2, R3: 10 kΩ x 1/8 W resistor, brown, black, orange

R4: 1 kΩ x 1/8 W resistor, brown, black, red

P1: 1 MΩ potentiometer, lin or log

C1: 1 μF electrolytic or polyester capacitor (see text)

```
C2: 100 µF x 12 V elec-
  trolytic capacitor
S1: On/off switch
B1: 6 or 12 V power sup-
  ply or cells (see
  text)
Other: PCB or solderless
  board, cell holder
  (optional), plastic
  box, wires, solder,
  etc.
```

## Using a Fluorescent Lamp

The next version of visual biofeedback will light up a small fluorescent lamp even when powered by common 6-volt cells (C or D). The effect obtained from the circuit shown in Figure 3.2.7 is quite different from the previous versions.

The lamp will flash in an intermittent mode with a stroboscopic effect. The frequency of the intermittence depends on what will be changed by the subject when pressing the electrodes. Figure 3.2.8 shows the waveshapes in different conditions for the circuit.

The circuit uses the 4093 IC as its base, as in the original basic version. But in this configuration two gates are used as oscillators.

One of them has its frequency dependent on the amount of resistance between the electrodes X1 and X2, that is, the resistance of the skin of the subject, if it is a person. This oscillator, running at a very low frequency (between 0.1 and 5 Hz) controls a second oscillator running also at a low frequency but a little higher, between 5 and 10 Hz.

The result is an intermittent signal applied to the base of a medium-power NPN transistor, as shown in Figure 3.2.8. This transistor has as a load the low-voltage coil of a transformer. The secondary is coupled to a small, fluorescent lamp. Lamps in the range from 4 to 20 watts can be used. Since the power generated by the circuit is only a few watts, the lamps will not light up with all their power.

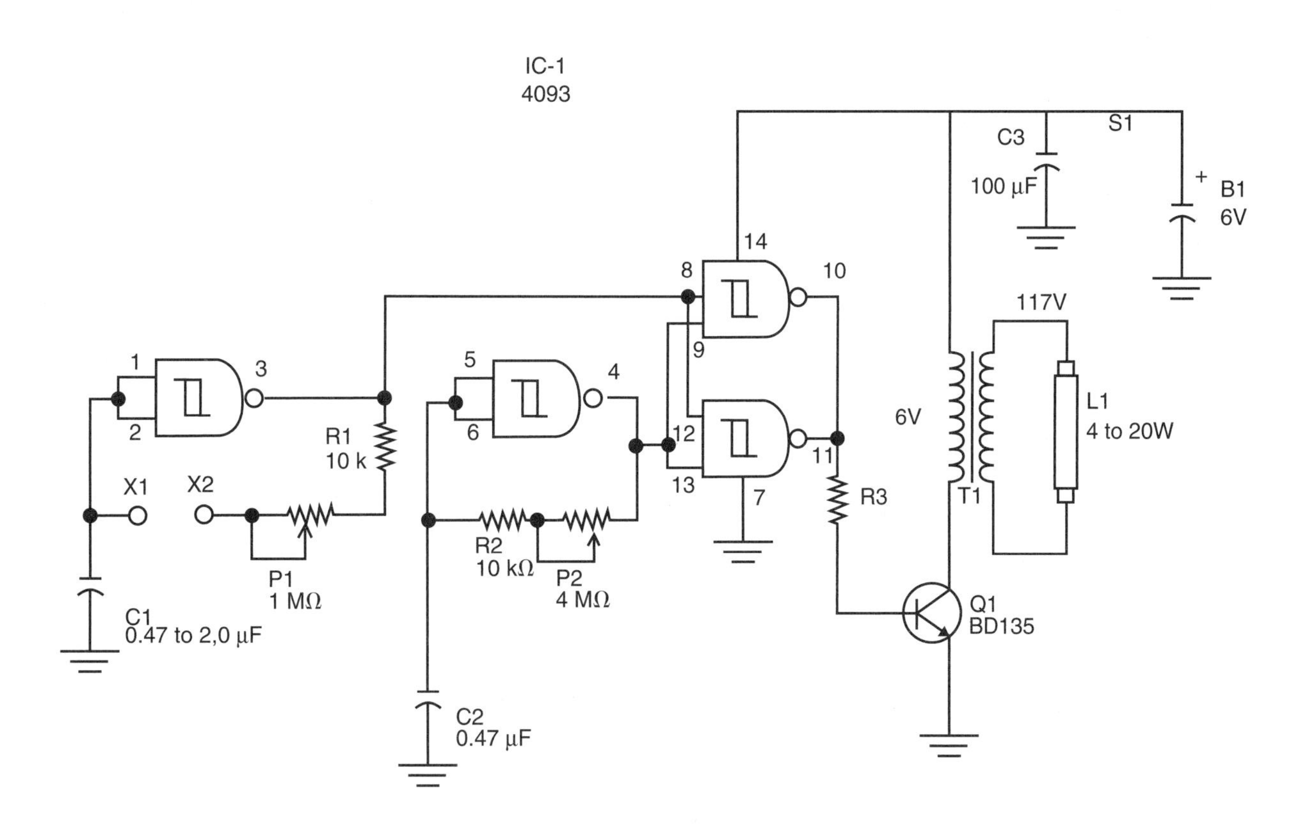


Figure 3.2.7 *Biofeedback using a fluorescent lamp*

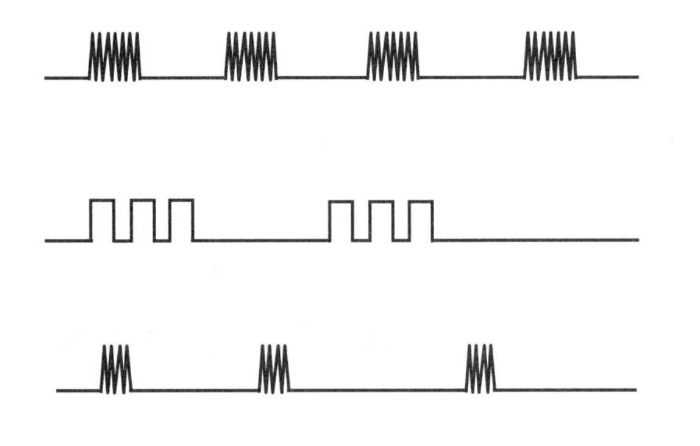

Figure 3.2.8 *The waveshapes of the voltage applied to the fluorescent lamp*

The second oscillator is important not only for preventing a stroboscopic effect from the light produced by the lamp but also because it is necessary for the pulsed current to induce high voltage in the secondary of the transformer.

As in the previous versions, do not power the circuit from transformerless power supplies. The circuit can be mounted on a PCB or solderless board. Q1 must be mounted on a heatsink.

T1 is any power transformer with a primary for the AC power line (117 VAC or 220/240 VAC) and a secondary of 6 volts with currents in the range from 200 to 500 milliamps. If you want more power to the lamp flashes, power the circuit from a 12-volt supply and replace Q1 by a TIP31.

## Testing and Using

Power the circuit on and place your fingers on the electrodes. The electrodes must be separated, but you must touch them at the same time. The flash rate of the lamp will change as you press and release your fingers, changing the pressure on the electrodes.

C1 and C2 can change the frequency of both the stroboscopic effect and the intermittence. Try some experiments with these components according to the applications.

A sensitivity control can also be added. It is a 4.7 MΩ potentiometer between pin 1 and 2 of the IC and the 0 volt line. The electrodes are the same as in the basic project.

## Parts List

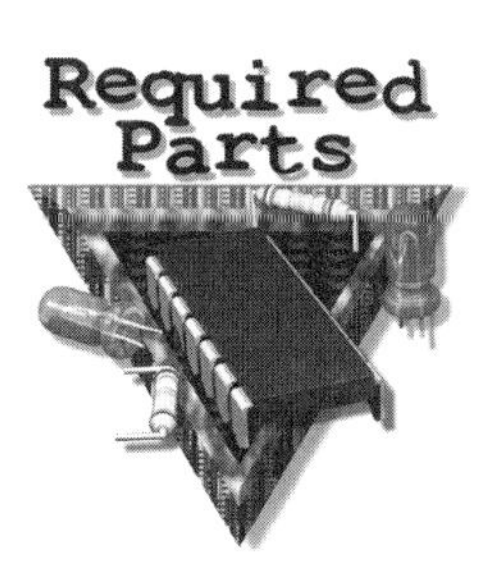


- IC1: 4093 CMOS integrated circuit
- Q1: BD135 NPN medium-power silicon transistor
- R1: 10 kΩ x 1/8 W resistor, brown, black, orange
- R2: 100 kΩ x 1/8 W resistor, brown, black, yellow
- R3: 1 kΩ x 1/8 W resistor, brown, black, red
- P1, P2: 1 MΩ potentiometers, lin or log
- C1: 0.47 or 1 μF polyester or electrolytic capacitor
- C2: 0.047 ceramic or polyester capacitor
- C3: 100 μF x 12 V electrolytic capacitor
- T1: Transformer (see text)
- L1: Any 4 to 20 W fluorescent lamp
- S1: On/off switch
- B1: 6 V power supply or cells
- X1, X2: Electrodes (see text)
- Other: PCB or solderless board, heatsink for Q1, cell holder or power supply, plastic box, wires, solder, etc.

## Sequential Biofeedback

A sophisticated biofeedback can be mounted using a sequence of 10 LEDs, as shown by the circuit in Figure 3.2.9.

The LEDs will run in sequence at a speed determined by the resistance of the subject connected to the electrodes. If the subject is a person, pressing one's fingers harder or becoming mentally stressed can possibly control the speed of the LEDs.

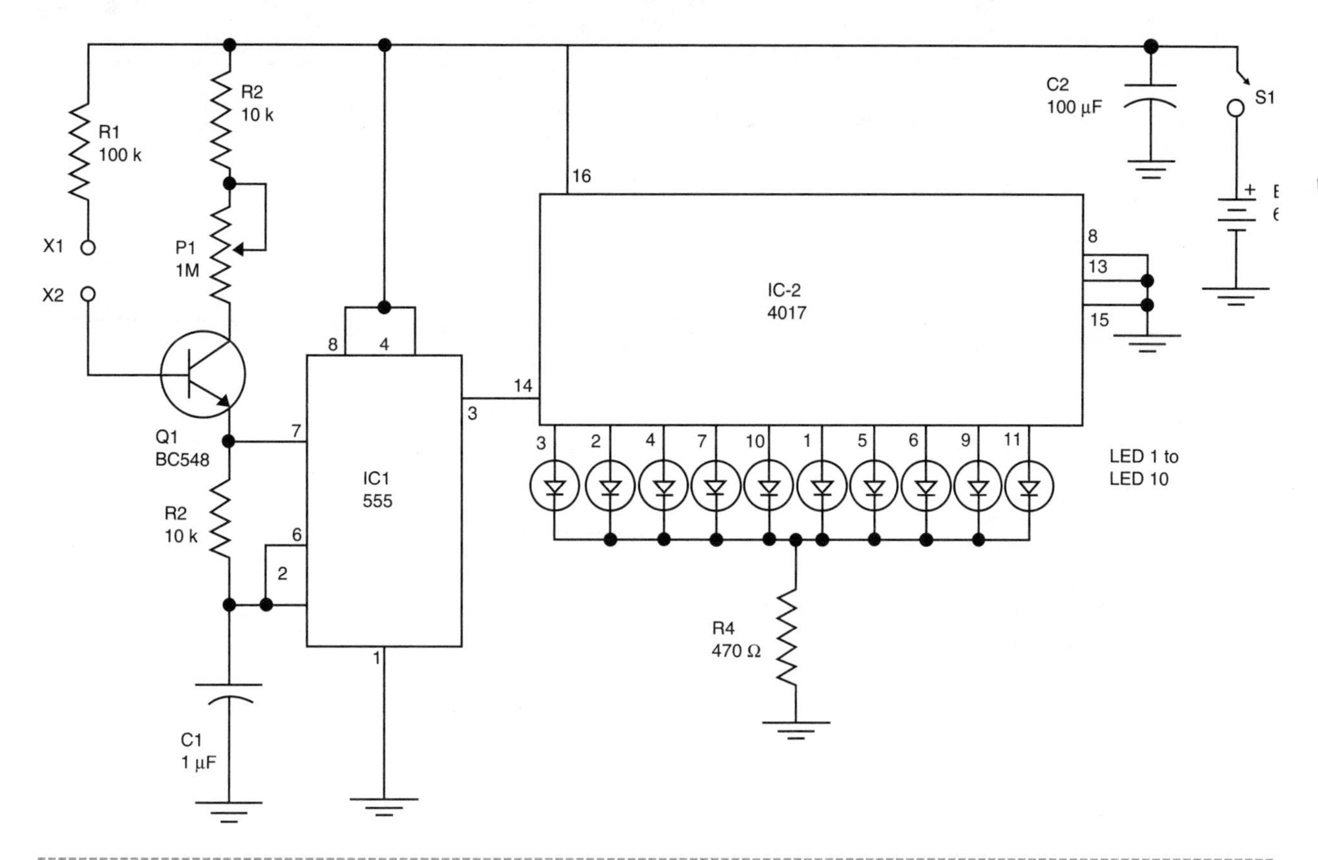


Figure 3.2.9 *Sequential biofeedback*

The basic speed of the LEDs is determined by C1. According to the experiment, this component can be changed. Any values between 0.1 and 1 µF can be used.

P1 adjusts the speed according to the resistance of the subject, but an additional sensitivity control can be added. It consists of a 4.7 MΩ connected between the base of Q1 and the 0-volt supply line.

The circuit can be mounted on a PCB or a solderless board for experimental purposes.

## Testing and Using

Power the circuit on and touch the electrodes with your fingers. The electrodes are the same as those used in the basic project.

Adjust P1 to the desired speed of the running LEDs. Check to see that pressing the electrodes will alter the speed. The device is now ready for use.

## Parts List

- IC-1: 555 integrated circuit timer
- IC-2: 4017 CMOS integrated circuit
- Q1: BC548 or equivalent NPN general-purpose silicon transistor
- LED1 to LED10: Common red LEDs (or other color)
- R1: 100 kΩ x 1/8 W resistor, brown, black, yellow
- R2, R3: 10 kΩ x 1/8 W resistors, brown, black, orange
- R4: 470 Ω x 1/8 W resistor, yellow, violet, brown
- P1: 1 MΩ potentiometer, lin or log
- C1: 1 µF polyester or electrolytic capacitor

C2: 100 μF x 12 V electrolytic capacitor

X1, X2: Electrodes (see text)

S1: On/off switch

B1: Four 6 V AA cells and holder

Other: PCB or solderless board, knob for P1, wires, solder, plastic box, etc.

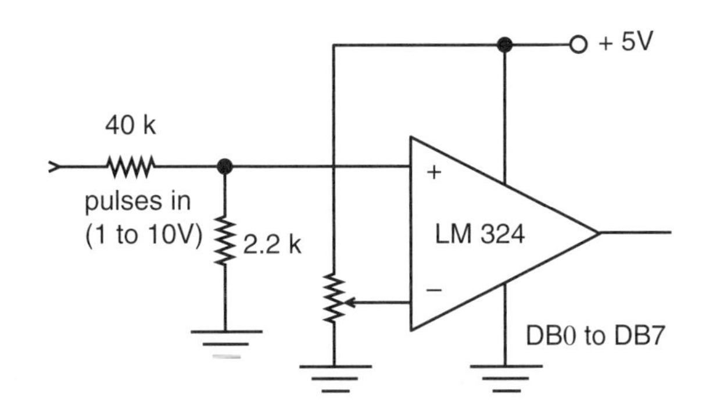


Figure 3.2.10 *Acquisition interface for connecting a living being with a computer*

## Other Ideas

Many other circuits can be controlled by the resistance of the skin or even electric signals generated by a living being. Some suggestions for experiments or other projects are given in the following sections.

## Using the Computer

Figure 3.2.10 shows an acquisition interface that can be used to transfer pulses to the parallel port of a computer. The pulses can be measured by software and displayed onscreen as visual effects (colors, shapes, etc.). It is up to the reader to create the software for this task.

## Other Types of Sensors

Other types of sensors can be used, replacing the metal plates in X1 and X2. Figure 3.2.11 shows how a *light-dependent resistor* (LDR) can be used to sense changes in the transparence of one's fingers. The amount of light passing across the fingers depends on the blood pressure and the heartbeat.

Adjusting the trigger point of the circuits (any version) can implement a circuit controlled by the heartbeat. This configuration can also be used with animals and plants.

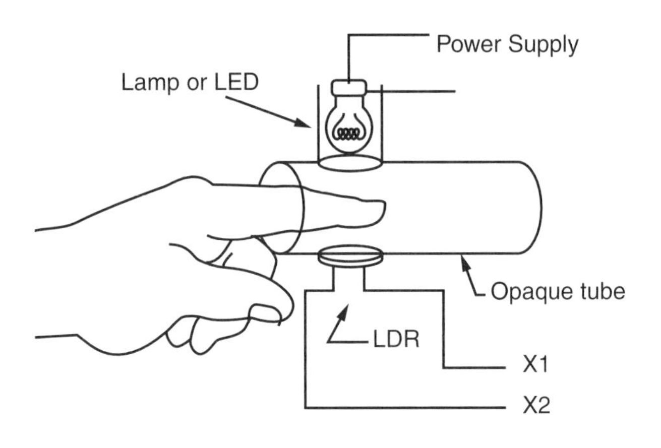


Figure 3.2.11 *Using a LDR as sensor*

Another kind of sensor is a diode placed to monitor temperature changes, as shown in Figure 3.2.12.

Any common general-purpose diode, such as the 1N4148 or 1N914, can be used as a temperature sensor.

## Powering a High-Power AC Lamp

Since high-power incandescent lamps are powered from the AC power line, they would be dangerous when used with the circuits described here. To perform experiments using high-voltage lamps, an isolating stage must be added to the circuits.

Figure 3.2.13 shows how to trigger an SCR from the pulses produced by the circuits described in this chapter.

Any incandescent lamp up to 100 watts can be used. The trigger point is adjusted by P1.

The SCR must be mounted on a heatsink. Don't touch any part of the circuit wired to the AC power line.

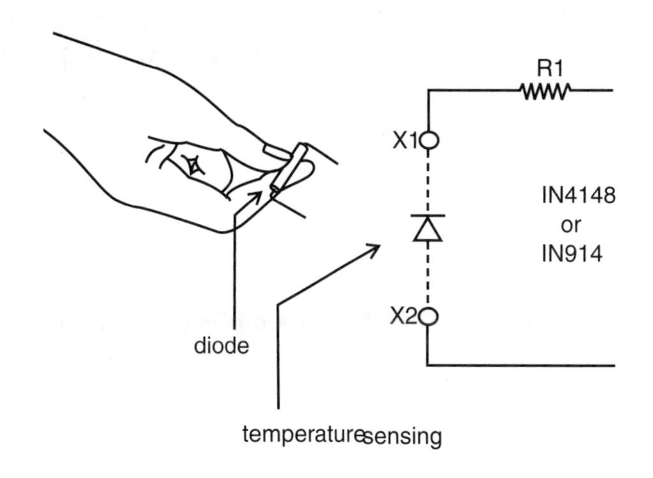


Figure 3.2.12 *Using a diode as a temperature sensor*

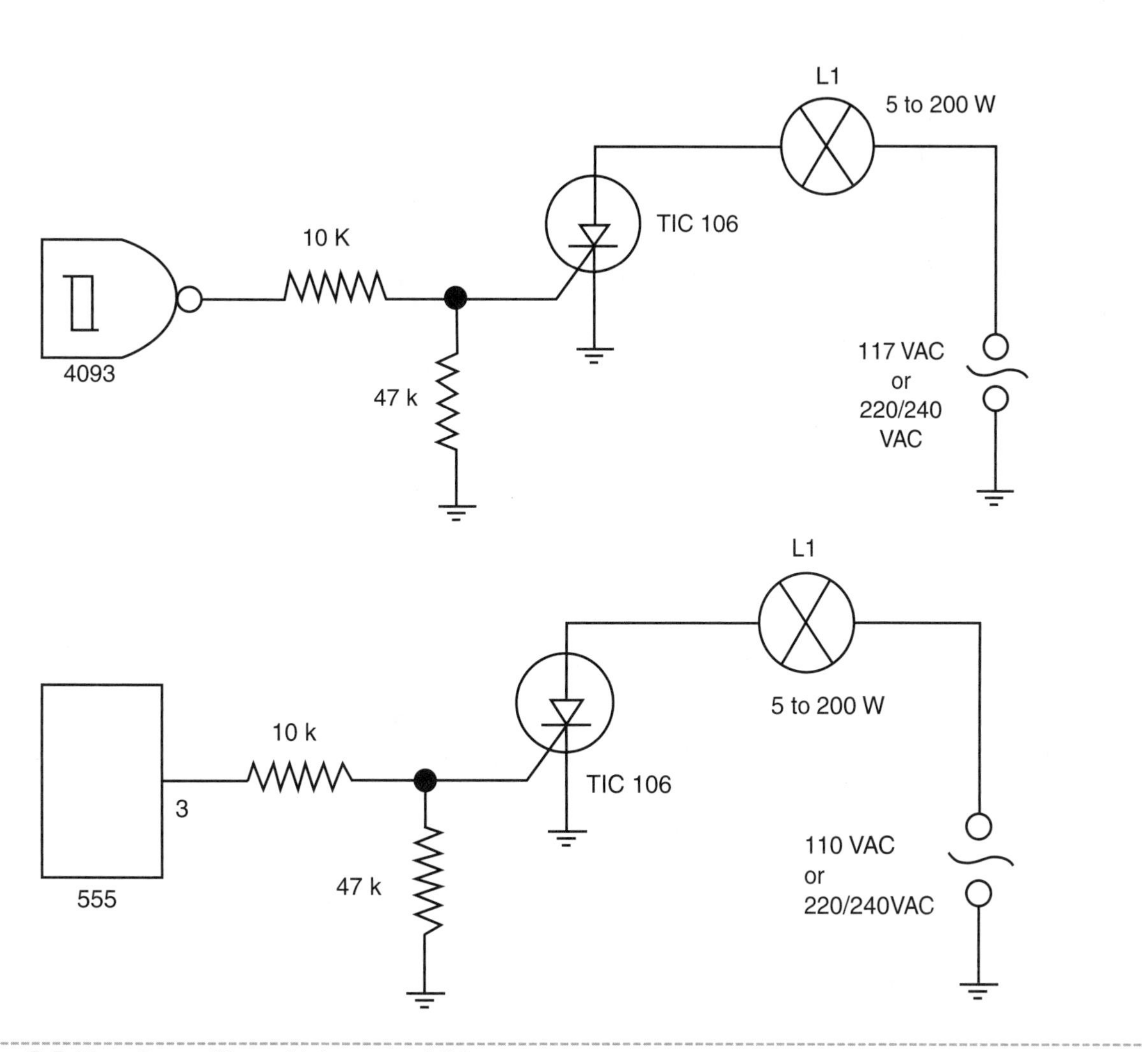


Figure 3.2.13 *Controlling a high-power AC lamp*

# Project 3—Audio Biofeedback

Interaction between living beings and electronic equipment can easily be performed with the aid of devices that use the biofeedback principle. As we saw in the previous project on visual biofeedback, experiments in human or animal conditioning can also be conducted using audio biofeedback.

In this project, the subject will touch the electrodes with his or her fingers, controlling the pressure of contact in order to make the circuit generate a tone with a desired pitch or to cancel a tone produced by the circuit.

Any change to the pressure exerted on the electrodes will make the oscillator alter its frequency. Thus, the subject can vary the tone pitch, controlling the strength of the electrodes in this biofeedback process.

Biofeedback means that the control of an electronic circuit is made by a loop that includes the subject's body, which could be a human being's body. It is a *biologic* or *bio* loop, carrying the control signal back to the input.

## How It Works

The circuit has the same configuration as in the previous project. It is formed by two audio oscillators, and each one of them is made from one of the four gates of a 4093 *complementary metal oxide semiconductor* (CMOS) *integrated circuit* (IC). The 4093 consists of four NAND two-input Schmitt gates that can be used in their original function or as inverters or digital buffers.

The first oscillator operates at a frequency determined by C1 and adjusted by P1, and it drives a piezoelectric transducer (BZ) via the digital buffer formed by one of the four gates at the IC. The second oscillator operates at a frequency determined by C2 and the body of the subject, which is pressing on the electrodes. As the amount of pressure on the electrodes changes, so does the resistance and the frequency generated by the oscillator, as we have explained previously. This second oscillator also drives the same output transducer via a digital buffer formed by one of the four gates in the 4093 IC.

If the frequencies of the generated signals are the same, no sound is produced by the transducer (beat zero), because the signals are canceled. This is the *point of balance* or *zero point* of the subject, and it can be reached by controlling the amount of pressure on the electrodes.

## How to Build

In Figure 3.3.1, we have the schematic diagram of the audio biofeedback. The experimental version can be mounted on a solderless board, as shown in Figure 3.3.2.

The circuit is powered from four AA cells or a 9-volt battery. The electrodes are made with two metal rods as in the previous project on visual biofeedback. Do not use a transformerless power supply.

The transducer BZ is any piezoelectric type or a piezoelectric tweeter without the internal transformer that should be removed.

The circuit can easily be housed in a small plastic box with two pieces of wire, 30 or more inches long, that are connected to the electrodes via X1 and X2. X1 and X2 are optional as you can wire the electrodes directly to the circuit or use terminals.

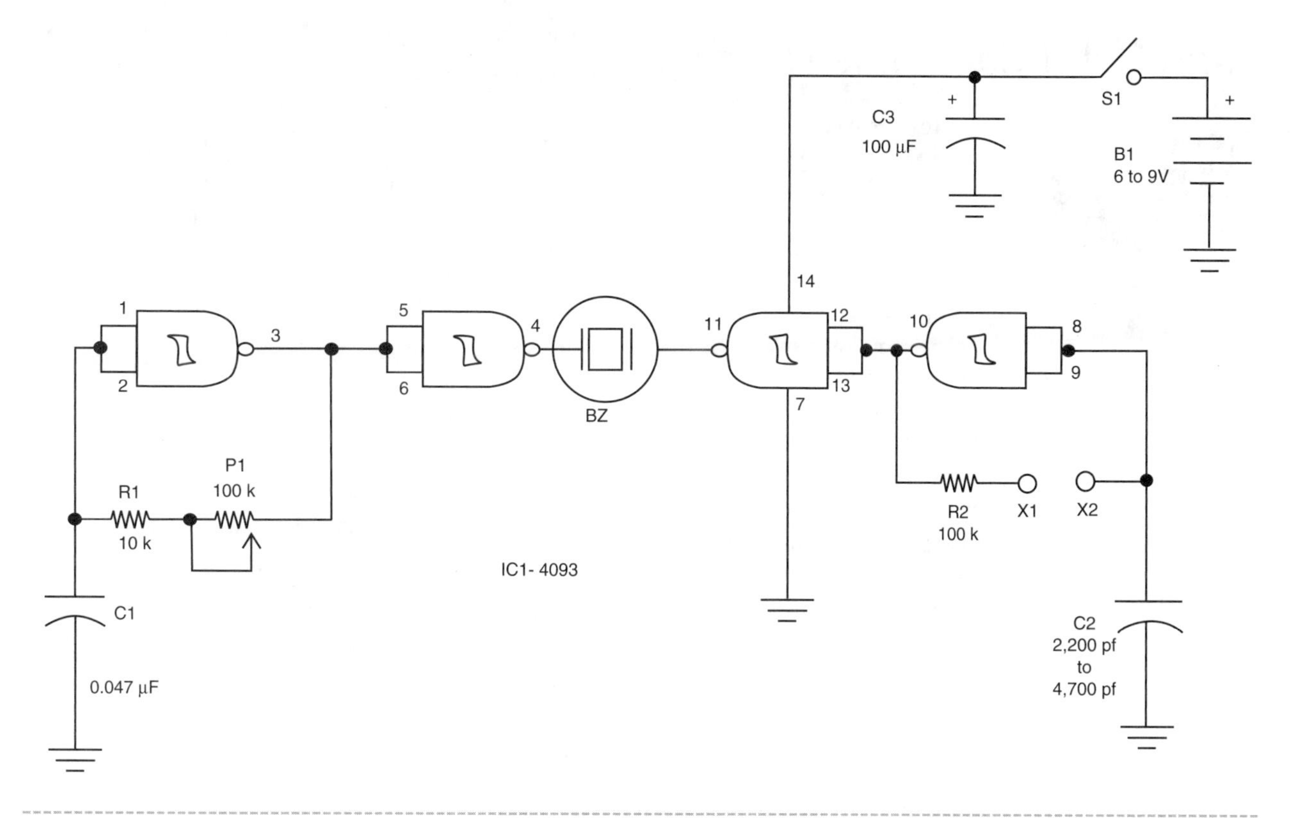


Figure 3.3.1 *Schematic diagram for the audio biofeedback*

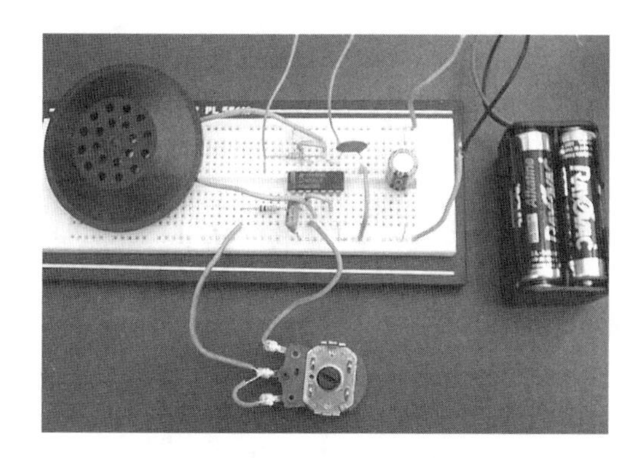

Figure 3.3.2 *The circuit mounted on a solderless board*

## Testing and Using

Apply pressure to both the electrodes at the same time in one hand and adjust P1 to a low-frequency tone using the other hand. Then grasp the electrodes, one in each hand, and try to cancel the tone or keep the pitch constant. When using the device, the experimenter should control the tone by pressing the electrodes. Experiments with animals or plants can also be made, connecting the electrodes to different subjects.

## Parts List: Audio Biofeedback

- IC1: 4093 CMOS integrated circuit, four Schmitt NAND gates
- R1: 10 kΩ x 1/8 W - resistor, brown, black, orange
- R2: 100 kΩ x 1/8 W resistor, brown, black, yellow
- P1: 100 kΩ potentiometer, lin or log
- C1: 0.047 μF ceramic or metal film capacitor
- C2: 2,200 to 4,700 pF ceramic capacitor
- C3: 100 μF x 12 V electrolytic capacitor
- S1: Single pole, single throw (SPST) toggle or slide switch (on/off switch)
- B1: 6 to 9 V AA cells or battery
- BZ: Piezoelectric transducer (see text)

## Additional Circuits and Ideas

Many different methods can be used to control the tone produced from an oscillator detecting skin resistance, from monitoring sensors, or from electrodes used with animals and plants. Besides the one described in the basic version, others are given in this section, using different techniques and components.

## High-Power Audio Biofeedback Using the 555

Figure 3.3.3 shows an audio feedback setup that can drive a small loudspeaker, which is much more powerful than the piezoelectric transducer used in the basic version.

This circuit is formed by only one oscillator, so the subject (the person or animal) must control the tone generated by the circuit, not cancel it. This circuit uses a 555 IC configured as an astable multivibrator. The frequency is controlled by the resistance between the positive rail of the power supply and pin 7.

P1 adjusts the basic frequency, and Q1 is used as a variable resistor that changes the resistance according to the amount of base current. Because the base current depends on the resistance between the electrodes (X1 and X2) and it is given by the skin of the subject, it is clear that the frequency reproduced by the loudspeaker depends on the amount of feedback of the person connected between the electrodes.

The circuit is very sensitive due to the presence of the transistor, and even resistances of many megohms will cause the circuit to produce audible sounds or clicks. Of course, not just people can be connected to the electrodes. Animals, plants, and even aquariums can be used.

To control the sensitivity, a 4.7 MΩ potentiometer can be wired between the base of Q1 and the 0-volt rail of the circuit.

C1 determines the center frequency of the oscillator and can be changed according to the application. The reader is free to experiment using values between 0.470 and 100 μF.

The circuit can be powered from cells or a power supply with voltages between 6 and 12 volts. When using power supplies above 9 volts, the transistor Q2 must be mounted on a heatsink. If a power supply is used, it must have a transformer. Do not use transformerless power supplies because they are not isolated from the AC power line, and the electrodes could cause severe shock hazards.

The circuit can be mounted using a *printed circuit board* (PCB) or a solderless board. The electrodes are the same as in the basic project, depending on the application.

A Darlington power transistor, such as the TIP120, can be used, replacing Q2. It is capable of handling a high-power loudspeaker. For that reason, it is necessary to increase R4 to 4.7 k ohms.

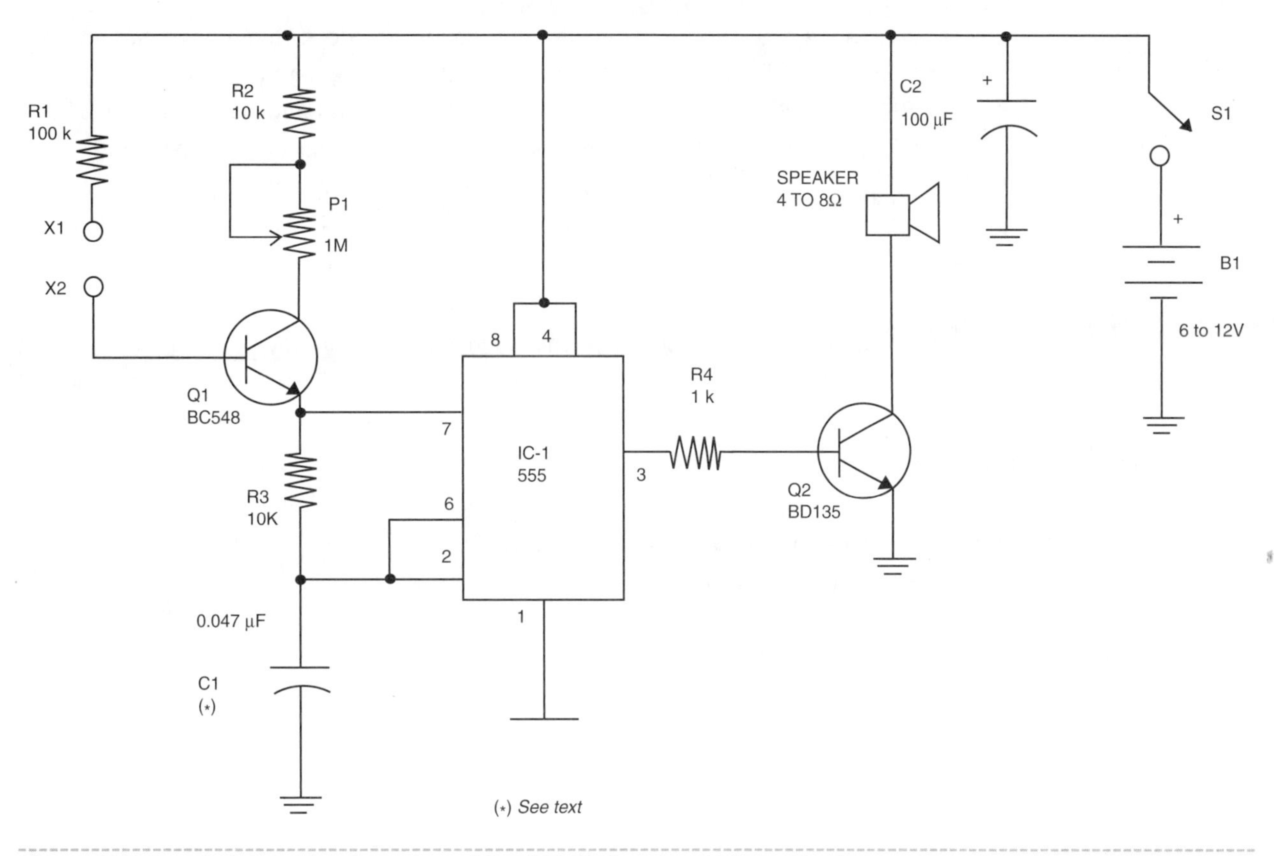


Figure 3.3.3 *Driving a loudspeaker with a 555 IC for visual biofeedback*

## Testing and Using

The first step in testing is to power on the circuit. If the electrodes are separated, the loudspeaker will not produce any sound. Maybe some random clicks will be heard due to any current loss in the transistor.

If you place your fingers on the two electrodes at the same time, the loudspeaker will produce a tone, acting on P1. You will see that the tone will change when you apply pressure to the electrodes.

## Parts List

IC1: 555 integrated circuit

Q1: BC548 or equivalent NPN general-purpose silicon transistor

Q2: BD135 or equivalent NPN medium-power silicon transistor

X1, X2: Electrodes

SPKR: 4 to 8 Ω, 5 to 15 cm loudspeaker

R1: 100 kΩ x 1/8 W resistor, brown, black, yellow

R2, R3: 10 kΩ x 1/8 W resistor, brown, black, orange

R4: 1 kΩ x 1/8 W resistor, brown, black, red

P1: 1 MΩ potentiometer, lin or log

C1: 0.047 µF ceramic or polyester capacitor (see text)

C2: 100 µF x 12 V electrolytic capacitor

S1: On/off switch

B1: 6 or 12 V power supply, battery or cells (see text)

Other: PCB or solderless board, cell holder (optional), plastic box, wires, solder, etc.

## An Intermittent Circuit

The next version of audio biofeedback produces an intermittent sound like a siren, but it is controlled by the electrodes. The effect obtained from the circuit shown in Figure 3.3.4 is quite different from the previous versions.

The loudspeaker will produce bips at a rate that is dependent on the pressure exerted on the electrodes. Figure 3.3.5 shows the wave shapes in different conditions of operation for the circuit.

The circuit uses the 4093 IC as its base, as in the original basic version, but in this configuration two gates are used as oscillators. One of them has its frequency dependent on the amount of resistance between the electrodes X1 and X2, that is, the resistance of the skin of the subject if it is a person. This oscillator, running at a very low frequency (between 0.1 and 5 Hz), controls a second oscillator running also at a low frequency but a little higher, between 500 and 1,000 Hz.

The result is an intermittent signal, applied to the base of a medium-power NPN transistor. This transistor has the loudspeaker as its load.

The sound is loud depending on the power supply voltage and loudspeaker. Power that ranges from 200 milliwatts (mW) to 1 watt will be produced from the supply. As in the previous versions, do not power the circuit from transformerless power supplies.

The circuit can be mounted on a PCB or solderless board. Q1 must be mounted on a heatsink.

If you want to supply more power to the sound, power the circuit from a 12-volt supply and replace Q1 by a TIP31.

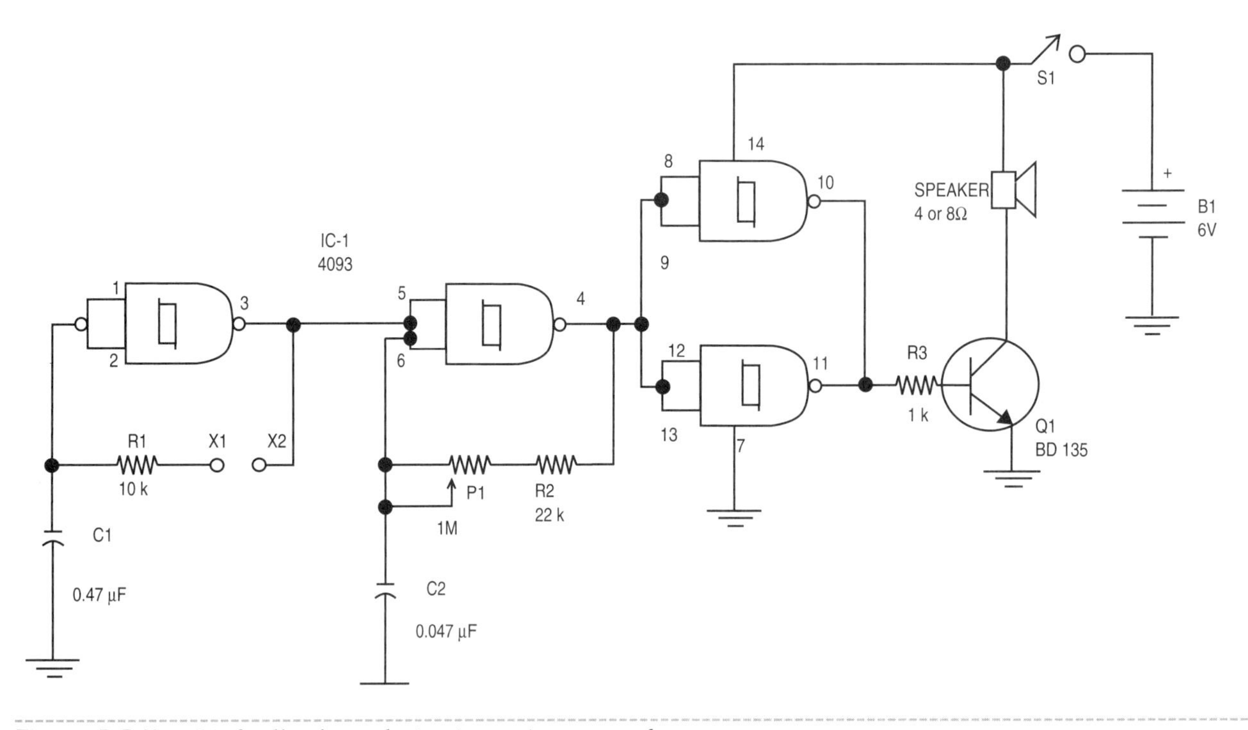


Figure 3.3.4 *Biofeedback producing intermittent sounds*

Low resistance between X1, X2

High resistance between X1, X2

Figure 3.3.5 *The wave shapes of the voltage applied to the transducer*

## Testing and Using

Power the circuit on and place your fingers on the electrodes. The electrodes must be separated, but you must touch them at the same time. The sound will change as long you press and release your fingers, changing the pressure on the electrodes.

The frequency of both the tone and the intermittence can be changed by replacing C1 and C2. Try experimenting with these components according to the applications.

A sensitivity control can also be added. It is a 4.7 MΩ potentiometer between pin 1 and 2 of the IC and the 0-volt line. The electrodes are the same as those described in the basic project.

## Parts List

IC1: 4093 CMOS integrated circuit

Q1: BD135 NPN medium-power silicon transistor

R1: 10 kΩ x 1/8 W resistor, brown, black, orange

R2: 22 kΩ x 1/8 W resistor, red, red, orange

R3: 1 kΩ x 1/8 W resistor, brown, black, red

P1: 1 MΩ potentiometer, lin or log

C1: 0.47 µF or 1 µF polyester or electrolytic capacitor

C2: 0.047 µF x 12 V polyester or electrolytic capacitor

SPKR: 4- or 8-ohm x 5 to 10 cm small loudspeaker

S1: On/off switch

B1: 6 V power supply or cells

X1, X2: Electrodes (see text)

Other: PCB or solderless board, heatsink for Q1, cell holder or power supply, plastic box, wires, solder, etc.

## Other Ideas

The interaction between a living being (a human, for instance) and an electronic circuit can be done in several different ways. Additional suggestions for interaction are provided in this section.

### Using the Computer

Figure 3.3.6 shows an acquisition interface that can be used to transfer pulses to the parallel port of a computer. The pulses can be measured by software and displayed onscreen as visual effects (colors, shapes, etc.). It is up to the reader to create the software for this task.

### Other Types of Sensors

Other types of sensors can also be used, replacing the metal plates in X1 and X2.

Figure 3.3.7 shows how a *light-dependent resistor* (LDR) can be used to sense changes in the transparence of your fingers. The amount of light passing across one's fingers depends on one's blood pressure and heartbeat.

Adjusting the trigger point of the circuits (any of the versions) can allow the circuit to be controlled by the heartbeat. This configuration can also be used with animals and plants.

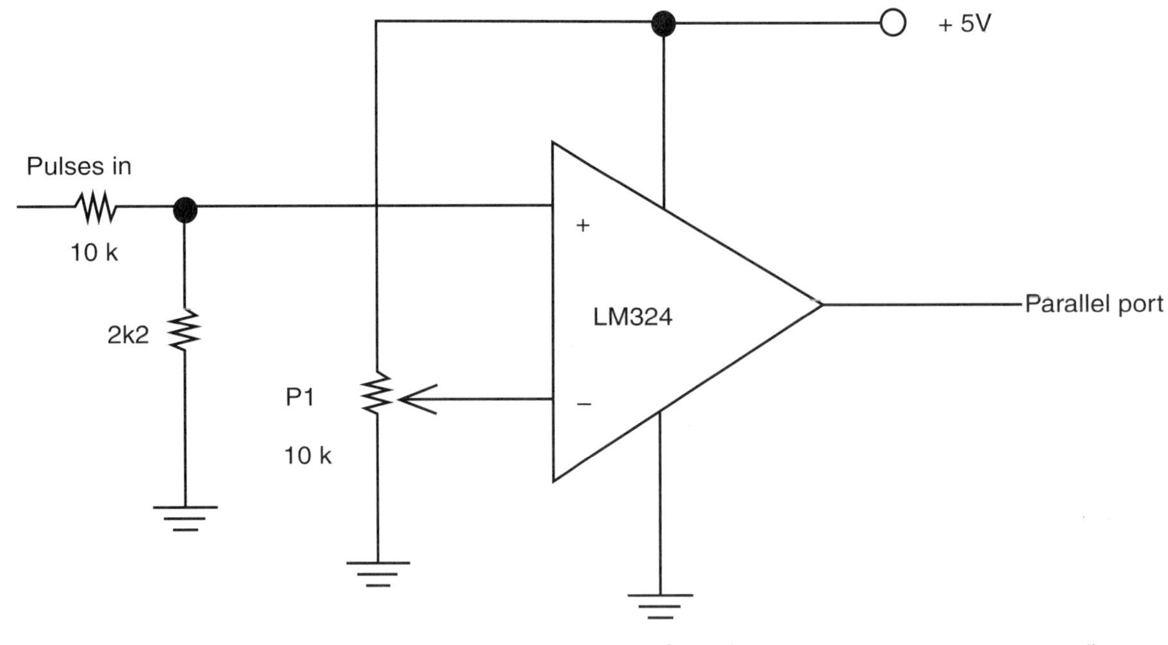


Figure 3.3.6 *Interface to get effects into a PC*

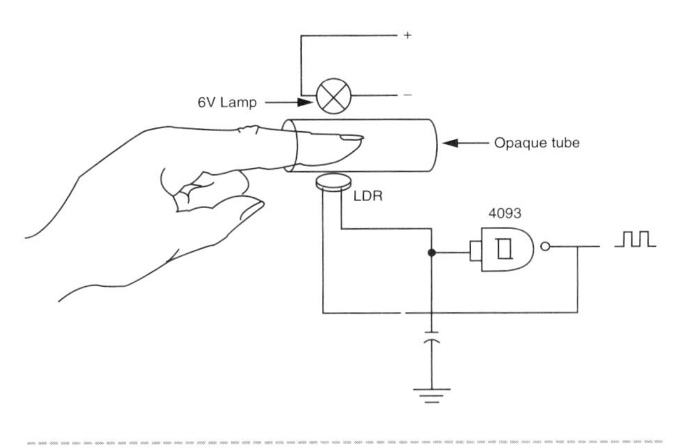


Figure 3.3.7 *Using an LDR as a sensor*

## Getting Better Performance and Learning More About the Circuit

Here are a few ideas for improving upon the circuit's performance and for helping you get to know the circuit's characteristics more closely:

- Connect X1 and X2 to electrodes to experiment with living beings, such as plants, and detect physiological changes in their behavior.
- Replace the electrodes with copper wires without isolation and insert them into an aquarium. Will the fish alter the frequency of the sound? Perform an experiment involving this subject.
- Explain why skin resistance changes with the pressure on the electrodes.
- Program experiments in concentration, yoga, and hypnosis using this circuit. Remember that this kind of experiment must be made with specialized assistance.
- Respiration and heart rates, along with blood pressure, tend to increase after a stressful activity. You can determine the influence of stress on the skin's electric resistance from the produced audio tone. Perform experiments involving stress, and use a stressed subject to produce the audio tone.
- The circuits discussed can also be used for polygraphs or lie detectors. Any alteration of the skin resistance due to stress caused by a compromising question will change the tone pitch of the circuit.
- Using two wires as electrodes, you can detect modifications in the conductance of a liquid in experiments involving chemistry.

- Output to BZ can be connected to an oscilloscope to get visual feedback or allow a visual analysis of the phenomena. Wire IC pin 4 to the horizontal input (external sync) and pin 10 to the vertical input of the oscilloscope.
- Readers more experienced in electronics can connect a circuit to a powerful audio amplifier to perform experiments with loud tones.



# Project 4—Nerves Stimulator

One of the pillars of bionics is that living matter can conduct an electric current. This means that many experiments can be done in which living beings are stimulated with electric currents to monitor their reactions.

The project described in this section focuses on experiments with electrically stimulating living beings, in particular those with a nervous system, which means mammals (including humans), fish, insects, and many others. Of course, the circuit can also be used to stimulate other beings, such as plants, microorganisms, and so on. The use of the circuit in bionics is limited only by the reader's imagination.

The nerves stimulator consists of a high-voltage generator powered from simple cells. The circuit itself is quite simple and has no critical parts or adjustments. It is very important that the reader avoid the danger of connecting the device directly to the AC power line. This means that, although the circuit can produce peaks of voltage reaching more than 200 volts, in the case of shock, the reader will experience a hard shock, but not a fatal one.

Of course, it is recommended that the reader never use this device with others without the presence of an adult or a teacher who could limit the effects of electric shocks. The circuit mustn't be used on persons with pacemakers or heart problems.

## The Bionics Experiments

Many experiments can be suggested using this project:

- You can use the circuit to show how the human nervous system senses different voltages.
- Experiments with living beings, such as fish in an aquarium, can be conducted to show how an alternating electric field can alter their behavior.
- Plants and other beings can be stimulated in experiments at a lab to see how their circadian rhythms and general behavior are altered.
- Experiments can be conducted to test if pain can be controlled by nerve stimulation (under the supervision of specialists, of course).
- A trap for insects can be created using this circuit.

## The Electronic Circuit

It is not possible to directly convert a low DC voltage into a high DC or AC voltage without of the use of special circuits. These circuits are called inverters and can be found in common applications, as in fluorescent lamps driven by common cells, or in car batteries.

To convert a low *direct current* (DC) voltage into a high *alternating current* (AC) voltage, the recommended configuration is to use a transformer. But to drive a transformer, it is necessary to change the pure DC current into a pulsed current or AC. Transformers can't operate with pure DC currents.

Consequently, an electronic circuit is necessary to connect the transformer with the DC power supply, as shown by Figure 3.4.1.

The oscillator produces current pulses, a square signal that, when applied to the base of the power transistor, is amplified and transferred to the low-voltage coil of a small transformer.

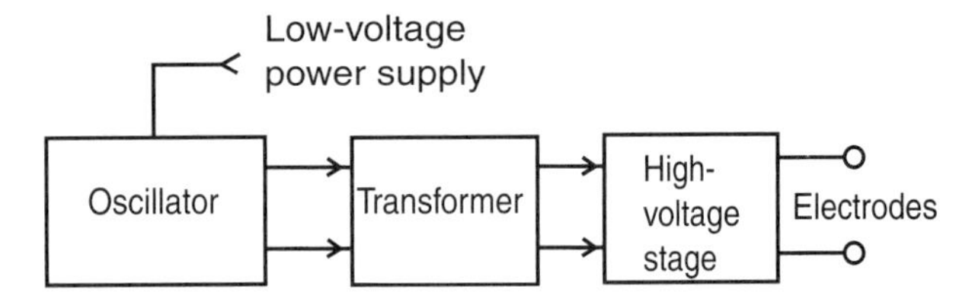


Figure 3.4.1 *Block diagram representing the nerves stimulator*

In the secondary coil of this transformer, a high voltage appears, with peaks reaching more than 200 volts, enough to stimulate any living being, including humans. To control the voltage applied to the specimen, a potentiometer is used. In parallel, a neon lamp indicates that a high voltage is being produced.

The frequency of the oscillator can be changed both by a potentiometer and selecting the timing capacitor (C2 or C3). This means that the bionics researcher can select the type of signal to be applied to the specimen under analysis.

When S2 connects the 0.047 $\mu$F capacitor, the circuit operates with a high-frequency signal between 100 Hz and 3 kHz approximately, adjusted by P1.

When S2 uses the 1 $\mu$F capacitor, the circuit produces low-frequency pulses, ranging from 0.1 to 3 Hz. The neon lamp will blink when the pulses are produced. The reader is free to change the capacitor if different ranges of frequencies are needed for the experiments.

The neon lamp is the best high-voltage indicator, because it only glows when the voltage across its terminals is more than 80 volts (typ, or typical). When glowing, the neon lamp only consumes a low amount of power, which does not affect the amount of high voltage produced by the circuit.

Four cells make up the power supply, and for the best autonomy it is recommended that C or D cells be used. The circuit will drain between 100 and 300 milliamps, according to the characteristics of the transformer used in the project.

## How to Build

Figure 3.4.2 shows the basic version of the nerves stimulator.

A *printed circuit board* (PCB) or a solderless board can be used to implement the project. A suggestion for a PCB pattern is shown in Figure 3.4.3.

Figure 3.4.4 shows the circuit mounted on a solderless board. For experimental purposes, this is the best way to build the circuit.

The transformer is the only critical part of the project. Any small transformer having a primary rated for the AC power line and a secondary rated from 5 to 6 volts with currents ranging from 100 to 500 milliamps can be used. The final results, or the maximum output voltage, will depend on these components, so we recommend that the reader experiment with more than one transformer.

The probes depend on the experiment. Figure 3.4.5 shows some probes, which are used according to the experiment.

## Testing and Using

Testing is very simple, especially if you are a courageous reader.

First, place the cells in the cell holder and turn on S1. Place S2 in position to connect C3, and after adjusting P1 you will see the neon lamp glow and flash. If the neon lamp shows that a high voltage is being produced, you can pass to the next part of the test.

Put the potentiometer P2 in the minimum voltage output position and then hold the two wires connected to the output between your fingers, as shown in Figure 3.4.6.

At first, the reader will feel a mild sensation, then discomfort, and then a painful shock. Of course, this is only a sample of what this circuit can do!

When using the circuit, start every time from the minimum of P2. Do not turn the circuit on before adjusting the output.

## Parts List

IC-1: 555 - Integrated circuit, timer

Q1: TIP31 (A, B, or C) NPN silicon power transistor

R1: 22 kΩ x 1/8 W resistor, red, red, orange

R2: 10 kΩ x 1/8 W resistor, brown, black, orange

R3: 1 kΩ x 1/8 W resistor, brown, black, red

R4: 220 kΩ x 1/8 W resistor, red, red, yellow

P1: 1 MΩ, lin or log potentiometer

P2: 10 kΩ lin potentiometer

C1: 470 μF x 12 V electrolytic capacitor

C2: 0.047 μF ceramic or polyester capacitor

C3: 1 μF polyester or electrolytic capacitor

S1: On/off switch

S2: 1 pole x 2 positions SPDT

T1: Transformer (see text)

B1: A 6 V or four C or D cells and holder

NE-1: NE-2H neon lamp

J1, J2: Output jacks

Other: Plastic box, PCB or solderless board, knobs for P1 and P2, probes, etc.

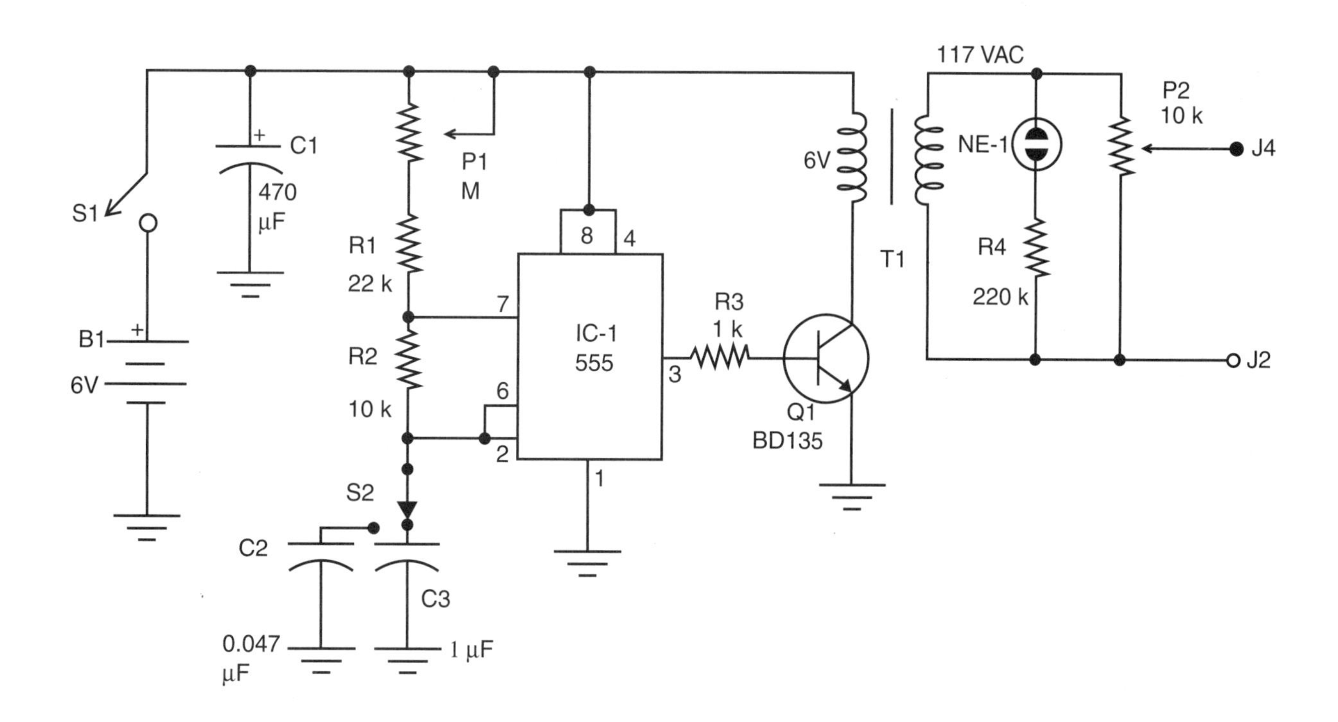


Figure 3.4.2 *The complete schematic diagram for the nerves stimulator*

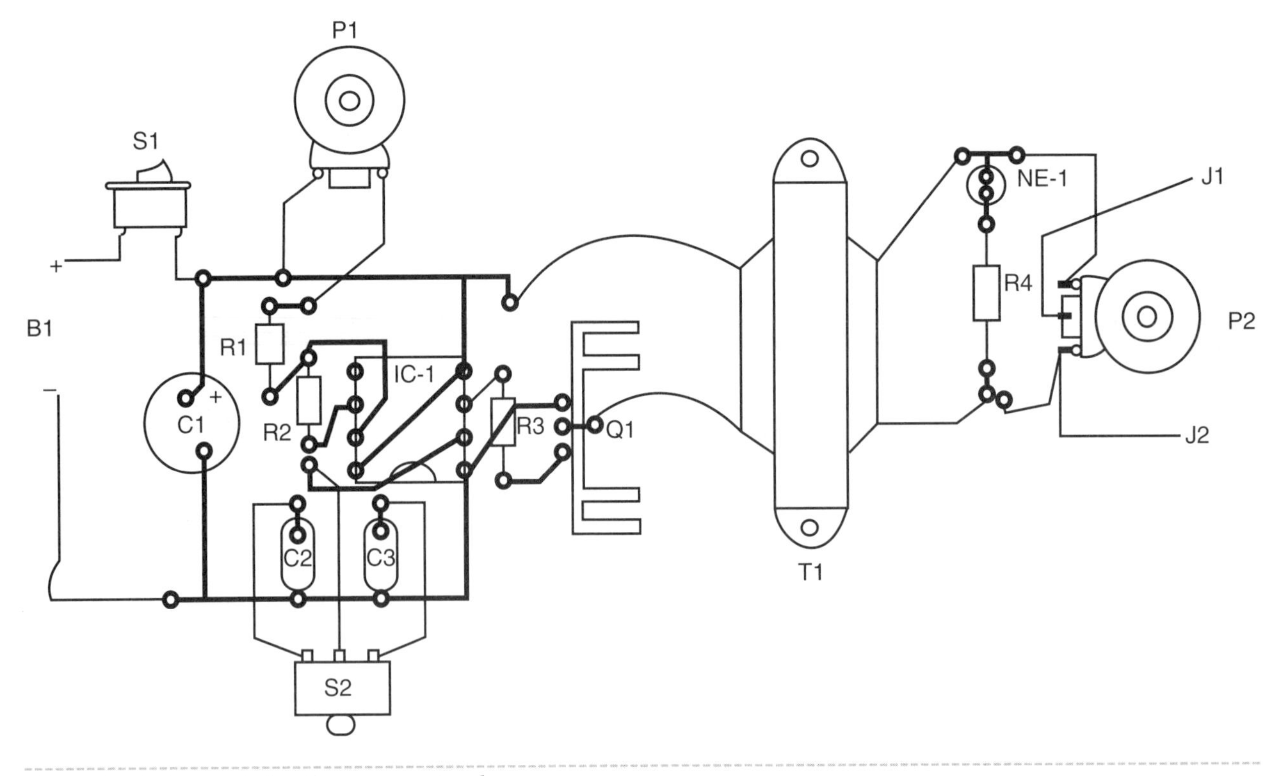


Figure 3.4.3 *PCB for the nerves stimulator*



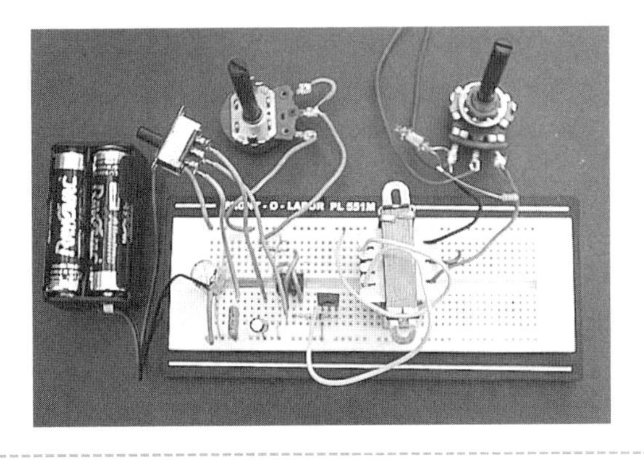

Figure 3.4.4 *The circuit mounted on a solderless board*

## Additional Circuits and Ideas

The circuit suggested here is not the only configuration you can use to generate a high voltage for experiments starting from cells or a low-voltage supply. Other versions are given in the following sections.

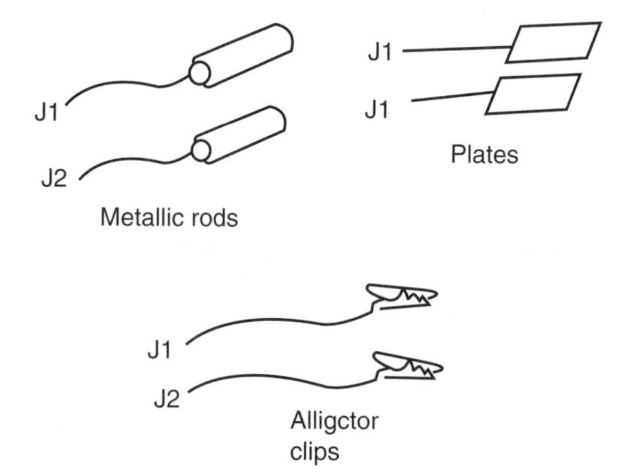


Figure 3.4.5 *Some probes for the nerves stimulator*

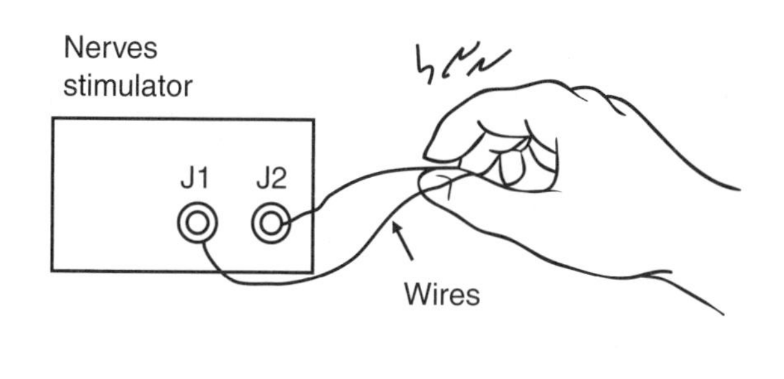


Figure 3.4.6 *Testing the high voltage*

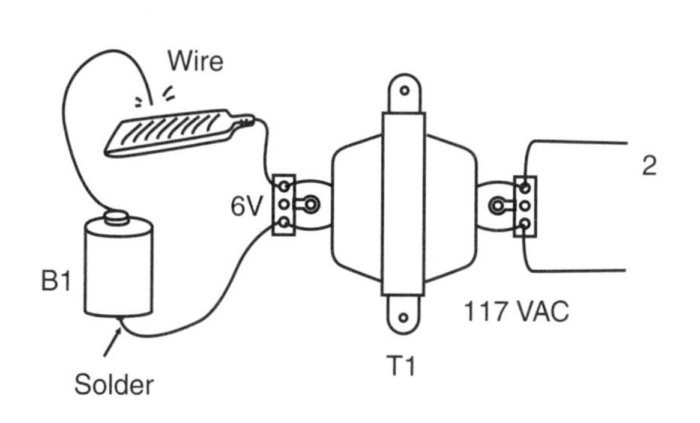


Figure 3.4.7 *The simplest nerves stimulator*

## The Simplest High-Voltage Generator

Using only four components, the circuit shown in Figure 3.4.7 can also be used to demonstrate the nerves stimulation or to make some experiments with electric shocks.

A pulsed current is produced by scraping the wire on a file, with the current flowing across the low voltage of the transformer. Thus, a high voltage is induced in the secondary.

Of course, this is the only way to produce a big voltage. If you keep the wire in contact with the file after scraping, no high voltage will be induced because no changes will occur in the current flowing across the circuit. A potentiometer and a neon lamp, as in the basic voltage, can be added to monitor and control the output voltage.

## Parts List

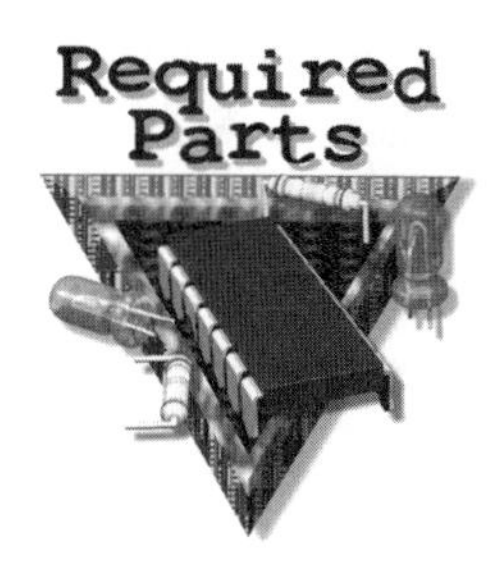


T1: Transformer, the same as in the basic version

B1: 1 to 4 C or D cells

X1: Flat file

X2: Electrodes, as in the basic version

Other: Wires, cell holder (optional), solder, etc.

## High Voltage Using One Transistor

Figure 3.4.8 shows another circuit that generates high voltages from common cells. The circuit is a Hartley oscillator built around the transistor Q1. The frequency is determined by the LC-tuned circuit where L is the secondary of T1 and C is C2.

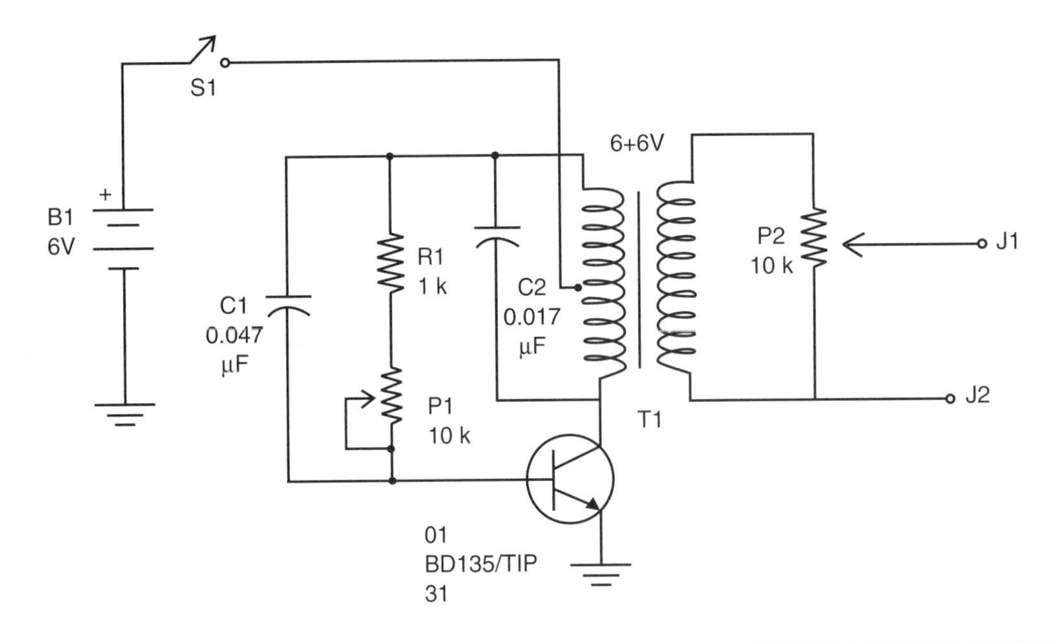


Figure 3.4.8 *Circuit using a transistor*

C1 and R1/P1 form the feedback network that keeps the circuit in oscillation. This network also has some influence on the frequency of the circuit.

P1 can be used to adjust the frequency in a narrow range. This adjustment helps the reader find the point of best performance for the circuit.

The transformer is the same one used in the basic version. According to the transformer, it could be necessary to find the best values for C1 and C2 that provide the highest voltages in the output.

The transistor must be mounted on a heatsink and, since the current drain is high, C and D cells are recommended to serve as the power supply. Do not use transformerless power supplies in this circuit or in any other that has direct contact with living beings (including humans).

The testing and using processes are the same as in the basic project. A neon lamp can be added to indicate the high-voltage generation.

Caution: Do not power this circuit from transformerless power supplies!

## Parts List

Q1: TIP31 (A, B, or C) or BD135 NPN silicon medium-power transistor

T1: Transformer, as in the basic version but with a center tap

P1, P2: 10 kΩ lin or log potentiometers

R1: 1 kΩ x 1/8 W resistor, brown, black, red

C1: 0.047 µF ceramic or polyester capacitor

C2: 0.1 µF ceramic or polyester capacitor

S1: On/off switch

B1: 6 V or four C or D cells and holder

J1, J2: Output jacks

Other: PCB or solderless board, heatsink for the transistor, knobs for P1 and P2, plastic boxes, wires, solder, etc.

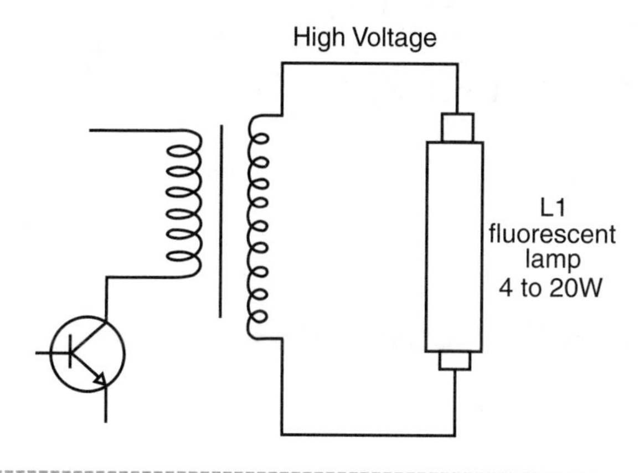


Figure 3.4.9 *Powering a fluorescent lamp*

## Powering a Fluorescent Lamp

The high-voltage generators used to make experiments with nerves stimulation, can also be used as experimental inverters. A small, common 4- to 8-watt visible or *ultraviolet* (UV) fluorescent lamp can be powered by the circuits shown here.

Remove the potentiometer from the output and the neon lamp. Then plug the fluorescent lamp to the output, as shown by Figure 3.4.9.

Powering on the circuit, in the versions where P1 is present, and adjust this component to the best performance (the highest power).

The UV version can be used to find fluorescence in many different substances, including organic materials, and it can be used to conduct experiments with living beings. How plants, insects, and other animals sense UV light is another interesting bionics experiment.

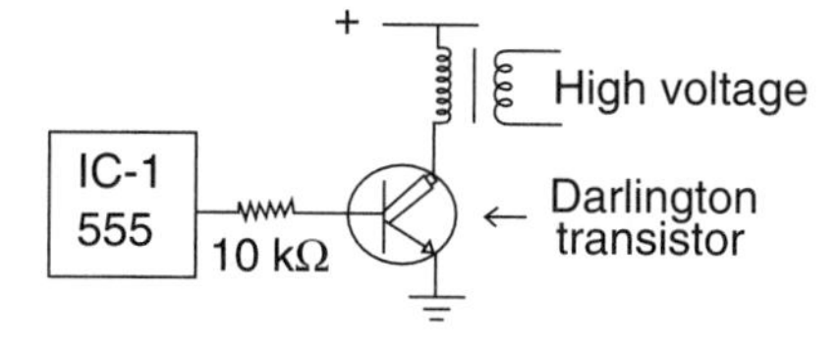


Figure 3.4.10 *Using Darlington and power MOSFET transistors*

## Using a Darlington Transistor or Power MOSFET

Darlington transistors and power *metal-oxide-semiconductor field effect transistors* (MOSFETs) can both be used in the basic version of the nerves stimulator. Figure 3.4.10 shows how these transistors can be wired to the circuit.

The Darlington transistor types that can be used are the TIP110, 111, 112, 120, 121, and 122.

As for the Power MOSFETs, any type with a drain current of 1 amp and up can be used. With some types, it will be necessary to increase the power supply voltage to 9 volts for the best results.

## A Collective Shock

An interesting experiment involving many people can be performed using the nerves stimulator. A group of people would be arranged in a circle with two individuals on one side each holding onto an electrode. When the group joined hands, they would experience a collective shock, as shown in Figure 3.4.11.

When the circle is ready for the experiment, turn on the S1 switch, keeping P2 open.

Some interesting questions can be made after the experiment (or before):

- Who experiences the strongest shock: the ones at one side of the circle or everyone? Explain.
- Why are all the people affected?
- What is a closed circuit?
- Why are even those wearing shoes that provide isolation affected by the shock?

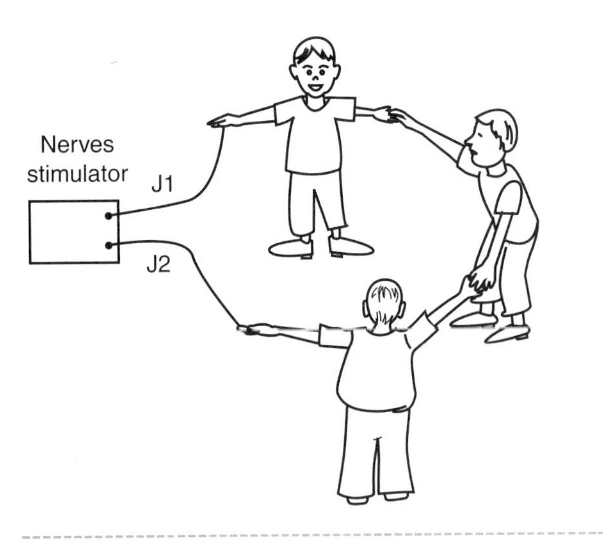


Figure 3.4.11 *A collective shock*



# Project 5—Stroboscopic Lamp

Common light sources, such as the sun, an incandescent light, or a flame, produce a continuous flux of radiation, which is received without any problem by many living beings. The common means for detecting a light source is through sight, but some creatures, such as some species of snakes and insects, have the ability to see *infrared* (IR) light. Also, plants and many microorganisms use the light of the sun to store energy through the use of chlorophyll.

Animals and humans use any source of light to see objects. The light reflected in the ambient media allows our eyes to form pictures of the objects we see.

Of course, any disturbance in the light that fills our surroundings or the presence of a light source with special properties can affect our behavior and the behavior of many other living beings.

One form of light that can cause unexpected behavior changes is stroboscopic light. The interaction between living beings and a stroboscopic source can result in interesting experiments and projects involving bionics.

A stroboscopic light is a light source that flashes from 1 to 10 times by second. You can mount one using an electronic circuit that will drive a common light source such as an incandescent lamp, a fluorescent lamp, an LED, or a xenon lamp.

Xenon lamps are the best for this application because they can produce very strong short-duration light pulses, such as the ones seen on police cars and ambulances. The advantage to the xenon lamp is that when using them as stroboscopic light sources, the flashes will be produced faster.

To help the reader conduct some experiments using stroboscopic light, this section describes a basic project using a common incandescent lamp. In the section on additional projects and ideas, we are going to demonstrate other circuits of stroboscopic light sources using *light-emitting diodes* (LEDs), fluorescent lamps, and a powerful version using a xenon lamp.

## The Experiments in Bionics

When you see a flashing stroboscopic lamp at a party or a club, the interaction between you and that light source is very simple to explain: You have the sensation of a moment in time being "frozen" from that intermittent source of light. Why this phenomenon happens will be discussed later in this section, but the importance of a light source in bionics experiments is that you can use it to discover many things about the behavior of many different living beings.

The following are some experiments you can perform involving the interaction between an electronic circuit that flashes a lamp and humans, animals, and plants:

- What happens with a plant (even a tomato plant) if it grows under the illumination of a stroboscopic light source?
- What changes in the behavior of a small aquarium fish if it is illuminated by a stroboscopic lamp? You can use this experiment to see how the frequency of the electric fish changes with a stroboscopic illumination. Illuminating the aquarium where the electric fish is, the frequency can be changed after some delay due to the induction of an uncomfortable situation.
- Can you induce panic or other discomfort in others under the illumination of a stroboscopic light source? Yes. Placing persons in a room where the only light source is a stroboscopic lamp, a panic situation can be produced.
- Can you use the modulation of a light source to transmit information, serving as a visual link between humans and a computer? For instance, you can modulate pulses produced by an LED to transmit sound from one place to other, using a phototransistor as receiver.
- What is the limit of perception of frequency changes or the frequency of the flashes from a stroboscopic lamp sensed by a human being or animal?

## What Is the Stroboscopic Effect?

When a rotating electric fan is illuminated by a flashing light source so that a flash arrives as the fan blades are running, the blades will seem to stand still at a fixed position, as suggested by Figure 3.5.1.

This is a useful way of observing fast-moving objects such as machinery or insect wings. If the

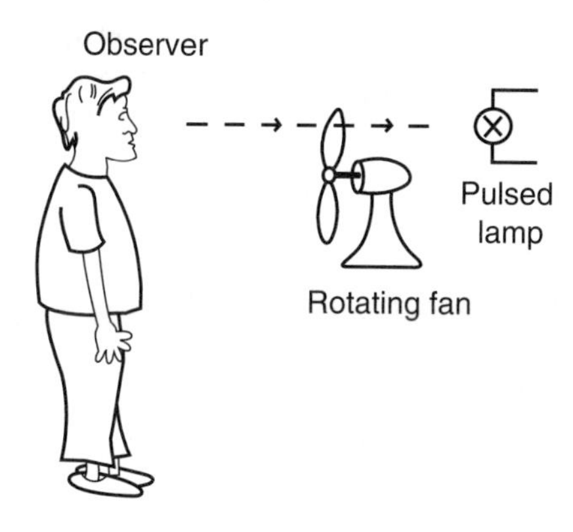


**Figure 3.5.1** *The stroboscopic effect can be observed using a rotating fan.*

flashes occur less frequently, the object will seem to move more slowly than its actual movement. You can observe this effect by rotating a small card fan in front of your TV, as shown by Figure 3.5.2.

Another way to observe the stroboscopic effect is to use a fluorescent lamp. This kind of lamp, operating on a 60 Hz alternating current, goes on and off 120 times per second. The light from the lamp would go out completely, except that the phosphors have some phosphorescent or "carry-over" action. That is, they continue to glow for a short time after the existing radiation is cut off.

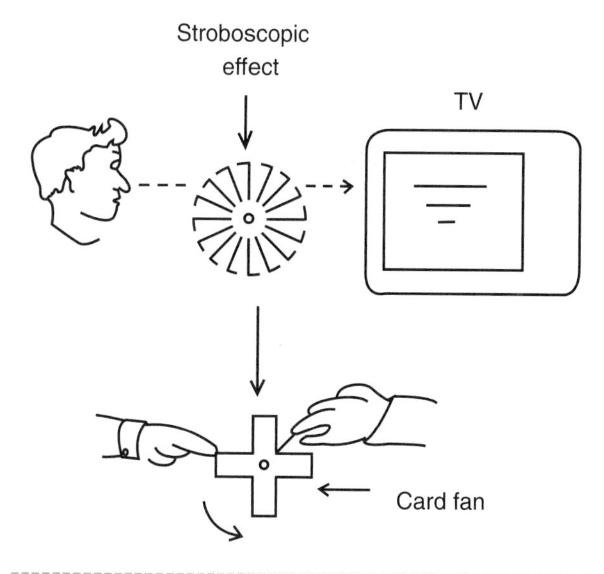


**Figure 3.5.2** *Observing the stroboscopic effect in front of a TV set*

This variation in light output is known as *flicker*. The flicker rate for a fluorescent lamp is 120 cycles per second. At the ends of the lamp, each alternate flash results in an effective rate of 60 flashes per second.

The 120-cycle flicker rate is too fast to be detected by human sight. The 60-cycle flicker can be detected, but only by the periphery of the retina. For this reason, lamp flicker is seldom noticed except when observing the ends of the lamps out of the corner of one's eye or when using related equipment.

The stroboscopic effect gives the illusion of motion or nonmotion as the case may be. For example, it was thought that the pulsating effect of an alternating current operating at 60 Hz produced a flicker, which when used on deteriorating machinery would appear to indicate that the machinery was slowing down or even stationary.

The danger of the stroboscopic effect is self-evident, but with modern fluorescent tubes the problem has been minimized by reduced flicker effect and should not cause problems. In some industrial situations, it is common for the ambient light source to be in opposition of phase and therefore to not have the presence of the stroboscopic effect.

## The Circuit

Common incandescent lamps are very slow when it comes to producing stroboscopic flashes. Such lamps cannot offer more than five flashes a second, because the inertia of the filament does not allow it enough time to return to the cold state, make small temperature changes, and create another flash, as shown by Figure 3.5.3.

If an incandescent lamp produces flashes at a rate higher than four or five per second, the lamp will begin to flicker. Crossing the 10-flashes barrier, the flicker will not have enough amplitude to be detected by our eyes.

So for our circuit, using incandescent lamps in the range between 0 and 4 or 5 Hz will be useful. The basic circuit consists of a relaxation oscillator using a neon lamp and a TRIAC circuit.

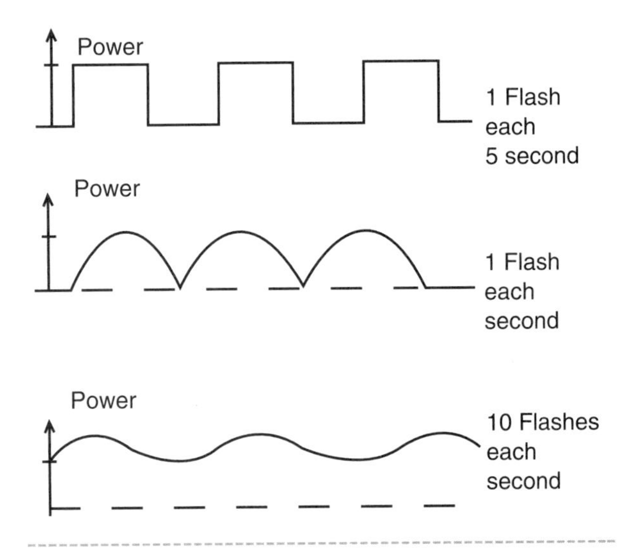


Figure 3.5.3 *The inertia of the filament of an incandescent lamp limits its ability to flash at a higher frequency.*

The load is an incandescent lamp with power capabilities of up to 150 watts. When the circuit is powered on, the capacitor C1 charges across D1, R1, and P1. It charges until it reaches the trigger voltage of the neon lamp, about 80 volts.

At this moment, the neon lamp conduces (or begins to let the electric current pass across it), and the capacitor C1 discharges through the gate of the TRIAC. The TRIAC conduces for a moment, powering the lamp used as a load.

The flash lasts until the charge of the capacitor reduces the gate voltage to a value that is enough to keep the TRIAC on. At this moment, the TRIAC turns off and a new cycle of operation begins.

R2 controls the speed of C1's discharge and thus the duration of the flash. The reader is free to find the best value for this component. Values that range from 2.2 to 10 kΩ must be tested if the flashes are not powerful enough for the application.

The circuit can also be operated from the 220/240 AC power line. It is enough to replace D1 by a 1N4007 and the TRIAC by a TIC226D.

## How to Build

Figure 3.5.4 shows the basic version of the schematic diagram for the stroboscopic lamp.

Since few components are used, the circuit can be mounted on a terminal strip, as shown in Figure 3.5.5. Of course, for readers with more resources for mounting electric devices, a *printed circuit board* (PCB) can be used.

The TRIAC must be mounted on a heatsink, mainly if the lamp is powered at higher than 60 watts. Remember that the circuit is not isolated from the power supply line, so any exposed part can cause severe shocks.

Place all the parts inside a plastic box. Only the lamp will be placed externally, and a long pair of wires can be used to wire the lamp to the circuit. Lengths of up to 10 meters will not cause any problems for the circuit's operation. You can use color lamps or wire many lamps in parallel because the sum of the power will not pass the maximum value allowed by the circuit.

## Testing and Using

First, plug the circuit into an AC power line. Adjusting P1, the lamp will flash at different rates, so find the best effect.

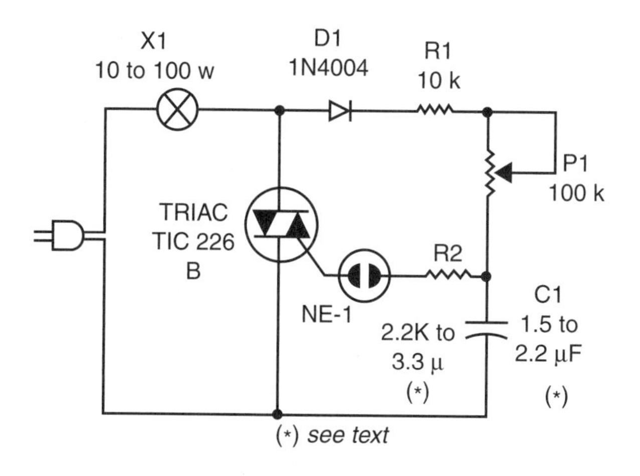


Figure 3.5.4 *Schematics for the basic version*

If the power of the flash is not high enough, change R2. Higher values will result in longer light pulses, but the higher limit of the frequency range will be reduced.

## Parts List

TRIAC: TIC226B TRIAC for 117 VAC or equivalent (TIC226D for 220/240 VAC)

D1: 1N4004 silicon rectifier diode (1N4007 for 220/240VAC)

NE-1: NE-2H or equivalent neon lamp

X1: 10 to 150 W incandescent lamp according to the AC power line

R1: 10 kΩ x 1/8 W resistor, brown, black, orange

R2: 2.2 kΩ or 3.3 kΩ x 1/8 W resistor, red, red, red or orange, orange, red

P1: 100 kΩ or 220 kΩ lin or log potentiometer

C1: 1.5 to 2.2 µF x 100 V or more polyester capacitor

Other parts: PCB or terminal strip, heatsink for the TRIAC, power cord, knob for P1, wires, solder, etc.

## Additional Circuits and Ideas

Short pulses of light can be produced using many types of sources, such as LEDs, fluorescent lamps, or xenon lamps. LEDs, for instance, can reach very high frequencies for produced pulses and are suitable for experiments and applications where other sources don't function.

The following sections cover some circuits using these sources. The reader is free to change compo-

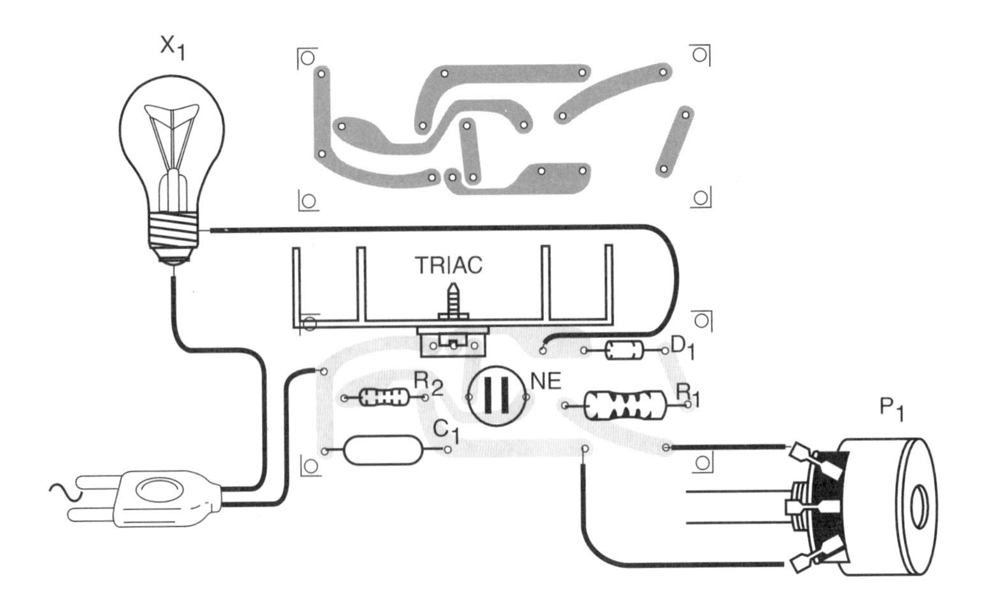


Figure 3.5.5 *PCB for the stroboscopic lamp*

nents to find the best performance according to the experiment being conducted.

## Using LEDs

The circuit in Figure 3.5.6 uses LEDs rather than lamps and is powered from a DC power supply or cells. The circuit uses a IC-555 (timer) configured as an astable multivibrator driving a power transistor.

The frequency is controlled by P1 in a range determined by C1, and the pulse duration is determined by R2. These components can be changed from 4.7 to 47 k. White bright LEDs are recommended, but depending on the application, you can use common LEDs of any color.

The power drain depends on the number of LEDs used in the project and also on the values of R4 and R5. These resistors can be increased if you don't need high power or if you want to use common LEDs.

## Parts List

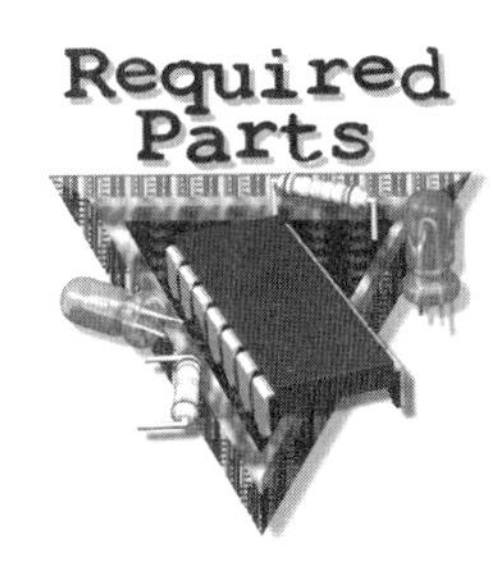


IC-1: 555 integrated circuit and timer

Q1: TIP32 PNP silicon power transistor

LED1 to LED4: White LEDs (see text)

P1: 1 MΩ lin or log potentiometer

R1, R2: -10 kΩ x 1/8 W resistors, brown, black, orange

R3: 1 kΩ x 1/8 W resistor, brown, black, red

R4, R5: 47 x 1/8 W resistors, yellow, violet, black

C1: 1 µF polyester or electrolytic capacitor

S1: On/off switch (SPST)

B1: A 6 V or four AA, C, or D cells with a holder

Other parts: PCB or solderless board, knob for P1, wires, solder, plastic box, etc.

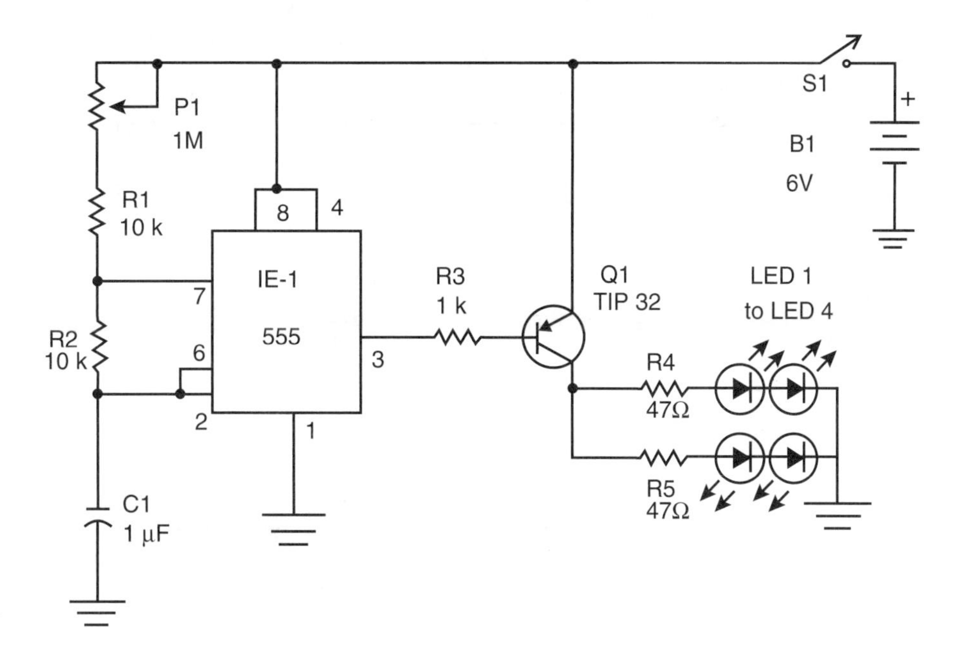


Figure 3.5.6 *Stroboscopic LEDs*

## Fluorescent Stroboscopic Light

Fluorescent lamps are faster than incandescent lamps, so they are better for producing fast flashes in a stroboscopic circuit.

The circuit shown in Figure 3.5.7 is not powerful, but it can be used in some experiments involving bionics such as animal conditioning, working with stress, and others. The circuit consists of a relaxation oscillator using a neon lamp to trigger a *silicon-controlled rectifier* (SCR).

The capacitor C1 charges through R1 and D1 until the voltage across the lamp rises enough to trigger the light. At this moment, the neon lamp lights up, and the capacitor C2 is discharged through the gate of the SCR. The result is that the SCR conducts the discharge current of C1, which flows across the low-voltage winding of the transformer.

The high-voltage pulse produced at the secondary is also applied to the lamp that flashes for a moment. The pulse rate can be controlled by P1, and the power of the flash depends on the capacitor C1.

The transformer can be any type used in power supplies with a primary rated to 117 VAC and a secondary of 9 to 12 volts and currents ranging between 250 and 600 milliamps.

Additionally, any 5- to 20-watt fluorescent lamp can be used. Even ones that do not function anymore because they are weak will glow using this circuit. The high-voltage pulses can easily be higher than 117 VAC, triggering the lamp on.

Depending on the components, to compensate tolerances C2 can be changed. Values that range from 0.1 to 0.47 $\mu$F can be tested.

To operate from the 220/240 VAC power line, change R1 to 1 kΩ, replace D1 with a 1N4007, and use the TIC106D for the SCR.

It is not necessary to mount the SCR on a heatsink because it functions at short time intervals without generating large amounts of heat.

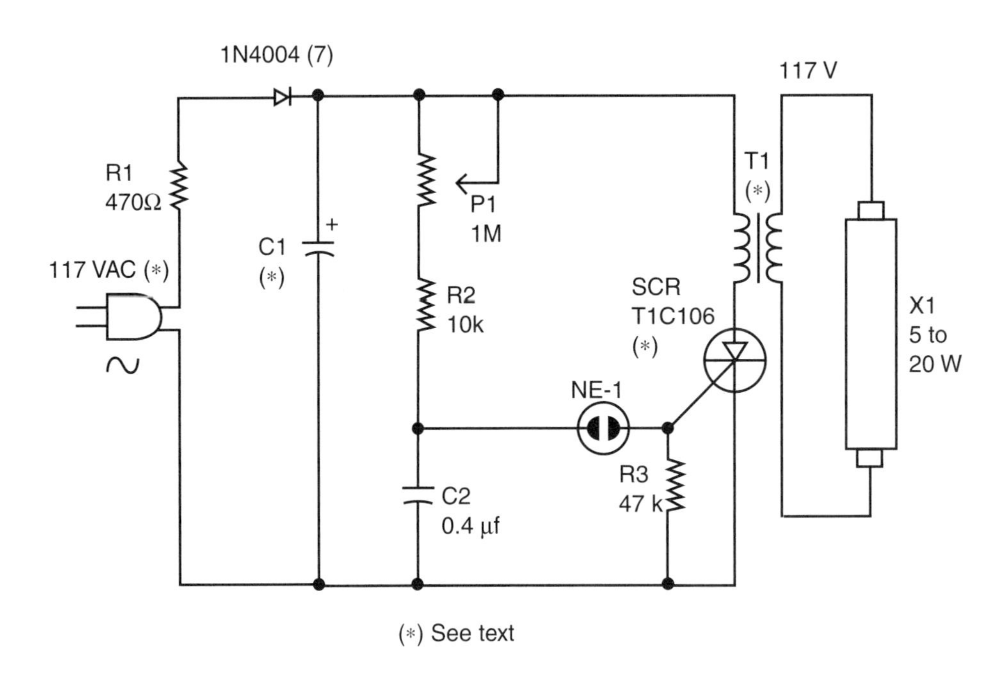


Figure 3.5.7 *Circuit using a fluorescent lamp*

## Parts List

**CAUTION**

SCR: TIC106B (117 VAC power line) or TIC106D (220/240)

D1: 1N4004 (117 VAC) or 1N4007 (220/240 VAC) silicon rectifier diode

NE-1: NE-2H or equivalent neon lamp

R1: 470 Ω x 10 W (117 VAC) or 1 kΩ x 10 W (220/240 VAC) wire-wound resistor

R2: 10 kΩ x 1/8 W resistor, brown, black, orange

R3: 47 kΩ x 1/8 W resistor, yellow, violet, orange

P1: 1 MΩ lin or log potentiometer

C1: 4.7 to 22 µF x 200 V (117 VAC) or 400 V (220/240 V) electrolytic capacitor

C2: 0.1 µF x 100 V or more polyester capacitor

T1: Transformer (see text)

X1: 5 to 20 W fluorescent lamp

Other: PCB or terminal strip, power cord, plastic box, wires, etc.

## Xenon Stroboscopic Light

The best way to generate high-power light pulses is to use a xenon lamp. When a xenon lamp is active, the gas resistance falls to a few ohms, allowing very high currents to pass through it. The result is the generation of very high power light pulses. Common xenon lamps are used in photographic cameras, beacons, and in stroboscopic light sources such as the one described here.

To trigger a xenon lamp, very high voltage pulses are necessary, up to 4 kilovolts or more, and to produce the light pulses a large capacitor is needed to provide the necessary energy.

In the circuit shown in Figure 3.5.8, a voltage doubler is used to charge the capacitor C2 with a voltage

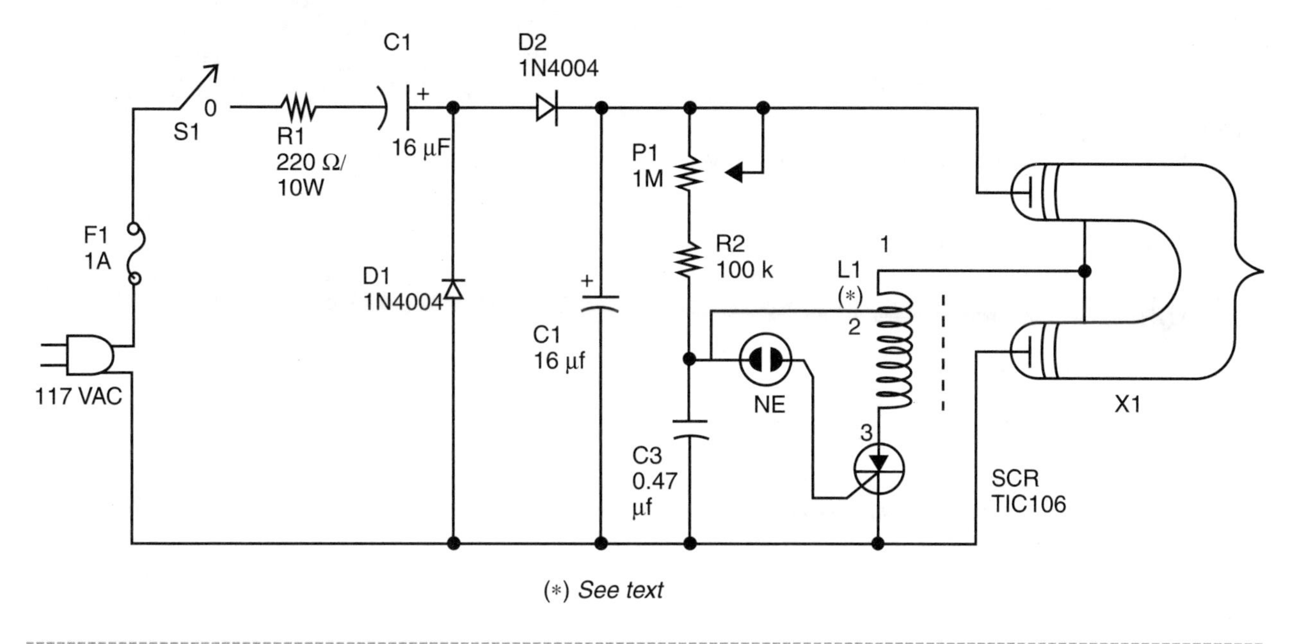


Figure 3.5.8 *Circuit with a doubler*

near 300 volts when plugged into the 117 VAC power line. The power will depend on the value of the capacitors. Values between 10 and 22 $\mu$F are recommended.

The pulse rate is determined by P1. This component controls the speed of the C3's charging process. When C3 reaches about 80 volts, the neon lamp turns on. Then C3 discharges across T1, generating a fast, high-voltage pulse of about 4 kilovolts, enough to trigger the xenon lamp.

When triggered, the gas inside the lamp becomes conductive, allowing the capacitor C2 to discharge across the lamp. The result is a powerful flash of light. When the capacitor C2 discharges, the lamp turns off and a new cycle of operation begins.

Any small xenon lamp powered by a few joules can be used. The trigger transformer can be bought with the lamp, but if you intend to use a lamp found in a old-style flash attached to a camera, the transformer can be wound as shown in Figure 3.5.9. According to the lamp, you can change the capacitors C1 and C2 to get more power.

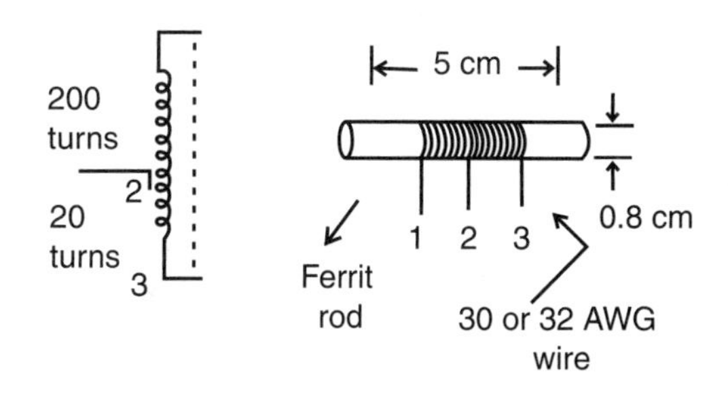


Figure 3.5.9 *The triggering transformer T1*

## Parts List

SCR: TIC106B SCR

D1, D2: 1N4004 (117 VAC) or 1N4007 (220/240 VAC) silicon rectifier diodes

NE: NE-2H or equivalent neon lamp

X1: Xenon lamp (see text)

R1: 220 Ω x 10 W wire-wound resistor

R2: 100 kΩ x 1/8 W resistor, brown, black, yellow

P1: 1 MΩ lin or log potentiometer

C1, C2: 16 μF x 200 V (117 VAC) or 400 V (220/240 VAC) electrolytic capacitors (see text)

C3: 0.47 μF x 100 V or more polyester capacitor

T1: 4 kV trigger transformer (see text)

F1: 1 A fuse and holder

Other: Power cord, PCB, plastic box, wires, knob for the potentiometer, solder, etc.

# Project 6—Bio-Amplifier

***Caution*: This circuit is not isolated from the power supply line, so take care protecting all the exposed parts. Even the wires to the lamp can cause a severe shock due to the high-voltage pulses.**

Any biological process is accomplished by the generation of electric signals. Most of these signals are very low, continuous, or alternating voltages, in the range of few millivolts, meaning that only very sensitive amplifiers can detect them.

Oscilloscopes are useful tools for viewing these signals, but they are very expensive instruments, and normally the common bionics evil genius doesn't have one at his or her lab.

Another tool for detecting signals produced by living beings uses a high gain amplifier, such as the one that will be described here. It is cheaper and can be built using a few common parts, all of which can be found at your nearest electronic component dealer.

Signals produced by the metabolism of plants, insects, fish, and even humans can be amplified to drive an indicator. In this project, the indicator is a moving coil galvanometer in order to keep the cost of the device low.

The circuit is very safe because it is not plugged into the AC power line. It can be powered from common cells or a battery, presenting a very low current drain and a very low input impedance. This is a very important characteristic if you do not have much electronics experience and intend to detect signals generated by your own body.

The project consists of a high-gain operational amplifier with an external gain control. The circuit can be used to detect very weak signals ranging between 10 microvolts and 1 volt. Remember that the nerve cells (the neurons) of a human generate pulses between 10 and 50 millivolts.

Because of the very high input impedance (many teraohms, or thousands or billions of ohms), it is important to not load the circuit when testing, and keep in mind that living beings are normally high-impedance sources of signals. A large number of projects and experiments in bionics can be performed using this amplifier.

## Experiments in Bionics

The low voltage generated by living beings can be used to drive many electronic circuits or be observed to learn more about the way the voltages are produced. Experiments can be performed to see how these voltages change when the specimens are placed under special conditions (the influence of light, magnetic fields, chemical substances, etc.).

The following list of experiments and applications is not a complete one, but the reader, using his or her imagination, can add many others:

- Detect voltages generated by plants. The Backster effect* can be studied using a bio-amplifier.
- Detect changes in the resistance of skin, using the amplifier as a lie detector or in experiments to detect stress.
- Observe changes in the electric potential of electrodes immersed in an aquarium due to the presence of fish or other beings.
- Detect voltages produced in biological processes such as fermentation or putrefaction.

Cleve Backster has conducted hundreds of experiments showing that plants respond to our emotions and intents, as do severed or crushed leaves, eggs (fertilized or not), yogurt, scrapings from the roof of a person's mouth, sperm, and so on. Backster's claims were refuted, however, by Horowitz, Lewis, and Gasteiger (1975) and Kmetz (1977). Kmetz denied Backster's ideas in an article for the *Skeptical Inquirer* in 1978. He said that Backster had not used proper controls in his experiments. When controls were used, no detection of the plant's reaction to thoughts or threats could be found. They also said that the cause of the polygraph contours could have been due to a number of factors, such as static electricity, movement in the room, and changes in light and humidity. For our readers who need to see to believe, why not try the experiments? To learn more about the subject, we recommend the book *The Secret Life of Plants* by Peter Tompkins and Christopher Bird.

*Cleve Backster published some research in the *International Journal of Parapsychology* ("Evidence of a Primary Perception in Plant Life," vol. 10, no. 4, Winter 1968, pp. 329-348). He tested his plants using a polygraph machine and found that plants react to thoughts and threats.

## How It Works

The voltages generated by biologic processes are very low, on the scale of few millivolts, and the impedance of the source is very high. This means that the total power involved in the processes is very low, needing high gain amplifiers to be used in any application, such as driving a meter.

The ideal device for the amplification of very low voltage signals is an operational amplifier (see Figure 3.6.1). An operational amplifier has a very high input impedance, a large voltage gain, and a low impedance output.

The amplifier has two inputs, an inverting input and a noninverting input. The output is a single line and the gain is fixed by the feedback resistor R2. The ratio between R2 and R1 determines the gain of the amplifier.

A common operational amplifier can be programmed to have gains between 1 and more than 100,000 times the signal, which necessitates replacing R2 with a potentiometer, as we are going to adopt in our project. Common operational amplifiers also have an input impedance of some megohms, but a special type of operational amplifier has very high input impedance, in the range of many teraohms. They use *field-effect transistors* (FETs) or *junction field-effect transistors* (JFETs).

High input impedance is very important in applications involving biological systems because it doesn't load the source. This means that the presence of the external amplifier causes a minimum of influence in

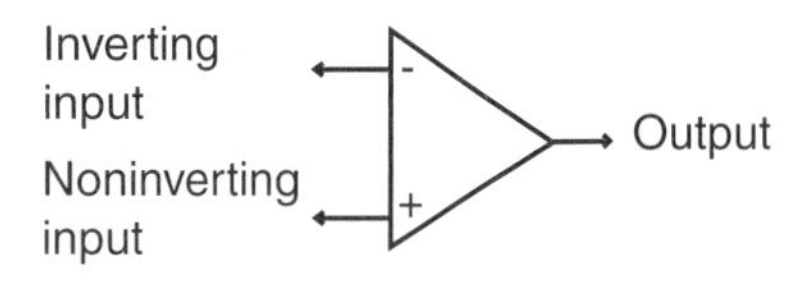


Figure 3.6.1 *The operational amplifier*

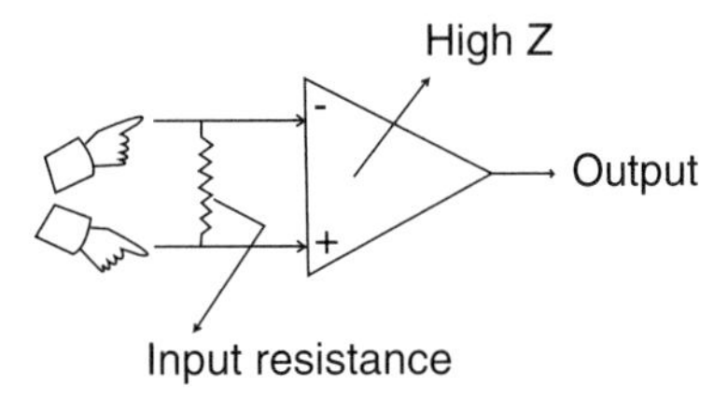


Figure 3.6.2 *High input impedance means less influence by the circuit in the sample to be studied.*

the system to be studied, as indicated by Figure 3.6.2. Loading the source occurs if a something attached to the circuit drains more current than is provided at the output of the circuit. For instance, if a 100 mA load is connected to the output of an amplifier with a maximum current of 50 mA, the amplifier will be loaded.

The basis of our circuit is the CA3140, a common JFET operational amplifier with an input impedance of many teraohms. The gain is programmed by a feedback resistor and is selected by a switch. In the high gain position, the circuit can present a gain of 100,000. The current drain is very low, so the cells can have their life extended for several weeks or even months.

The operation of the circuit is in the differential mode, meaning that the circuit amplifies the difference of the voltages applied to the inputs. This means that the output of the circuit when the voltage between the inputs is zero is half the power supply voltage. As the input 1 goes positive or negative when compared to input 2, the output voltage oscillates to more or less than Vcc/2 (half the power supply voltage).

Special care must be taken when connecting the inputs to a specimen. High impedance input also means sensitivity to noise. The wires must be short and shielded, and operation must be far from noise sources such as fluorescent lamps, transformers, and so on.

## How to Mount

Figure 3.6.3 shows the complete diagram of the basic version for the bio-amplifier. The reader can mount it using many common processes, but *printed circuit boards* (PCBs) and solderless boards are the best.

Figure 3.6.4 shows how to mount the circuit using a small PCB.

The components are housed inside a small plastic box, having as external controls the gain switch and the on/off switch, as shown in Figure 3.6.5.

Any galvanometer with full-scale currents between 50 and 200 $\mu$A can be used. A multimeter (analog or digital) can also be used, replacing the analog galvanometer.

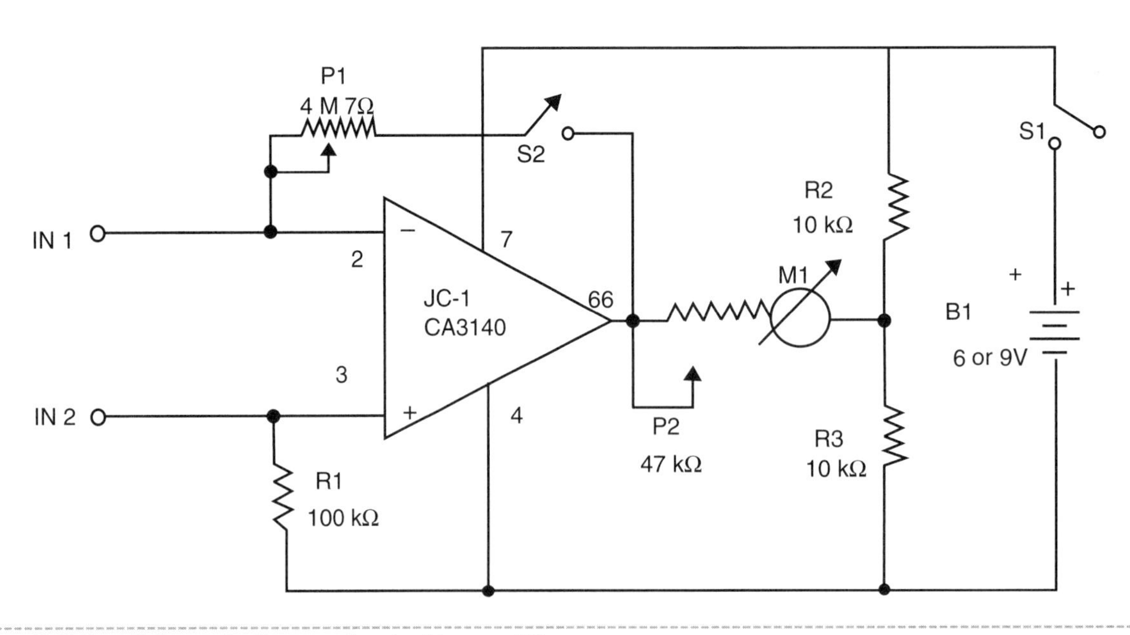


Figure 3.6.3 *Schematic diagram for the bio-amplifier*

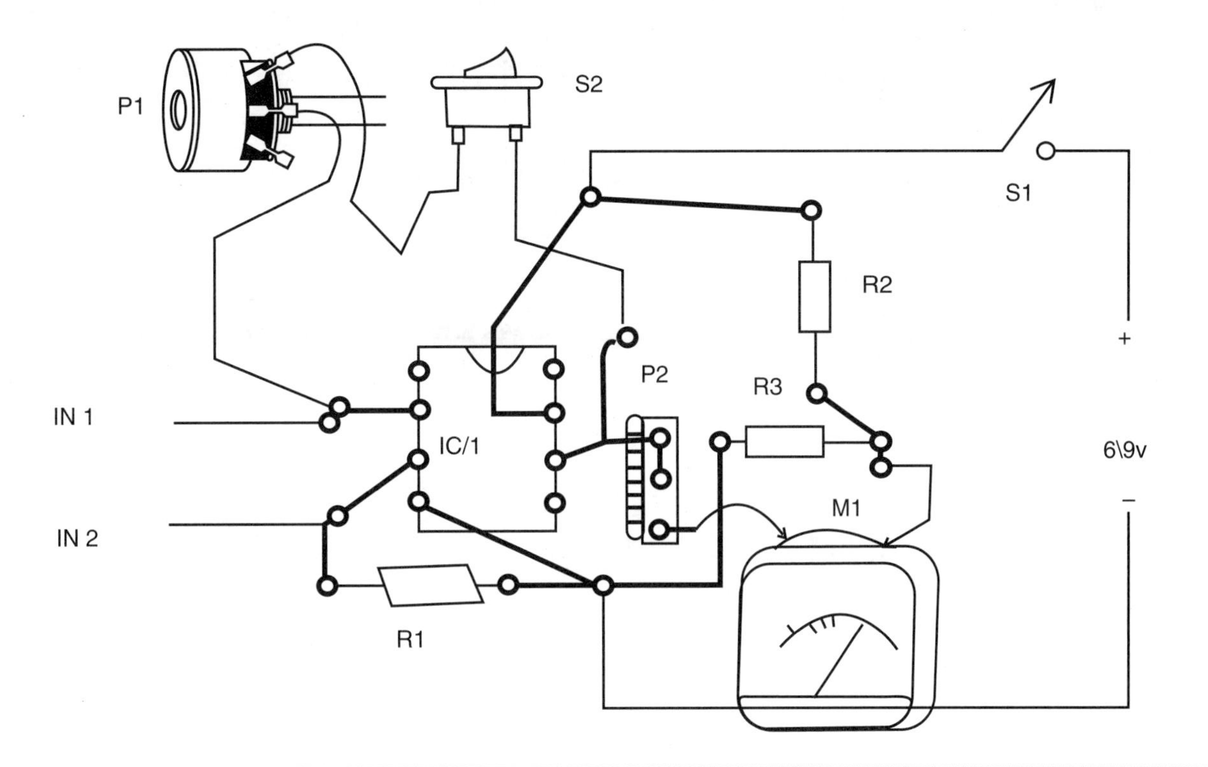


Figure 3.6.4 *PCB for the bio-amplifier*

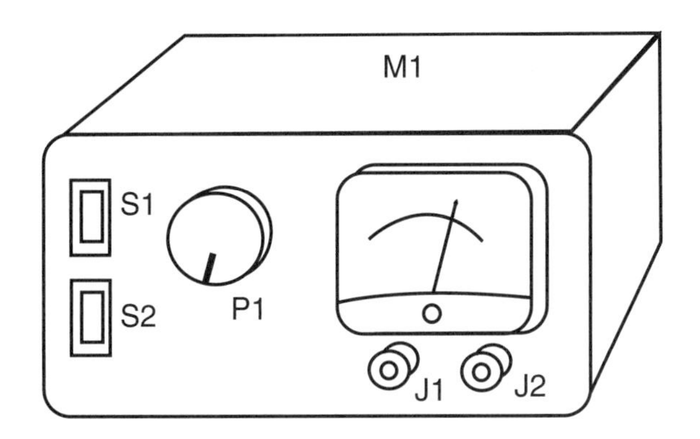


Figure 3.6.5 *The components are housed inside a plastic box.*

The electrodes depend on the application. To plug the electrodes into the circuit, plastic terminals can be used. Different colors are used to indicate the inverting (−) and noninverting (+) inputs of the circuit.

When handling the operational amplifier, avoid touching its terminals. The JFETs are very sensitive to *electrostatic discharges* (ESD). If your body is charged and you touch the component, you can burn it.

## Testing and Using

Place the power supply (cells or battery) in the circuit and turn on S1. Keep S2 closed for initial tests and open P1, adjusting this potentiometer to the highest resistance.

Touching the input terminals, adjust P2 to have visible indications that do not pass the end of the scale. Now open S2 and repeat the tests. The indicator will move even when you place your hand close to the input terminals without touching them.

The bio-amplifier can be used in two basic ways: as a voltage monitor or as a resistance monitor. When using the circuit as a voltage monitor, the signals generated by the specimen are amplified. The specimen acts as a voltage source or a generator, producing a *direct current* (DC) or an *alternating signal* (AC). Figure 3.6.6 shows how to place electrodes on a plant to observe the voltage generated by its biological activities.

The electrodes can be made by small metal plates. Use nonoxidable materials because the galvanic

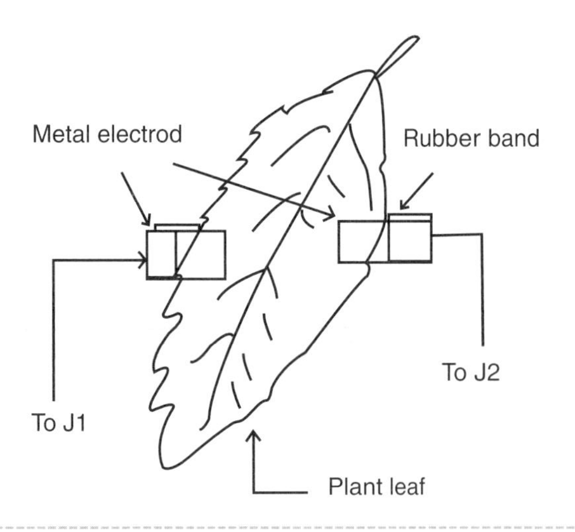


Figure 3.6.6 *Placing the electrodes in a plant*

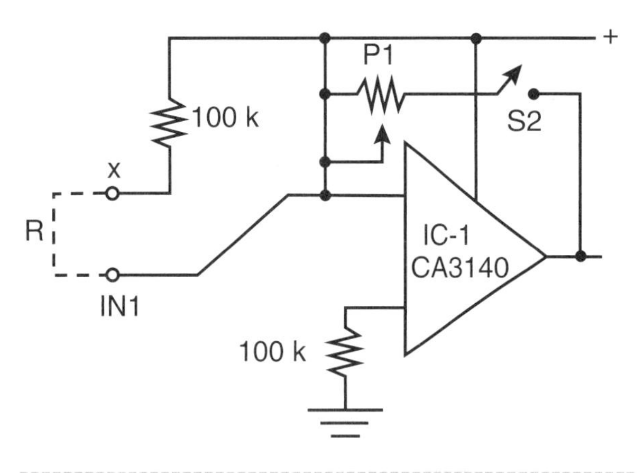


Figure 3.6.7 *Using the bio-amplifier as a resistance monitor*

action of the plant fluids can attack the electrodes, granting undesirable voltages and altering the indications. This galvanic effect can also kill the tissue where the electrode is placed in a few hours after the experiment has begun.

When using the bio-amplifier as a resistance monitor, the circuit will detect changes in the resistance of a specimen, as shown in Figure 3.6.7.

P1 and the resistance of a specimen act as bridge or voltage divider. Adjust P1 to get half the voltage of the power supply and then any change in the resistance will be amplified and shown by the instrument plugged into the output of the circuit.

## Parts List

IC-1: CA3140 integrated circuit and JFET operational amplifier

P1: 4.7 MΩ lin or log potentiometer

P2: 47 kΩ trimmer potentiometer

R1: 100 kΩ x 1/8 W resistor, brown, black, yellow

R2, R3: 10 kΩ x 1/8 W resistors, brown, black, orange

S1, S2: On/off switch

B1: A 6 or 9 V source or four AA cells and holder

M1: 0-50 μA to 0-200 μA galvanometer

Other: PCB or solderless board, cell holder or battery connector, wires, solder, electrodes

## Additional Circuits and Ideas

The circuit presented here is the simplest configuration you can mount using an operational amplifier. More elaborate configurations can be made using more than one operational amplifier or another type of output. The following are some of those configurations, and it is up to the reader to make experiments with those circuits.

## Audio Output (ADC)

Instead of driving a galvanometer, the circuit can drive an audio oscillator, as shown by the circuit in Figure 3.6.8.

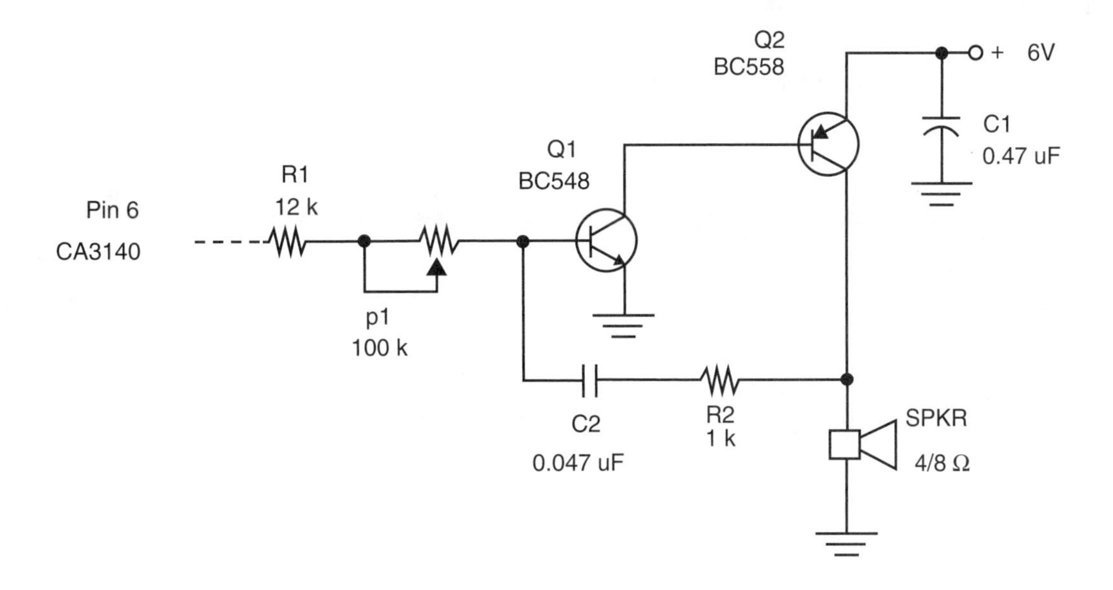


Figure 3.6.8 *Driving an audio oscillator*

The basic tone with no input signal is adjusted by P2 and determined by C1. You can change the values of C2 to alter the frequency of the oscillator; higher output frequencies correspond to higher input voltages. The circuit can also be used as a biofeedback device.

## Parts List

- Q1: -BC548 or equivalent general-purpose NPN transistor
- Q2: BC558 or equivalent general-purpose PNP transistor
- R1: 12 kΩ x 1/8 W resistor, brown, red, orange
- R2: 1 kΩx 1/8 W resistor, brown, black, red
- P1: 100 kΩ x 1/8 W trimmer potentiometer
- C1: 0.47 μF ceramic or polyester capacitor
- C2: 0.047 μF ceramic or polyester capacitor
- SPKR: Small 4- or 8-ohms loudspeaker
- Other: PCB or solderless board, wires, solder, etc.

## Using a 741 Operational Amplifier

The 741 is the most famous operational amplifier and due to its characteristics it can be used in a great number of practical circuits, including a bio-amplifier. A version using the 741 is shown in Figure 3.6.9.

This circuit uses a voltmeter or a multimeter with a scale of 0 to 1 volts as an indicator. Digital or analog multimeters can also be used.

The reader can also use a 50 to 200 $\mu$A galvanometer in series with a 10 kΩ resistor and a 100 kΩ trimmer potentiometer. The trimmer potentiometer can be adjusted to have 1 volt of the full scale. Choosing the appropriated value for R1, the circuit can reach end scales of 1 millivolt, 10 millivolts, 100 millivolts, or 1 volt.

The reader can add a switch to select the values for R1 according to the following table or simply use the best value for the application he or she has in mind.

| R1 (Ω) | Full Scale |
|---|---|
| 1 k | 1 mV |
| 10 k | 10 mV |
| 100 k | 100 mV |
| 1 M | 1 V |

As the reader can see, the 741 acts as a × 1,000 amplifier due to the ratio between P2/R3 and R1. P2 is used to adjust the correct gain of the circuit, compensating for the tolerances of the used components.

The circuit is powered from two 9-volt batteries, but because the current drain is very low, the life of these sources will last for several weeks. We don't recommend the use of power supplies plugged into the AC power line for safety considerations, as in other projects that involve the interaction with people.

P1 is used to adjust the zero offset and the electrodes, as in other versions of the bio-amplifier and in biofeedback projects.

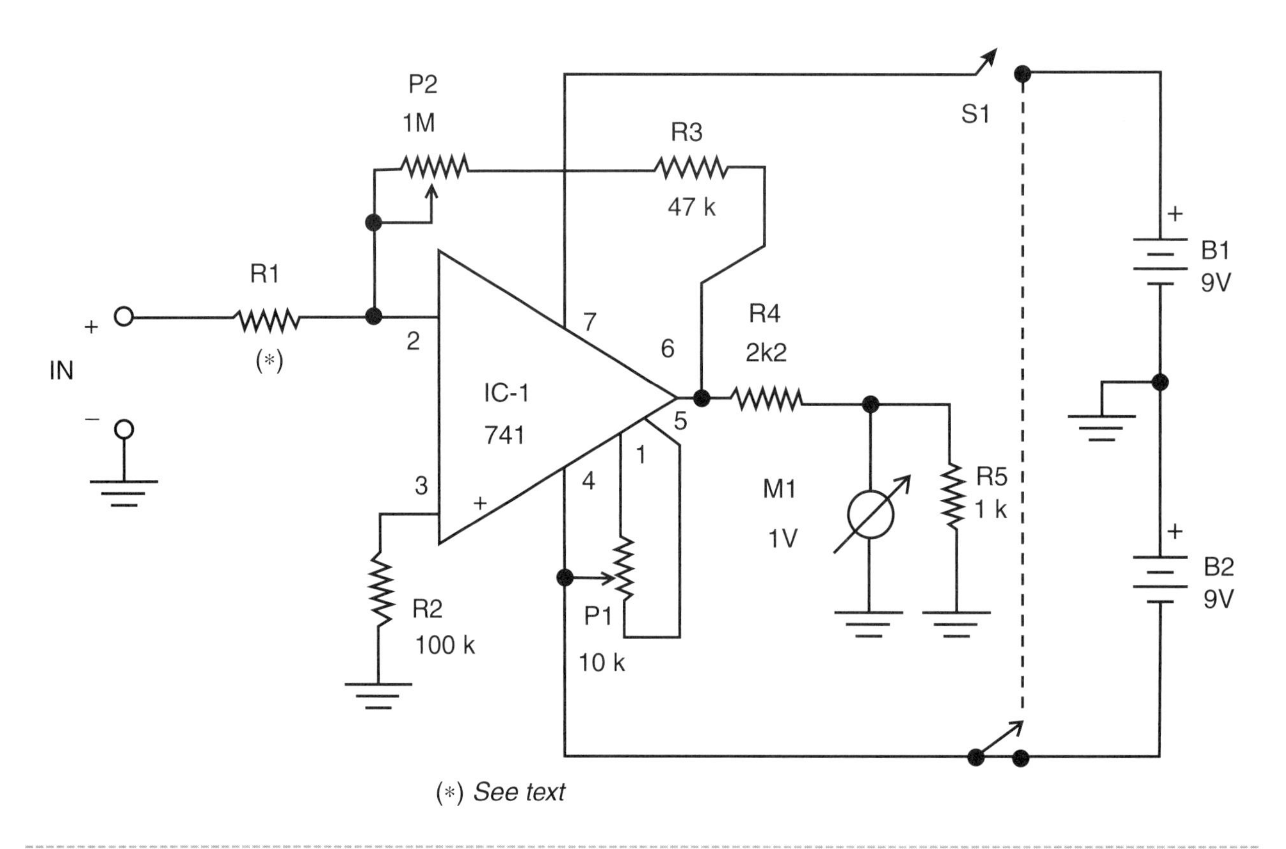


Figure 3.6.9 *Bio-amplifier using the 741*

## Parts List

IC-1: 741 integrated circuit operational amplifier

M1: 1 V voltmeter (see text)

R1: See table

R2: 100 kΩ x 1/8 W resistor, brown, black, yellow

R3: 47 kΩ x 1/8 W resistor, yellow, violet, orange

R4: 2.2 kΩ x 1/8 W resistor, red, red, red

R5: 1 kΩ x 1/8 W resistor, brown, black, red

P1: 10 kΩ trimmer potentiometer

P1: 10 kΩ potentiometer, lin or log

S1: DPDT switch

B1, B2: 9 V batteries and connectors

Other: PCB or solderless board, wires, electrodes, etc.

## Driving the PC

The bio-amplifier, when powered from a 5-volt source, can interface with a PC, as shown in Figure 3.6.10. This circuit uses an analog-to-digital converter ADC0831 (National Semiconductor), but equivalent configurations can be implemented using other analog-to-digital converters.

This 8-bit circuit has a serial input/output system that operates by a method called successive approximation, which produces an 8-bit digital representation of the analog input voltage. This means it has 256 possible digital levels for the selected input voltage range. The device can operate in a range between 0.0195 and 5 volts, and the trimpot adjustment depends on the scale to be measured.

More information about this device and how to use it in a data acquisition system such as the one suggested here can be found at the National Semiconductor Web site (www.national.com).

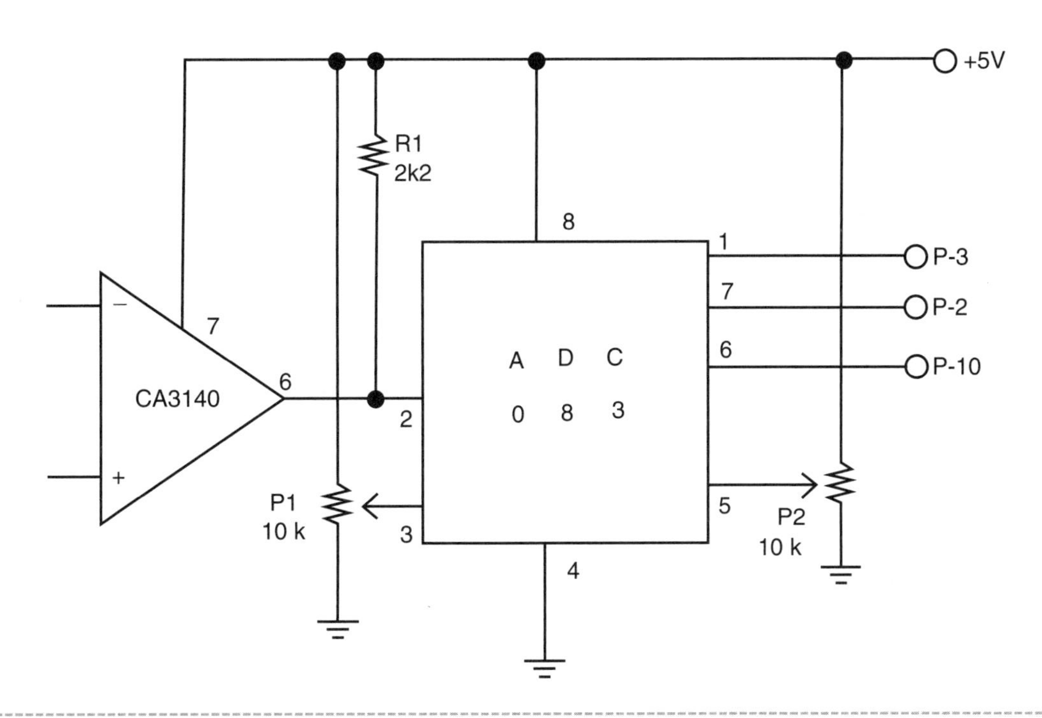


Figure 3.6.10 *Interfacing with the PC*

# Project 7—Panic Generator

What kind of device can cause a penetrating sound to resound inside your head? Of course, to know the answer for the question, you don't need to implant a loudspeaker inside someone's ear.

With the interesting circuit described here, you can make exciting experiments involving some strange acoustic phenomenon, such as interference, an ultrasound beat, and others that will be examined in this section. The circuit discussed here can produce sounds inside your ear without any physical connection or need for a device inside the head. But the most interesting aspect is that the person can't determine where the sounds are coming from, causing strange and frightened reactions, one of them being described in the name of the project: panic!

The panic generator was originally proposed by the author in a project published in *Electronica Total* (Brazil) in 1992. Since that time the circuit has been modified and revised to achieve a better performance using new devices that weren't available at the time of the original writing. The new version is presented here to the reader who wants to make some experiments involving an acoustic phenomenon that is not well known.

## Bionic Applications for the Project

The interactions between living beings and the sound source represented by the panic generator can be used in many bionic applications, some of them with unknown and imprevisible results. Experiments in this field can be considered very interesting.

### How It Works

The operation principle of the panic generator is a phenomenon called *beat*. So, to understand its usage, let's begin by explaining what *beat* is.

You may have learned about tuning forks in schoolbooks (see Figure 3.7.1). When they are struck, they vibrate at their natural frequency, which is determined by their size, material, and format.

An interesting phenomenon occurs when two tuning forks cut to different frequencies vibrate at the same time and in the same place. If you listen carefully, you not only hear the two basic tones (for which they are cut) emitted by the forks, but you may hear both a much lower and a much higher tone. This phenomenon occurs when the two tones are mixed inside your ear, or specifically in the timphanon membrane.

The two additional frequencies heard have the sum and the difference between the original frequencies produced by the tuning forks. If one fork vibrates at 800 Hz and the other at 300 Hz, it is very possible that you could hear a lower-frequency 500 Hz tone (800 − 300) and a higher-frequency 1,100 Hz tone (800 + 300), as suggested in Figure 3.7.2. This phenomenon, caused by the interference of the sound waves, is called *beat* and is also used by electronic circuits with high-frequency signals.

If you examine how the two additional frequencies are produced (1,100 and 500 Hz), you will find that they occur because each point of the eardrum (tympanic membrane) receives, at the same time, vibrations from the two sources. Each part has to vibrate at the same time from the two frequencies. The

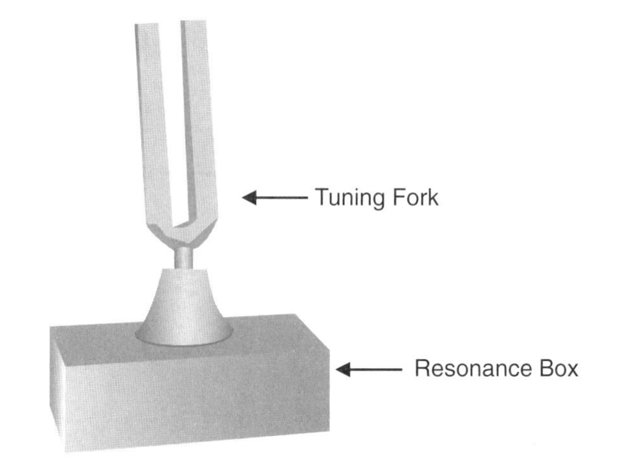


Figure 3.7.1 *A tuning fork*

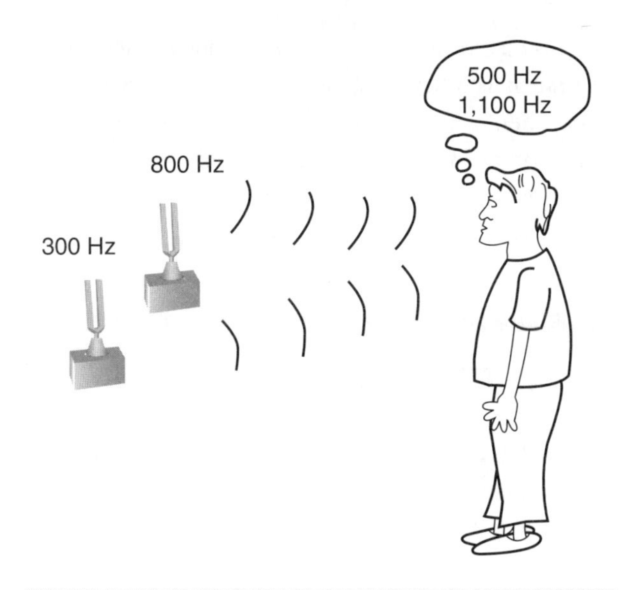


Figure 3.7.2 *The sum and difference of frequencies produced by two tuning forks*

movements caused by the two vibrations are combined and, as result of this mixing, two new frequencies are produced.

The important thing to notice in this phenomenon is that the two new frequencies (sum and difference) are not produced by the sound source but in the areas where the vibrations occur. So the two new frequencies take place inside your ear because the eardrum is generating the new vibrations. Thus, the 1,100 and 500 Hz vibrations are produced inside your ear!

Let's now go a step further and use the same experiments with ultrasonic sources, one of them operating at 19,000 Hz and the other at 20,000 Hz. (Remember that humans can't hear sounds above 18,000 Hz.) Of course, nobody can hear the two ultrasounds produced by the sources because they are operating above our hearing limit, but when the two tones are mixed inside the ear, they produce two new tones. One of them is the sum tone of 39,000 Hz, which is above the upper limit of our audible frequency range. But the other, the 1,000 Hz difference tone, can be heard, as it is in our audible range.

The interesting fact about this phenomenon is that the different tone is produced inside the ear, or precisely in the eardrum (tympanic membrane), so we have the strange sensation that the sound is generated inside our head or that it comes from nowhere, as shown in Figure 3.7.3.

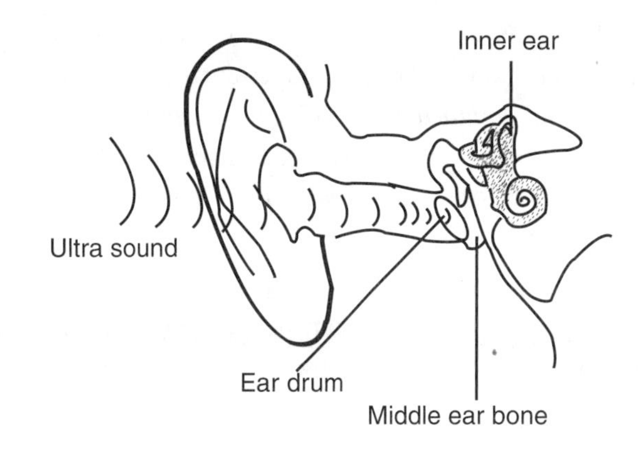


Figure 3.7.3 *Sounds from nowhere!*

Of course, if the ultrasounds are produced by high-power sources, the audible component generated inside our ear will cause a certain discomfort, and if the exposure to these effects takes place for a long period of time, it can even cause a sensation of panic. All of these factors are used in our project, which works as follows.

## How the Circuit Works

Two ultrasonic oscillators are mounted using two of the four gates in a 4093 *complementary metal oxide semiconductor* (CMOS) *integrated circuit* (IC), generating tones in the 20,000 Hz range. This frequency is chosen because small, common tweeters, such as the ones used in common audio equipment, can reproduce quality frequencies of up to 22,000 Hz.

The two remaining gates of the 4093 are used as low-frequency oscillators, modulating in frequency the high-frequency signals produced by the other oscillators. The two ultrasonic oscillators' frequencies change slowly to values under and above 20,000 Hz in a modulation process.

This process has an additional effect on the subjects. The beat changes the frequency, becoming less or more penetrating, like a siren. Because the modulations are not synchronized, the ultrasonic oscillations change their frequencies such that a random beat in the audible range is produced.

The signals of the two modulated oscillators are applied to powerful output stages using power *metal-oxide-semiconductor field effect transistors* (MOSFETs). These transistors can source some watts from ultrasounds to small, piezoelectric tweeters, as shown by Figure 3.7.4.

This random beat is in the audible range and, as we explained before, appears inside the ear, causing the strange effect of a deep sound coming from nowhere.

Figure 3.7.5 shows the waveshapes in different areas of the electronic circuit and also in the timphanon of a subject experiencing the vibrations produced by the circuit.

This electronic circuit can serve many practical uses, such as functioning as an alarm or as part of experiments involving human behavior, animal behavior, plant growth, and sound pollution. Other experiments related to music reproduction without loudspeakers are also suggested.

## How to Build

Figure 3.7.6 shows the schematic diagram of the panic generator.

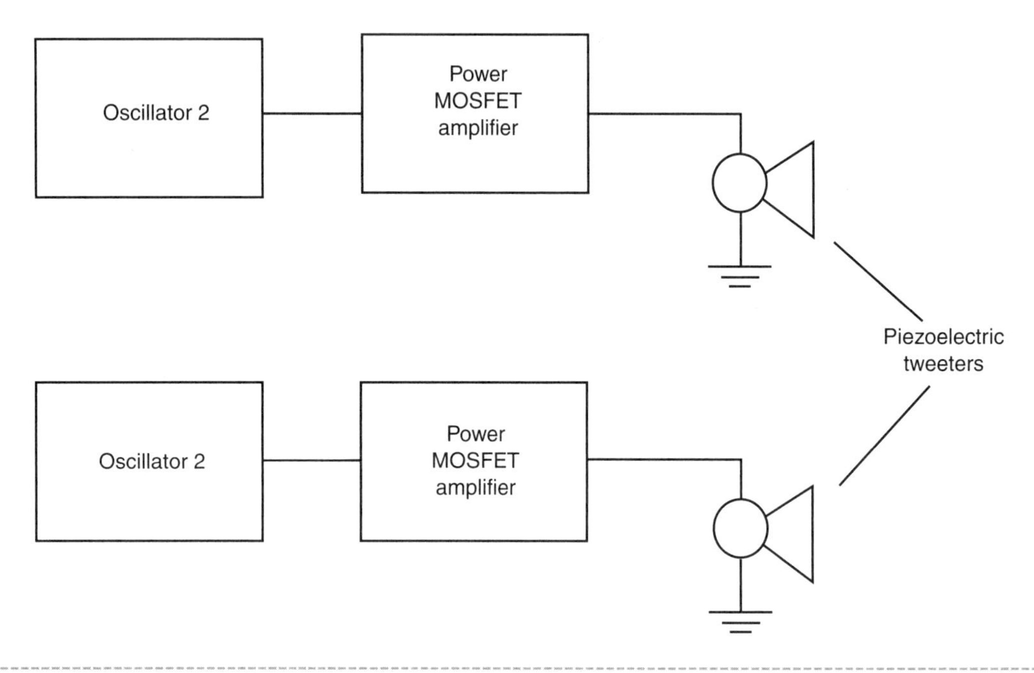


**Figure 3.7.4** *Block diagram for the panic generator*

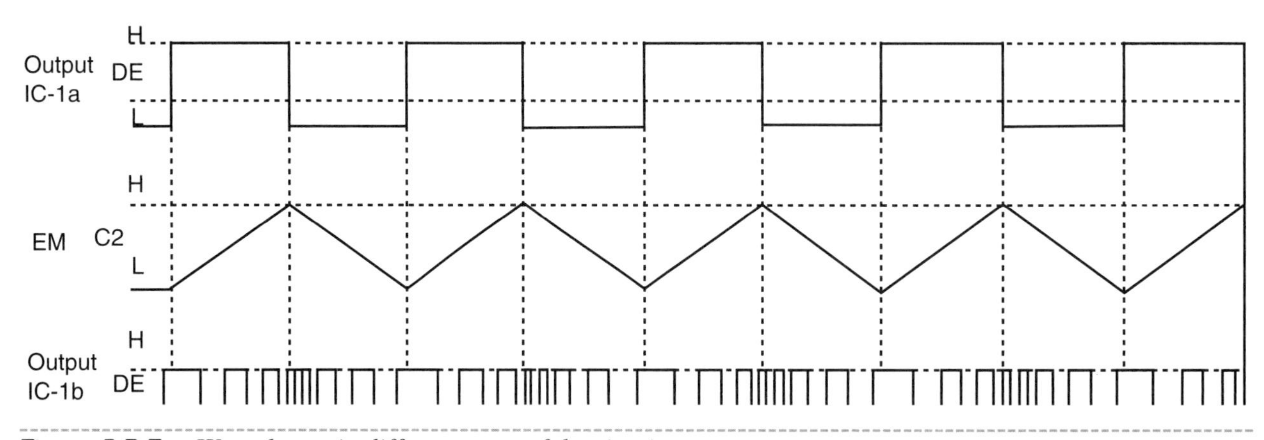


**Figure 3.7.5** *Waveshapes in different areas of the circuit*

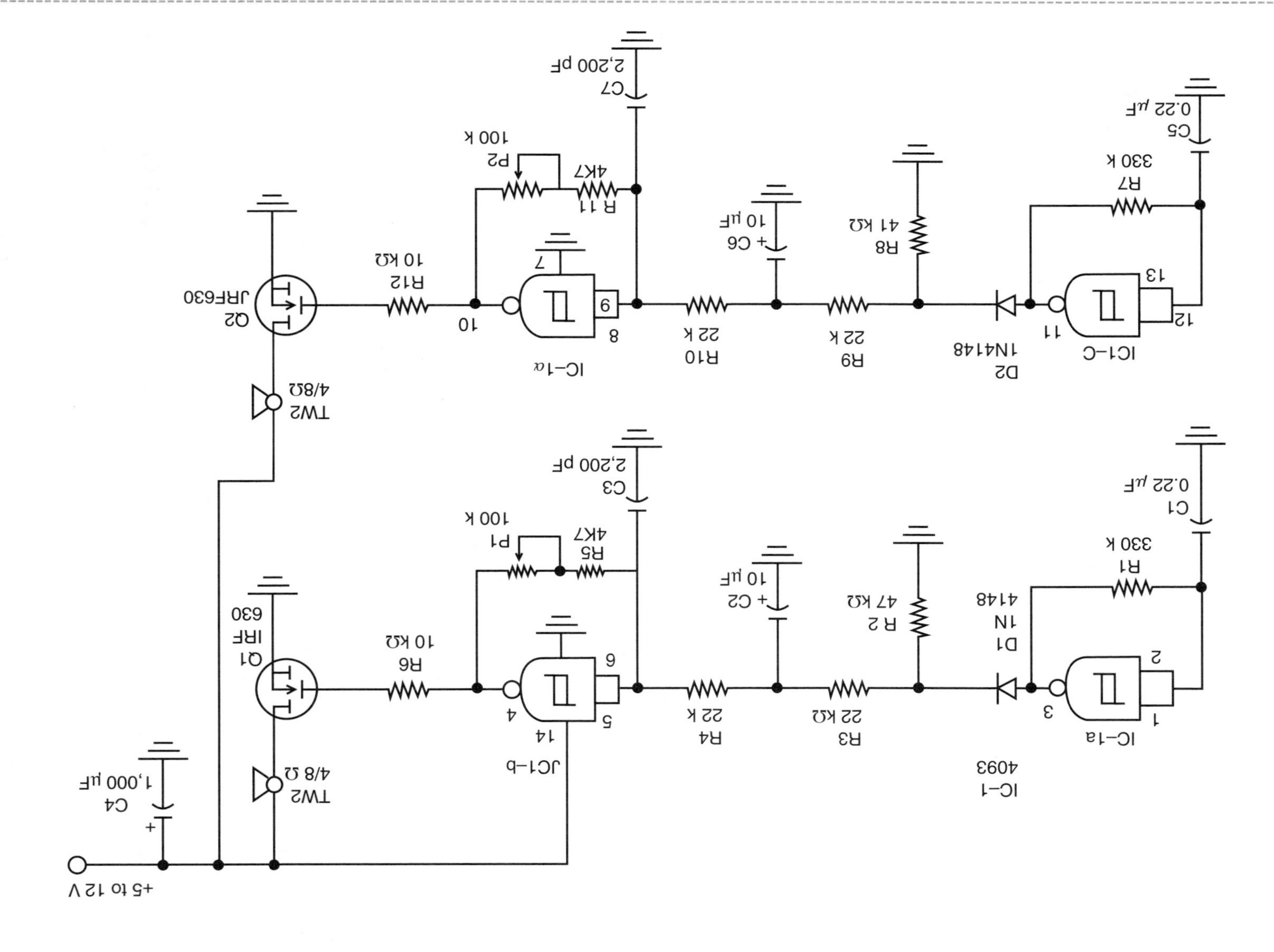


Figure 3.7.6 *Schematic diagram for the panic generator*

Figure 3.7.7 *Panic generator mounted on a PCB*

The circuit can be mounted on a *printed circuit board* (PCB), as shown in Figure 3.7.7.

In the figure, the transistors do not appear with their heatsinks, but they are necessary. Any power FET with a drain-source voltage rated to 200 volts or more and a drain current up to 2 amps can be used. The pattern for the PCB is suggested in Figure 3.7.8.

The tweeters are miniature types rated to 80 watts with a frequency response of up to 22,000 Hz or more. The circuit must be powered from a 9- to 12-volt × 3-amp source. A power supply for this circuit is suggested in Figure 3.7.9.

The transformer has a primary according to the power supply line voltage (117 VAC, for instance) and a secondary for 7.5 or 12 volts with currents ranging between 3 and 4 amps.

## Testing and Using

Place the tweeters 12 to 20 feet from each other. Wire one tweeter to the corresponding output, and turn on the circuit. Next, adjust the corresponding trimpot to cause the ultrasounds to occur (the sounds rise to the treble and disappear).

Then disconnect the tweeter and wire the second one, making the same adjustments. Disconnect the power supply and wire the two tweeters to their outputs. The effect should result in the sounds being produced inside your ear.

When conducting this experiment, take care not to cause harm to anyone who might be exposed to the sounds. Also, do not use the panic generator during long intervals where people or animals are around.

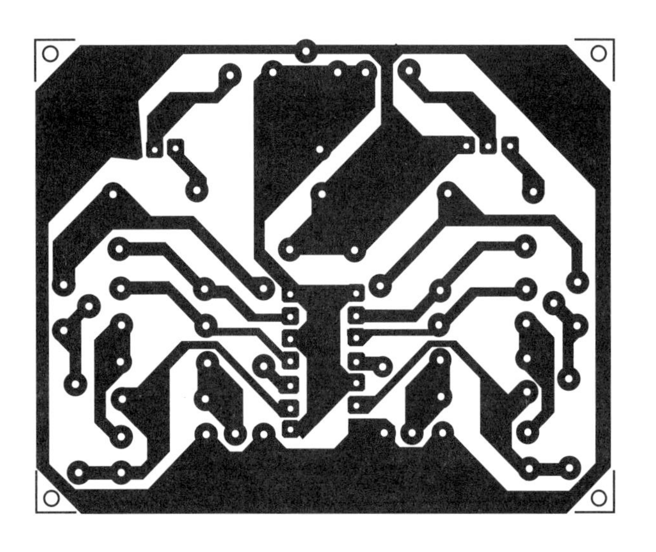

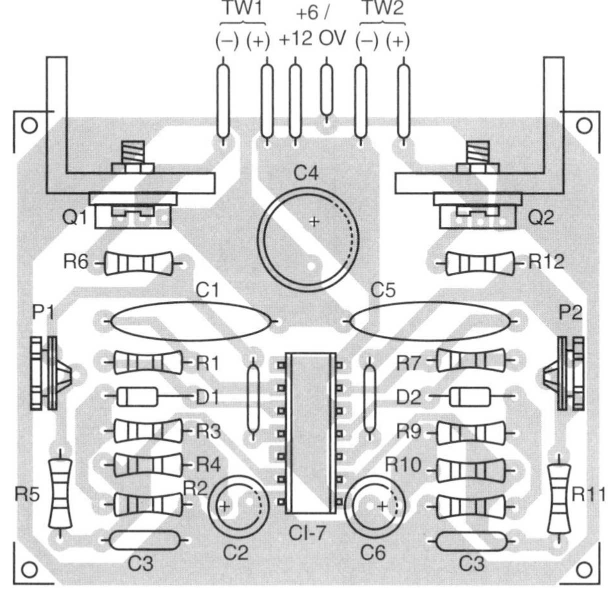


Figure 3.7.8 *The suggested PCB*

## Parts List: Panic Generator

IC1: 4093 CMOS integrated circuit

Q1, Q2: IRF630, IRF620, or equivalent power FET (see text)

D1, D2: 1N4148 general-purpose silicon diodes

R1, R7: 330 kΩ, 1/4-watt, 5% resistors, orange, orange, yellow

R2, R8: 47 kΩ, 1/4-watt, 5% resistors, yellow, violet, orange

R3, R4, R9, R10: 22 kΩ, 1/4-watt, 5% resistors, red, red, orange

R5, R11: 4.7 kΩ, 1/4-watt, 5% resistors, - yellow, violet, red

R6, R12: -10 kΩ, 1/4-watt, 5% resistors, - brown, black, orange

P1, P2: -100 kΩ - trimmer potentiometers or common potentiometers

C1. C5: -0.22 μF ceramic or metal film capacitors

C2, C6: - 10 μF/12 working voltage DC (WVDC) electrolytic film capacitor

C3, C7: 2,200 pF ceramic or metal film capacitor

C4: 1,000 μF/16 WVDC electrolytic capacitor

TW1, TW2: 4 or 8 Ω, mini-piezoelectric tweeters to 80 watts or more

Other parts: PCB or solderless board, wires, power supply (see text), heatsinks for the transistors, plastic box, etc.

## Additional Circuits and Ideas

The basic version of the panic generator can stand alone as a complete device, but the reader who wants to make further experiments has many options to explore. Some possibilities are given here.

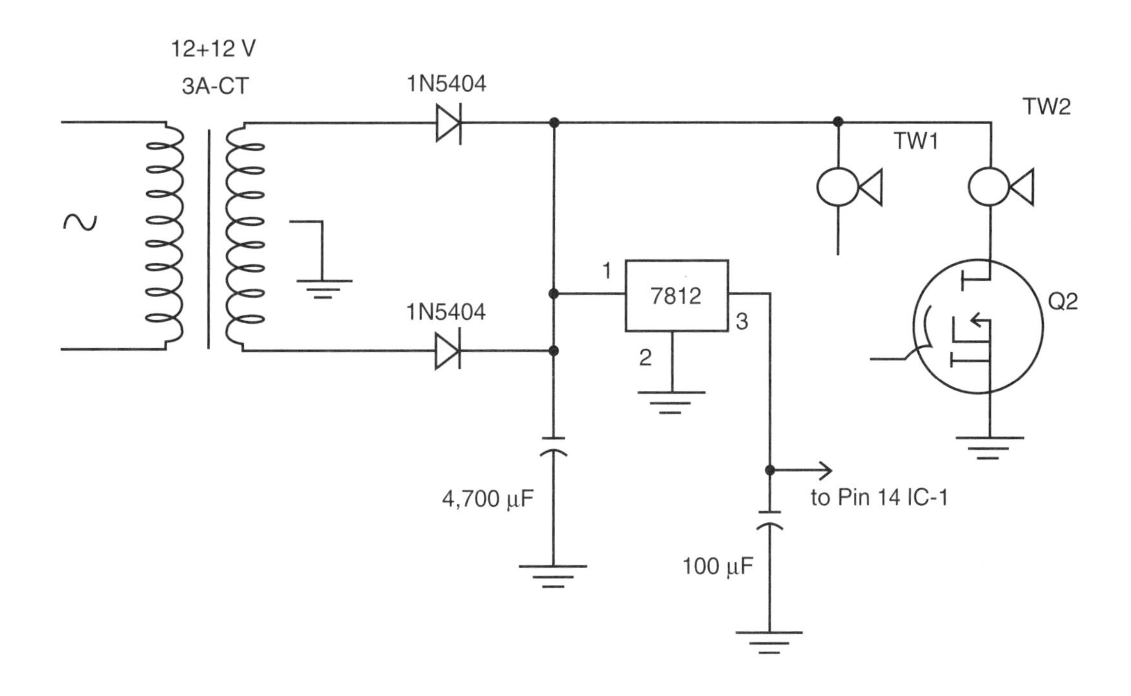


Figure 3.7.9 *Power supply for the circuit*

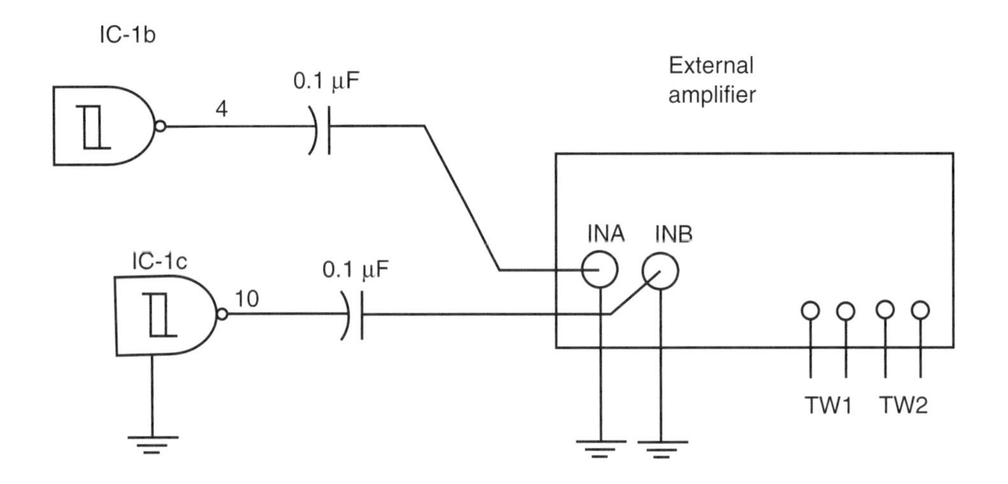


Figure 3.7.10 *Using an external amplifier*



## Using External Amplifiers

If the reader has a good, high-power amplifier, he or she can use it to power the tweeters, which must accept all the power of the amplifier.

It is only necessary that the amplifier have a good response to frequencies in the range between 15 and 25 kHz. Care must be taken not to overload the amplifier's output with these signals. Figure 3.7.10 shows how the external amplifier is connected to the oscillators, canceling the output transistors.

## Working with More Sources

The effects of the beat can be increased if more than two modulated ultrasound sources are used. If three sources are used, for instance, we will have six combinations of beats, with effects that even the author can preview. Figure 3.7.11 shows how two circuits like the ones described in the basic version can source four ultrasonic signals to fill an ambient space.

## A Circuit Using the 555 IC

Another circuit of a panic generator can be built using the famous 555 IC, as shown in Figure 3.7.12.

Four 555 ICs are used as astable multivibrators in this configuration. Two of them generate the high-frequency tones, applying them to powerful output transistors. Power MOSFETs are also mounted on heatsinks, and the frequency of these oscillators are adjusted by trimpots.

The other two 555s generate the modulation signals and are adjusted by the correspondent signals. The performance is the same as the basic version.

P3 and P4 adjust the amplitude of the modulation. Small changes in the components' values could be necessary to compensate for their tolerances and achieve the best performance.

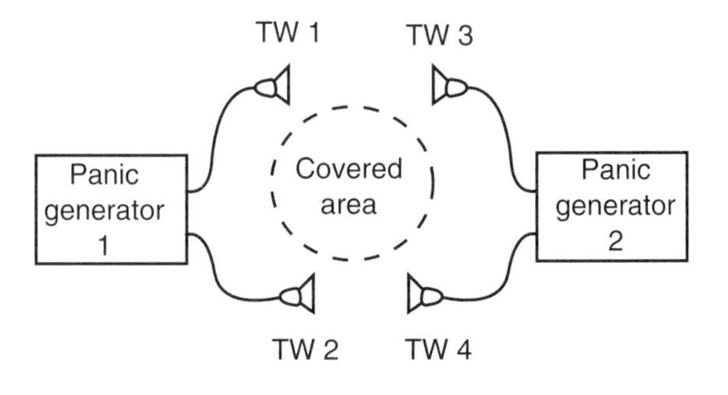


Figure 3.7.11 *Using four supersound sources*

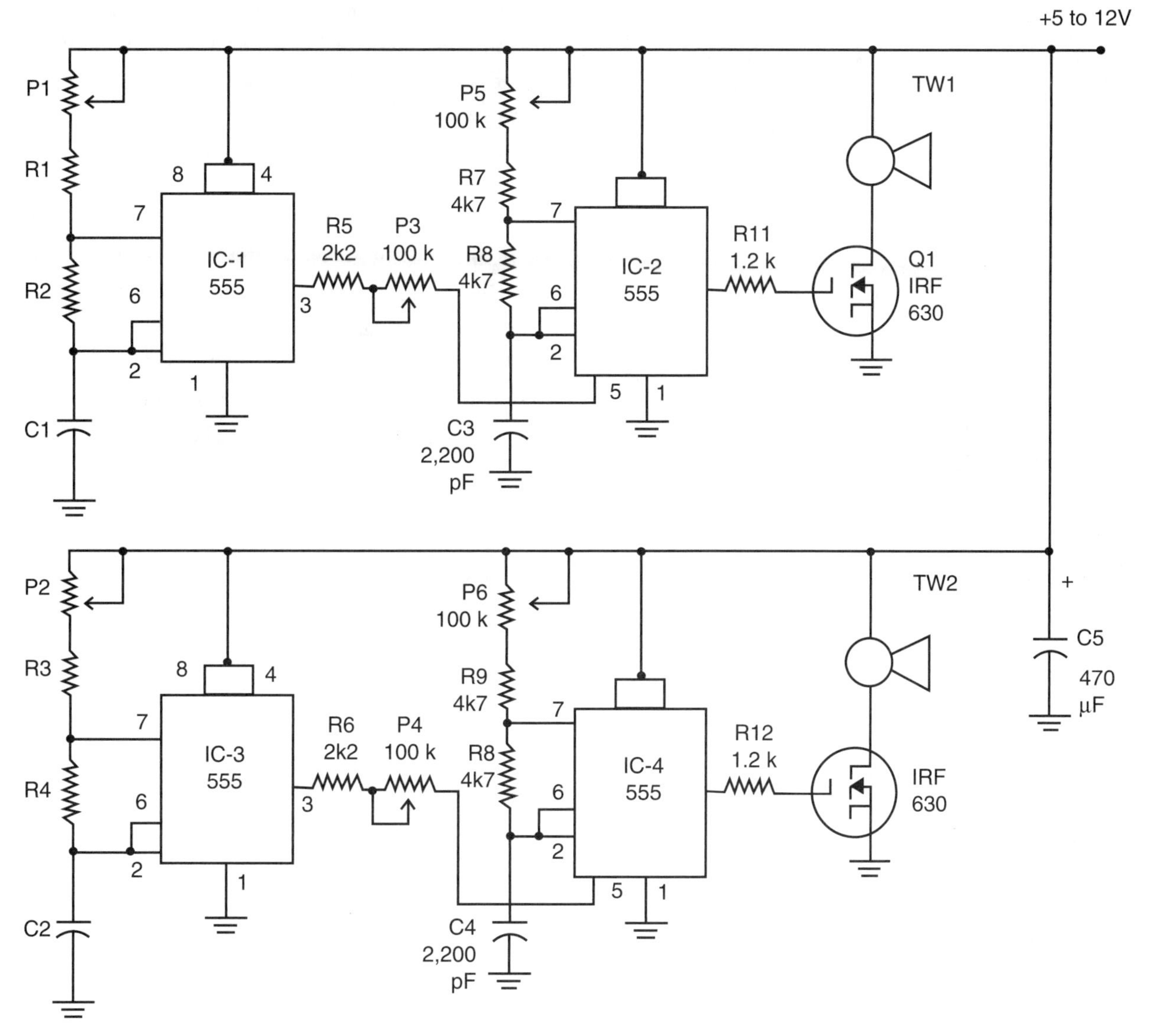


Figure 3.7.12 *A panic generator using 555 ICs*



## Parts List

IC-1 to IC-4: 555 integrated circuits

Q1 and Q2: IRF630 or any equivalent-power MOSFETs

R1, R2, R3, R4: 22 kΩ x 1/8 W resistors, red, red, orange

R5, R6: 2.2 kΩ x 1/8 W resistor, red, red, red

R7, R8, R9, R10: 4.7 kΩ x 1/8 W resistors, yellow, violet, red

R11, R12: 1.2 kΩ x 1/8 W resistors, brown, red, red

P1, P2, P3, P4, P5, P6: 100 kΩ x 1/8 W trimpots

C1, C2: 0.047 μF ceramic or polyester capacitors

C3, C4: 2.200 pF ceramic or polyester capacitors

C5: 470 μF x 16 V electrolytic capacitor

TW1, TW2: 4 or 8 Ω mini-piezoelectric tweeters to 80 watts or more

Other parts: PCB or solderless board, wires, power supply (see text), heatsinks for the transistors, plastic box, etc.

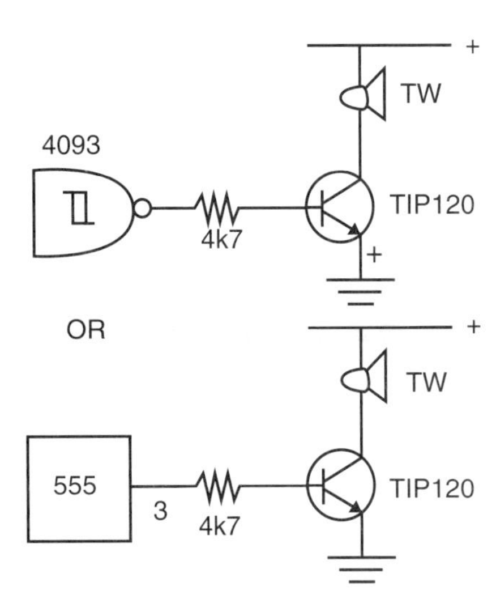


Figure 3.7.13 *Using Darlington transistors*

## Using Darlington Transistors

Power NPN Darlington transistors can be used, replacing the power MOSFETs. Figure 3.7.13 shows how to make this replacement.

No change is necessary in the PCB and in the basic circuit. Types such as the TIP120, TIP121, and TIP122 are suitable for this project. They also need to be mounted on heatsinks.

## Ideas to Explore

The effects of combining beat and ultrasounds are not well known within the science field. Recently, a program was shown on the Discovery Channel where the effects of many types of sounds, including low frequencies, were discussed. They didn't report anything especially dangerous or that any special effect affected internal organs, but we recommend care when working in this area.

Experiments have shown that exposure to ultrasounds and infrasounds (sounds below the audible limit or below 16 Hz) for long time intervals can cause digestive problems such as nausea and dysentery. We strongly recommend again to the reader to take care during experiments involving this circuit for long periods of time. Turn your device off if you feel something wrong with your stomach or intestines.

The author wrote some years ago an interesting comic story involving a mad professor who developed a powerful ultrasonic generator modulated by low frequencies (infrasounds) such as the one described here. Professor Ventura used it to scare birds from a particular square in his city. The birds were perched in the trees and anyone who tried to walk under them felt a splash on the head.

The professor was able to use his device to scare the birds away, but the side effect of the circuit was not anticipated by Professor Ventura: dysentery. After the mayor announced that the square was free from the birds, the effect exploded like a bomb: Any of the citizens who spent some time in the square left in a panic with a terrible dysentery crisis.

This idea was explored on the Discovered Channel where some documentary filmmakers tried to reproduce these effects using low-frequency, powerful amplifiers. Although the story exaggerates the supersound effects, we suggest that the reader try not to repeat this experience.

However, we'd like to present some ideas for some bionics experiments using your panic generator with friends and other living beings:

- Alter C2 and C6 to have a different modulation pattern.
- Use this circuit as part of an alarm, and control the power supply via a relay.
- Place the circuit in a location where you could attract the attention of people or animals.
- Expose volunteers to this circuit during time intervals between 10 and 15 minutes (but no more to prevent causing a panic). After the session, test the people's skill in some activities. The reader could determine how sonorous pollution can affect human reactions, such as causing stress.
- You can also use this circuit as an alternating method of pest control on farms. Investigate how can birds or insects can be affected by the vibrations produced by the panic generator. Be sure to keep domestic animals away from the device as exposure to the vibrations could be harmful.

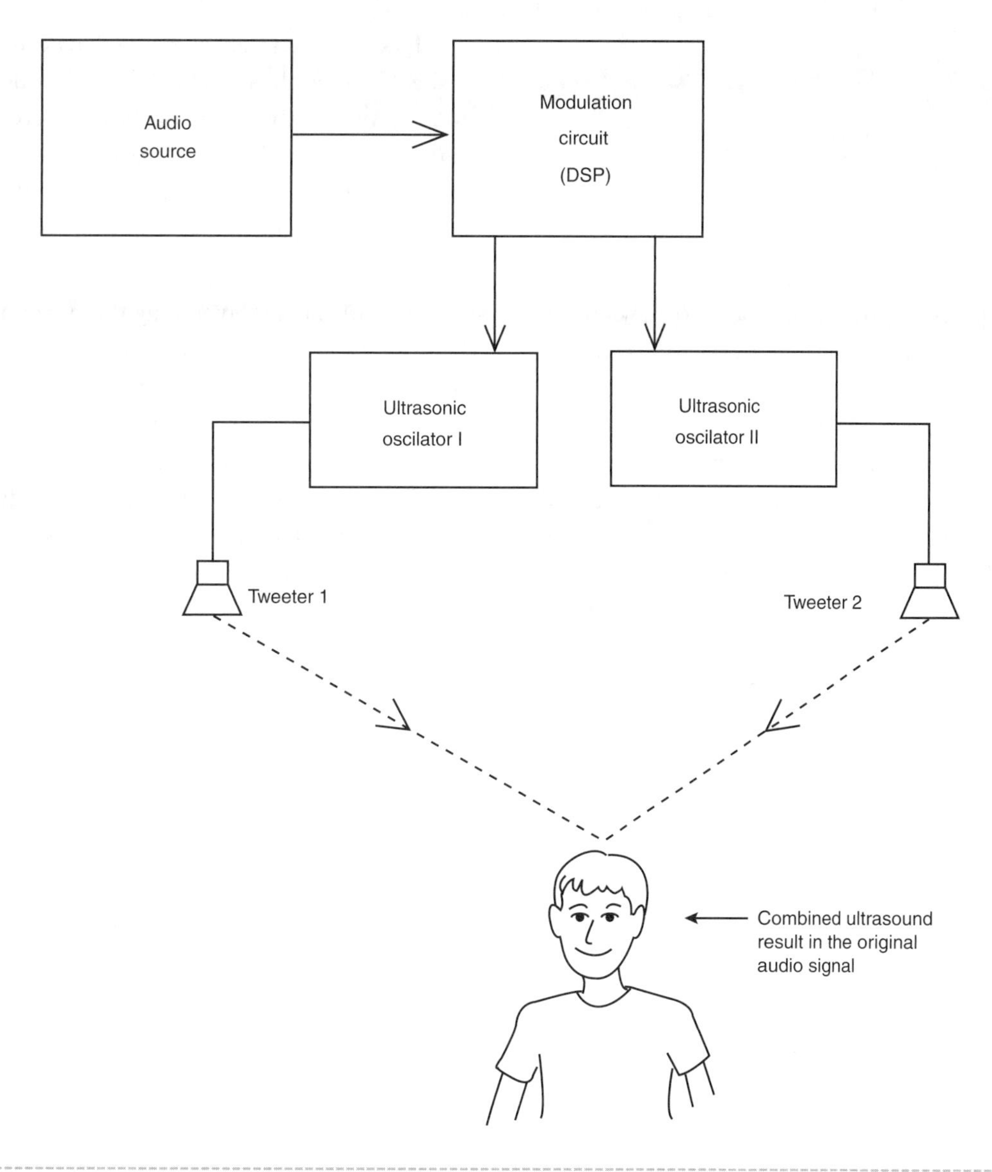


Figure 3.7.14 *A brontophonic* sound reproduction system, proposed by the author*

- Experiments in growing plants and germination can also be conduced using the panic generator. Select two groups of fast-growing plants, such as bean plants, and measure the rate of growth of each group: the one exposed to the panic circuit vibrations and the other that isn't. Then compare the growth.
- In an experiment involving the effect of magnetic fields on plant growth or germination, you can use coils instead of tweeters and apply the signals to the plants. The coils can be made by wiring 30 to 50 turns of common wire around cardboard squares (15 × 15 cm).
- Another possibility for this circuit is determining how ultrasounds can affect organic processes or interact with them. You can perform experiments with plant growth to see how noise and sounds affect common plants such as tomatoes and beans.

*Brontophonic comes from *brontus* a Latin word meaning thunder. This describes a kind of sound that produces the low frequencies that cause other vibrations in a higher frequency to occur, such as thunder vibrating a window. This can also occur when a heavy truck passes by in the street.

## An Idea to Explore: A New Sound-Reproducing System

You can use a new form of sound reproduction using the beat, as proposed by the author in 1992, where the sounds are generated inside the listener's ear without the need for loudspeakers or earphones. Two oscillators coupled to powerful ultrasonic transducers will be used. One of them operates in a fixed frequency and the other in the same frequency but modulated by a sound processor circuit, such as an audio amplifier. The beat resulting from the two super-sound signals inside the listener's ear will reproduce the sound applied to the audio amplifier, such as the sound from a CD player, microphone, or other source, as suggested in Figure 3.7.14.

Make sure that the listener will hear the sounds inside his or her head as the beat is produced in the timphanon membrane. he or she will have the strange sensation that the sound comes from nowhere. An American laboratory is now working with a commercial sound reproduction system using this effect, which is not quite so new as they claim.

The difficulty in developing this circuit is that the beat should correspond in frequency and intensity to the original sound, and a simple direct modulation is not enough for that. The signal to modulate the ultrasonic sound needs complex processing that would require some research. It is a very interesting subject to be explored by the reader who likes advanced work in bionics, electronics, and acoustics.

# Project 8—Magnetic Field Generator



Circadian rhythms are the natural changes that take place in living beings due to the day-night cycle. Important organic changes can be observed in many species of living beings, especially plants, due to this cycle.

The interaction of living beings with electronic devices that are controlled by circadian rhythms or circuits that influence circadian rhythms are interesting fields of bionics to explore. The reader can observe what happens with plants or other living beings, such as insects or even small mammals, when exposed to external influences such as electric fields, magnetic fields, or light of different wavelengths (colors). Experiments can induce changes in the animal or plant's behavior, and what is discovered can be used to create new bionic devices.

What we are proposing here is to mount a magnetic field generator for this kind of experiment. The circuit will generate high-current pulses that, flowing across a coil, will create a pulsed magnetic field. Figure 3.8.1 shows how the coil can be placed around a plant, applying the field to it.

The coil can be placed around cages of insects or other animals according to the experiment to be performed. The magnetic field can also be used to excite plants to grow faster or to produce larger fruit. We suggest experiments with tomato plants, which are easy to cultivate and observe.

In the project described here, the field strength can be controlled by a large range of values and also by the frequency. In special experiments with low-frequency fields, caution must be taken. The fields created by the low-frequency power lines are dangerous, causing cancer and other problems, and studies with plants and other living beings confirm this.

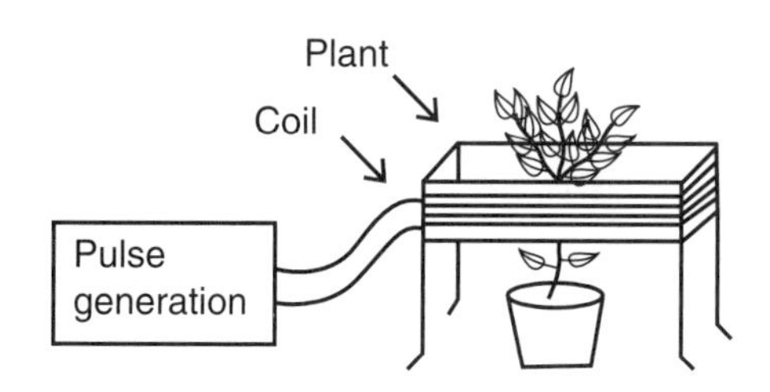


Figure 3.8.1 *Applying a magnetic field to a plant*

## Bionic Experiments and Projects

Many experiments and applications can be suggested based on this circuit:

- Study how circadian rhythms change when magnetic fields are applied to plants or insects.
- Conduct research to see how fruit develops under the influence of magnetic fields.
- Determine if magnetic fields can cause a change in the behavior of insects such as ants or bees.
- See how magnetic fields can influence the behavior of small aquarium fish.

It is important to know that the low-frequency fields produced by the device are very weak and are not harmful for humans, because they are concentrated only inside the coil where it is more intense, as shown in Figure 3.8.2. Otherwise, the device's strength is many times lower than the low-frequency magnetic fields produced by domestic appliances such as electric razors or appliances with motors that are much more dangerous.

## How It Works

The basic project consists of a relaxation oscillator plugged into the AC power line that has a *silicon-controlled rectifier* (SCR) as its main component. Figure 3.8.3 shows the block diagram for this circuit.

The AC power line voltage is rectified by D1, charging the capacitor C1 through R1, with the peak value of this voltage at about 150 volts in the 117

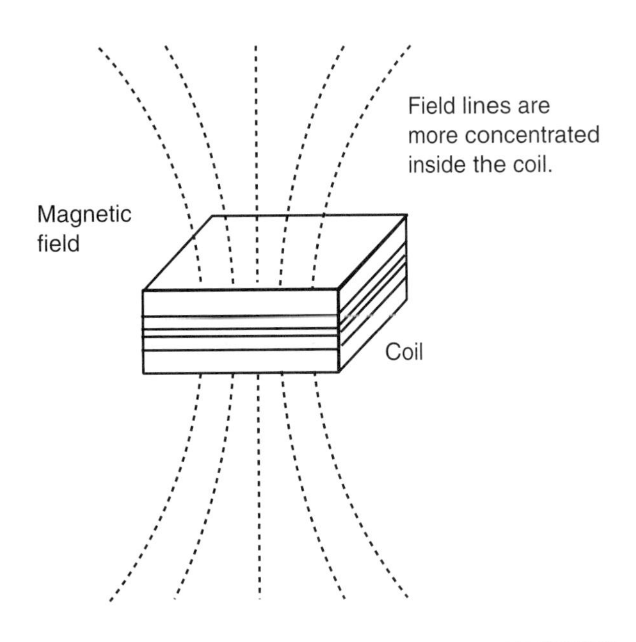


Figure 3.8.2 *The field is concentrated inside the coil.*

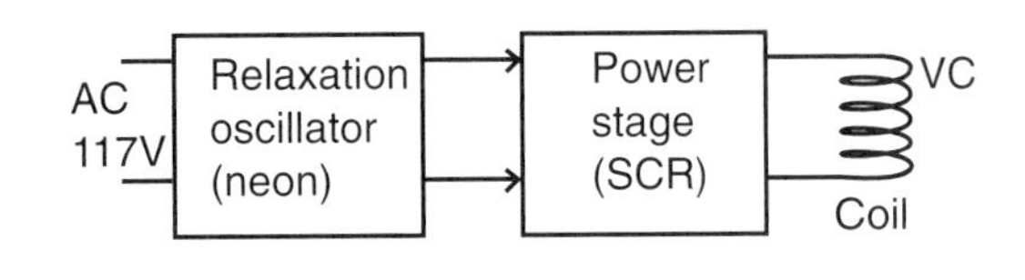


Figure 3.8.3 *Block diagram for this circuit*

VAC power line. The SCR, P1, R2, and C2 form a relaxation oscillator with a neon lamp. A coil wired in series with the anode of the SCR is the load.

This circuit works as follows: The capacitor C2 charges across the potentiometer P1 and R1 until the trigger voltage of the neon lamp is reached. This voltage is about 80 volts for common neon lamps, as recommended by the NE-2H for this circuit.

When the neon lamp is triggered, the capacitor C2 discharges to the gate of the SCR, triggering this component to the on state. In the on state, the SCR, C1, and the coil form a closed circuit. The capacitor C2 then discharges through the SCR, producing a powerful current pulse in the coil. This pulse produces the magnetic field for the experiment. As long as the voltage in C1 falls below the holding value, the SCR triggers off and a new charging cycle begins.

Triggering on and off, the relaxation oscillator produces high-power magnetic pulses in the coil. The strength of the pulses and the frequency depend on several factors, such as the following:

- The value of C1 determines the intensity of magnetic pulses; capacitors with a range between 1 and 32 $\mu$F can be used. This component also determines the frequency. Larger capacitors will produce stronger pulses, but the frequency is lower.
- The value of C2 and any adjustments to P1 determine the frequency.
- The number of turns of the coil determines how strong the magnetic field is.

Playing with these three parameters, the reader can create a version customized to the experiments or applications he or she has in mind.

## How to Build

Figure 3.8.4 shows the schematic diagram for the magnet field generator. The simplest way to mount this circuit is to use a terminal strip to fit the components, as shown in Figure 3.8.5. Of course, if the reader has the resources, he or she can mount the project using a *printed circuit board* (PCB).

The capacitor C1 is a high-voltage polyester or electrolytic type. If an electrolytic capacitor is used, the position must be observed because it is a polarized component. The minimum voltage for this component is given in the parts list. C2 is any polyester capacitor rated to 100 volts or more. R1 is a wire-wound resistor, and the values indicated between the brackets are for the 220/240-volt AC power line.

The coil is formed by 50 to 500 turns of any wire in a box that fits the assembly. You can use enameled wire (AWG) from 22 to 32 or common plastic wire of any gauge.

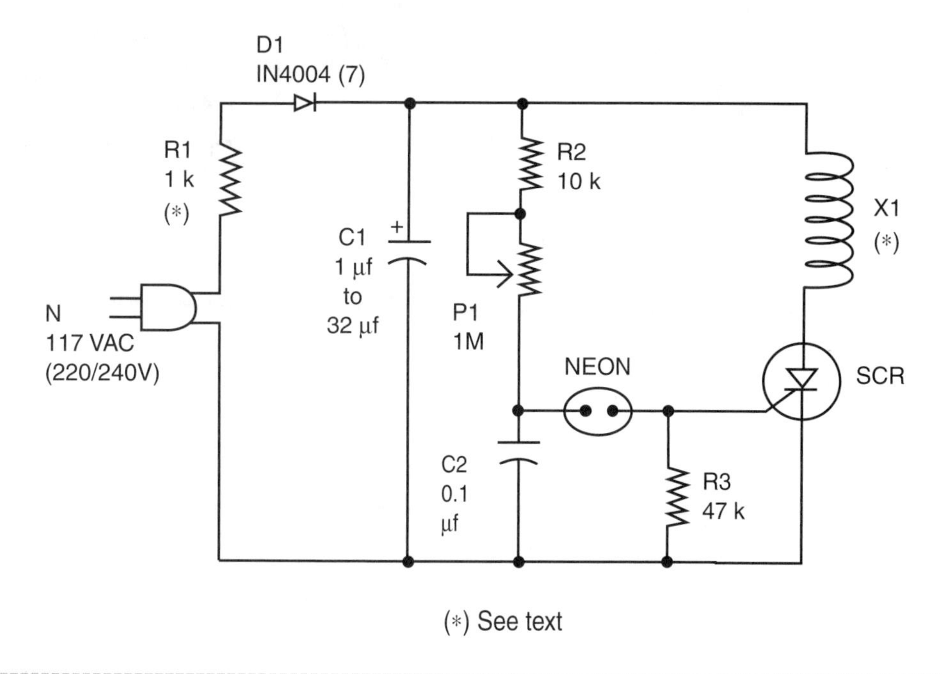


Figure 3.8.4 *Schematic for the magnetic field generator*

## Testing and Using

Plug the power cord into the AC power line. Adjusting P1, the neon lamp will flash. The flash rate depends on the value of C1 and the adjustment of P1.

The experiment is then conducted, placing the specimens inside the coil. Decide what the time frame will be and what you are intending to observe. Use additional resources to measure the changes in the specimen's behavior, such as a thermometer.

## Parts List

SCR: TIC106B(D) SCR

D1: 1N4004(7) silicon rectifier diode

NE: Neon lamp, NE-2H or equivalent

R1: 1 kΩ x 10 W (2.2 kΩ) wire-wound resistor

R2: 10 kΩ x 1/8 W resistor, brown, black, orange

R3: 47 kΩ x 1/8 W resistor, yellow, violet, orange

P1: 1 MΩ lin or log potentiometer

C1: 1 to 32 µF x 200 V (400 V) polyester or electrolytic capacitor (see text)

C2: -0.1 µF x 100 V polyester capacitor

X1: Coil (see text)

Other Parts: Terminal strip or PCB, power cord, wires, box, etc.

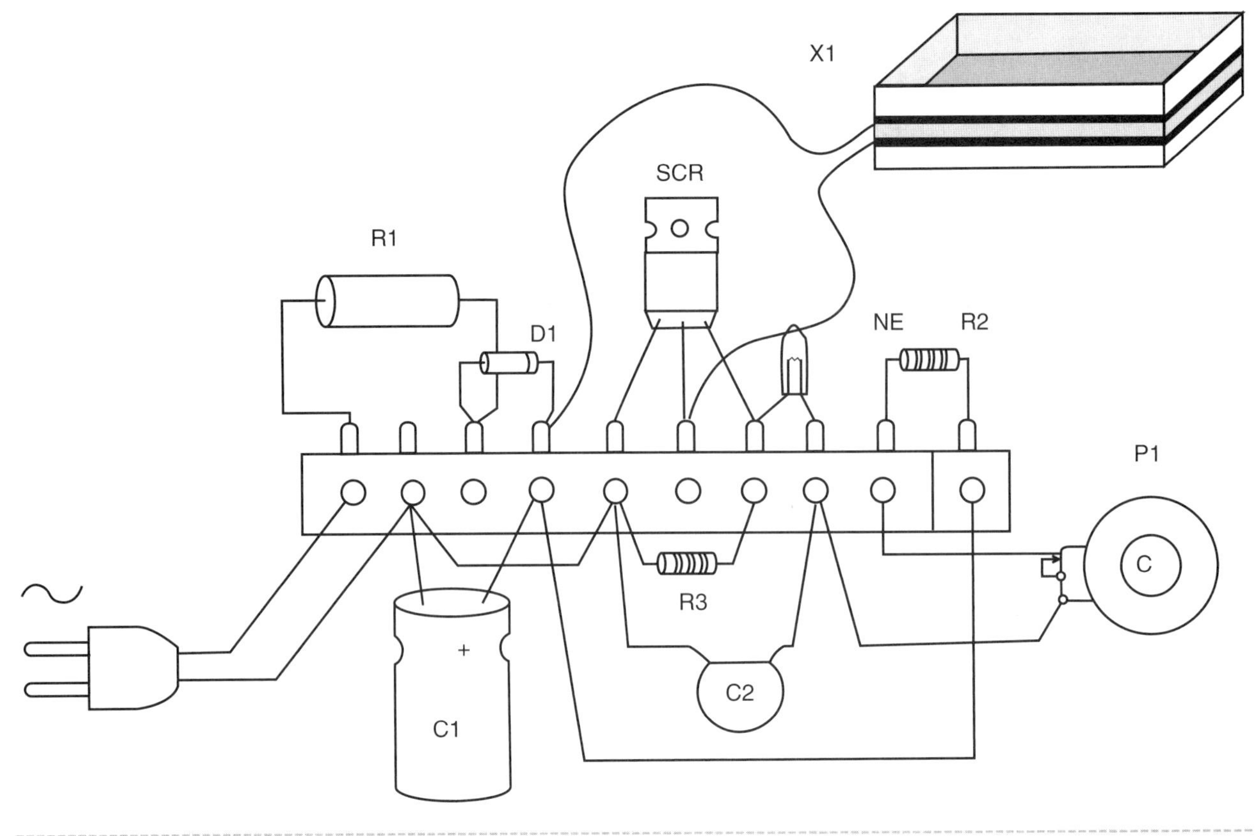


Figure 3.8.5 *Using a terminal strip to mount the circuit*

## Additional Circuits and Ideas

Whether they'll be used in experiments with living beings or for other means, producing low-frequency magnetic fields can be done in many different ways. Some circuits can also produce high-frequency fields, which are covered in this section.

### Using the AC Power Line

A very simple way to generate an alternating magnetic field with the frequency of the power line is shown by the circuit in Figure 3.8.6.

The advantage of this circuit is the reduced number of components. Otherwise, the main disadvantage is that it has a fixed frequency: 60 Hz.

The strength of the field can be adjusted, selecting the resistor in series with the coil. The resistor has an important role in limiting the current to secure values. A current larger than that supported by the transformer can burn this component.

The coil is the same used in the basic experiment. This circuit has the advantage of not being plugged directly to the AC power line. The transformer adds isolation, meaning that anyone can touch live parts of the circuit without the danger of shock hazards.

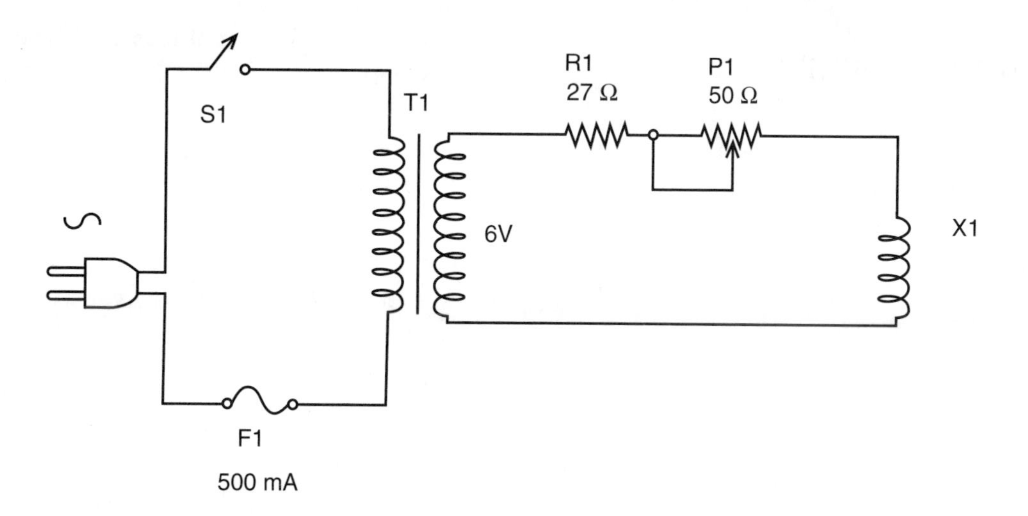


Figure 3.8.6 *Using the AC power line*

## Parts List

**Required Parts**

T1: 117 VAC (or 220/240 VAC) primary and 6 V x 300 mA to 500 mA secondary transformer

R1: 27 ohms x 5 W wire-wound resistor

P1: 50 ohms wire-wound potentiometer

X1: Coil (see text)

S1: SPST on/off switch

F1: 500 mA fuse and holder

Other: Power cord, terminal strip or PCB, box, solder, wires, etc.

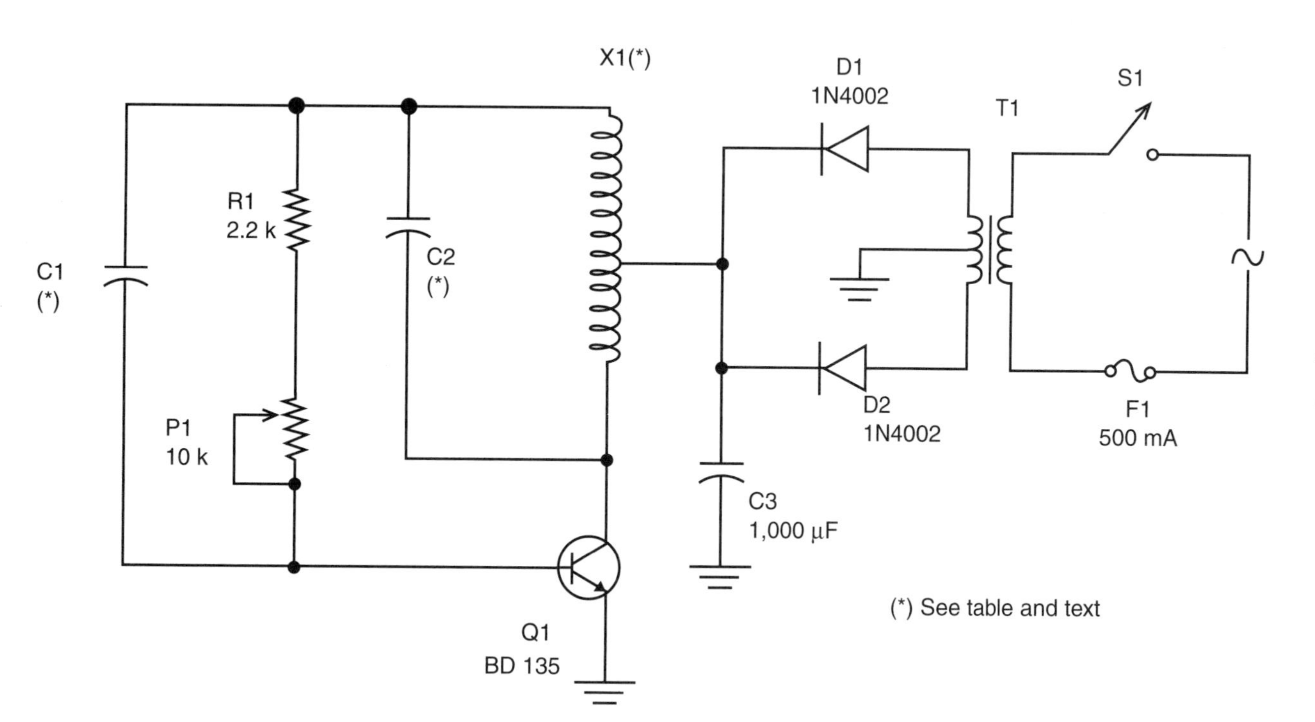


Figure 3.8.7 *Another magnetic field generator suitable for experiments in bionics*

## A Hartley Oscillator

The circuit shown in Figure 3.8.7 can be used to generate low- and medium-frequency magnetic fields. Depending on the components, the circuit can operate in the range from 1 to 10 kHz or from 1 to 5 MHz. The reader can conduct experiments using several circuits like this, covering a wide range of frequencies.

The frequency depends on the coil and can be adjusted in a narrow band of values by the potentiometer. The following table lists the number of turns used to cover the main frequency ranges.

| Frequency Range | C1/C2 | Coil |
|---|---|---|
| 1 kHz to 10 kHz | 0.022/0 .47 μF | 200 + 200 turns |
| 10 kHz to 50 kHz | 0.033/0 .22 μF | 150 + 150 turns |
| 50 kHz to 250 kHz | 0.01/0 .1 μF | 100 + 100 turns |
| 250 kHz to 1 MHz | 0.047/0 .047 μF | 60 + 60 turns |
| 1 MHz to 5 MHz | 0.022/0 .001 μF | 20 + 30 turns |

The coil can be wound around a piece of cardboard or plastic with its diameter, or its sides if a square, being 10 to 40 cm.

The transistor must be attached to a heatsink. This circuit is also isolated from the power supply line due to the presence of the transformer. This adds safety to the project.

## Parts List

- Q1: BD135 medium-power NPN silicon transistor
- D1, D2: 1N4002 silicon rectifier diodes
- C1, C2: See table
- C3: 1,000 μF x 12 V electrolytic capacitor
- R1: 2.2 kΩ x 1/2 W resistor, red, red, red
- P1: 10 kΩ potentiometer, lin or log
- X1: Coil (see table)
- T1: Transformer, with the primary according to the AC power line and the secondary rated at 6 to 7.5 V with the current at 300 to 500 mA
- F1: 500 mA fuse and holder
- S1: On/off switch
- Other: Power cord, terminal strip or PCB, wires, solder, etc.

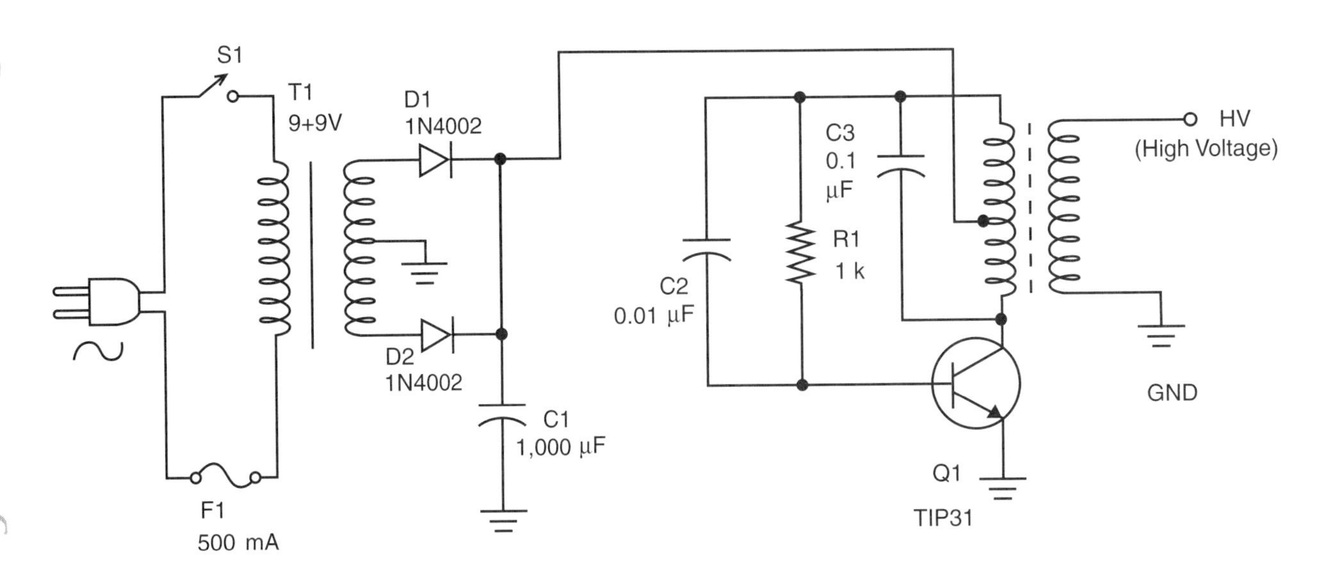


Figure 3.8.8 *High-voltage generator*

## Experimenting with Electric Fields

Electrostatic or electric fields, which are produced by a high voltage, can also be used in studies of circadian rhythms and external influences on living beings. The circuit shown in Figure 3.8.8 can be used to produce a voltage around 10,000 to 30,000 volts between two plates, where the specimens under observation are placed.

The strong electric field applied to the specimen can cause changes in its behavior. The reader can perform experiments to see how some fruit or a plant's flower is altered under the influence of this field, such as if it grows faster.

The circuit consists of a Hartley oscillator, having a high-voltage transformer as its load. This high-voltage transformer is a fly-back one, like the ones used to generate a high voltage in computer video monitors and television sets. Figure 3.8.9 shows this component and how the coil is wound in its core. This coil consists of two sets of 15 turns of common #22 or #24 wire placed in the low part of the core, as shown by the figure.

The high voltage in the secondary of the coil can be used directly if the reader wants to perform experiments with alternating electric fields, as shown in Figure 3.8.10.

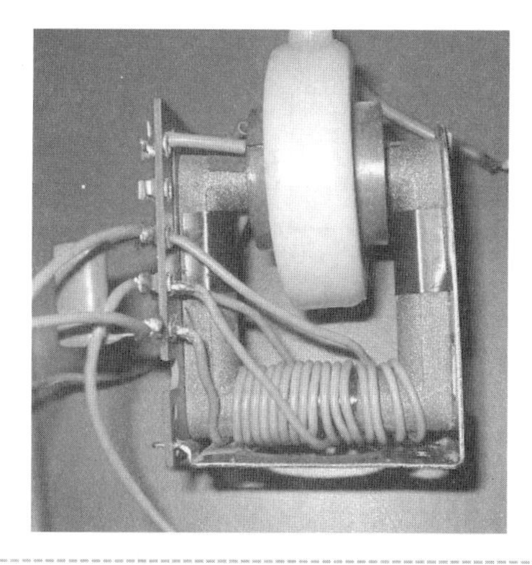

Figure 3.8.9 *The high-voltage transformer*

But if the reader wants to apply static (continuous) fields to the plant, the rectifier and filter shown in Figure 3.8.11 must be added.

The electrodes consist of two metallic plates measuring from 10 × 10 cm to 50 × 50 cm. The reader must be careful to not touch these plates when the circuit is in operation due to the high voltage, which could cause a severe but nonfatal shock.

The homemade capacitor C3 consists of two metal plates, with a glass plate between them as a dielectric that is 0.3 to 0.5 cm thick. The circuit can also be used to perform electrostatics experiments in a physics lab.

## Parts List

Q1: TIP31 silicon NPN power transistor

D1, D2: 1N4002 silicon rectifier diodes

C1: 1,000 μF x 25 V electrolytic capacitor

C2: 0.01 μF ceramic or polyester capacitor

C3: 0.1 μF ceramic or polyester capacitor

R1: 1 kΩ x 1/2 W resistor, brown, black, red

T1: Transformer, with the primary according to the AC power line and the secondary at 9 + 9 V or 12 + 12 V x 500 mA

T2: Fly-back transformer (see text)

S1: On/off switch

F1: 500 mA fuse and holder

For the high-voltage rectifier:

D3: High-voltage rectifier at 10 kV or higher

C4: Homemade capacitor

Other: Power cord, plastic box, solder, wires, etc.

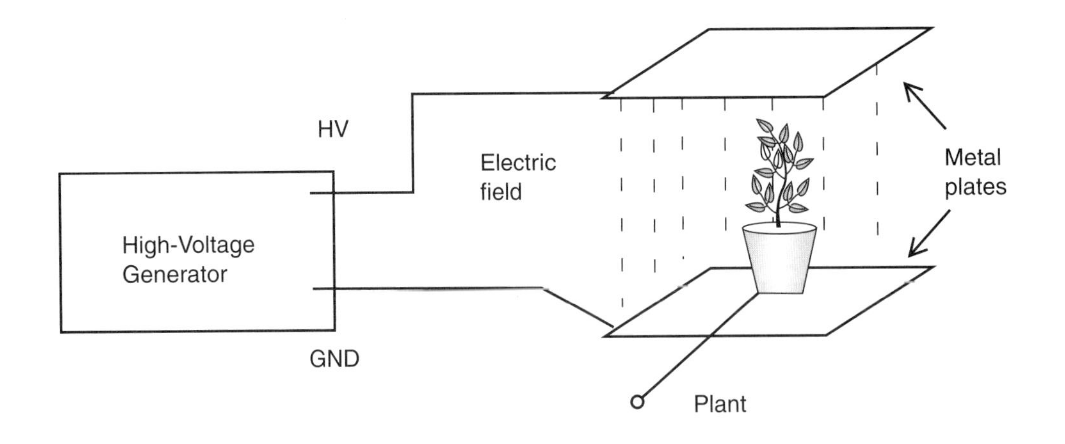


Figure 3.8.10 *Performing experiments with alternating fields*

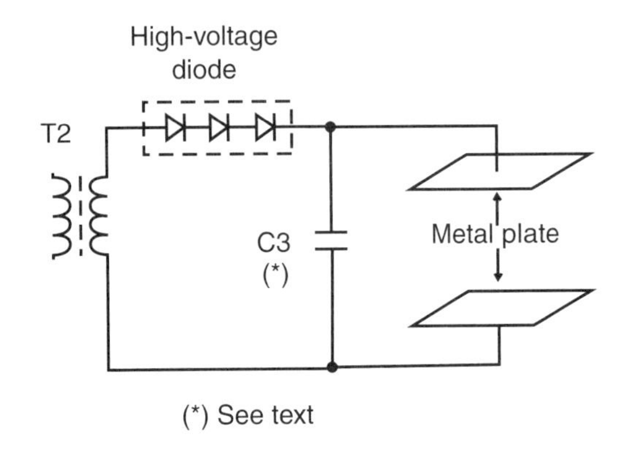


Figure 3.8.11 *Add rectifiers and filters*

## Constant Current Source

Experiments with static magnetic fields, or fields like the ones produced by magnets, can be made using the circuit shown in Figure 3.8.12. The strength of the produced magnetic field depends on the amount of current flowing across the coil and the number of turns of the coils.

P1 adjusts the amount of current, and values between 100 milliamps and 1 amp are recommended. Large values will lead to heat generation, mainly in

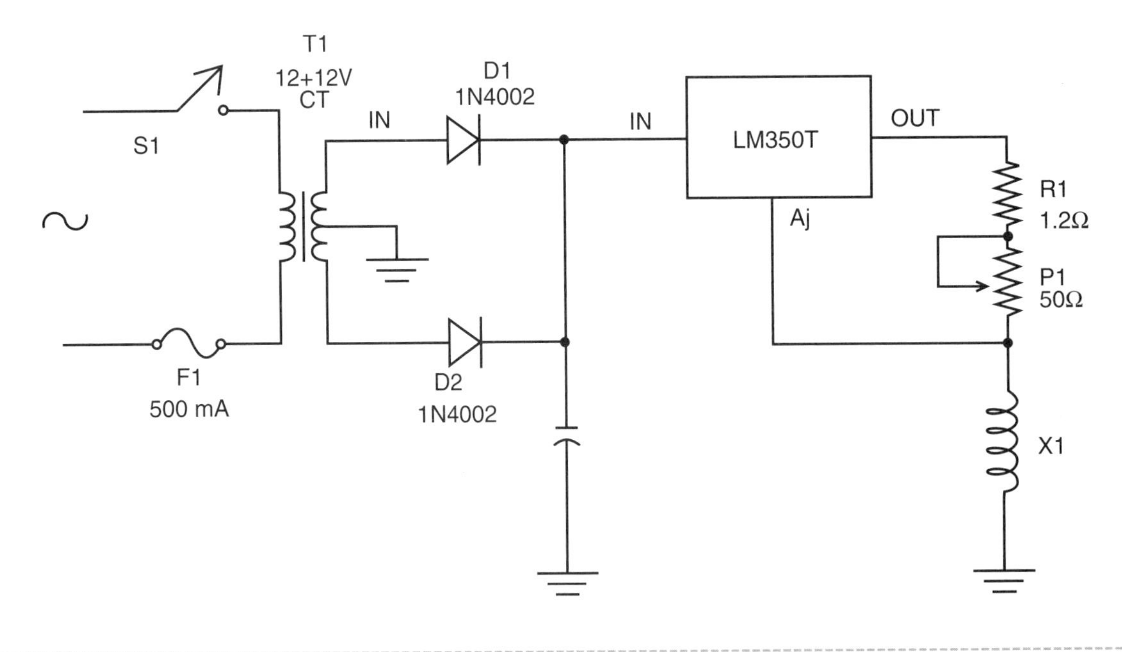


Figure 3.8.12 *A constant current source to produce static magnetic fields*

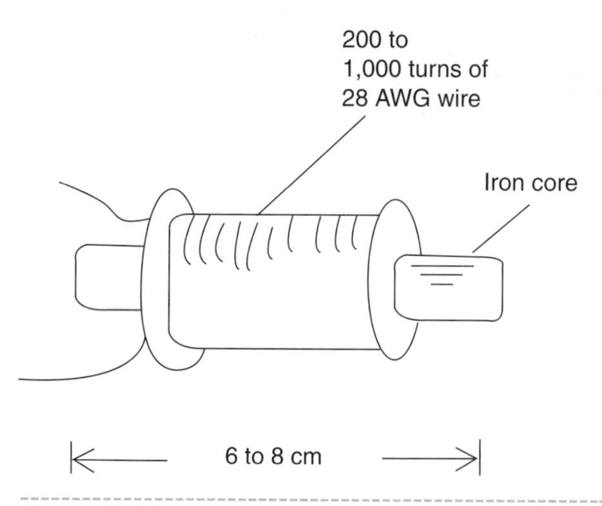


Figure 3.8.13 *Using an electromagnet to produce concentrated magnetic fields*

the coil, which is the same used in the previous experiments, and the *integrated circuit* (IC) must be mounted on a heatsink.

Experiments with concentrated magnetic fields can be made using an electromagnet, such as the one shown in Figure 3.8.13. The advantage of this electromagnet is that you can control the field strength, which is not possible with common magnets. The magnet is formed by 100 to 500 turns of 28 to 32 AWG wire in a plastic tube or an iron core (a screw can be used for this task).

## Parts List

IC-1: LM350T integrated circuit voltage regulator

D1, D2: 1N4002 silicon rectifier diodes

C1: 2,200 μF x 25 V electrolytic capacitor

P1: 50 Ω wire-wound potentiometer

R1: 1.2 Ω x 1 W wire-wound resistor

T1: Transformer, with the primary according to the power supply line and the secondary at 12 + 12 V x 1.5 A

S1: On/off switch

F1: 500 mA fuse and holder

X1: Coil (see text)

Other: PCB or solderless board, power cord, heatsink for the IC, wires, solder, etc.

## Magnetic Fields and Health

For many decades, research has been done to determine the influence of low-frequency magnetic fields on human health. The main focus of this research has been on the fields created by high-voltage power lines, computer monitors, and TV sets. The result of that research has revealed the dangers of such strong fields, possibly causing cancer and leukemia.

What happens is that many cells in our body have natural vibration frequencies (due to resonance) near the frequency of the AC power line, which is 50 or 60 Hz. Under the influence of the low-frequency fields, the atoms inside the cell tend to vibrate strongly, destroying or altering the structure of the cell.

In many countries, laws have been established to prevent homes from being built within 100 meters of any high-voltage power lines. Experiments with insects, plants, and other living things can reveal more about the influence of these fields.

# Project 9—Hypnotic LEDs

One of the most curious interactions between an external source of energy and a living being is one that causes the living being to undergo a hypnotic state. Not only humans, but many animals, can be hypnotized. In this state, they lose control of their movement and can even be subject to commands that, in normal conditions, they would not respond to.

Many techniques exist for hypnotizing a living thing, including humans. The simplest form, which is used by many, utilizes the continuous movement of a pendulum or repetitive sounds such as phrases, instrumental music, or fixed-frequency sounds. If the hypnosis uses an electronic device or interaction between a living being and an electronic device occurs, we can categorize this technique as a bionic process.

Such a hypnotic process is just what we are going to describe now: a project in which a living being (including a human) is put in a relaxed, hypnotic state. Such a frame of mind would be akin to the intermediate state between being awake and asleep, such as the "alpha" state used in transcendental meditation and other Eastern practices.

Of course, the device described here is not recommended for use with people without specialized aid. The states experienced during the hypnotic processes can be dangerous, but you can conduct some interesting experiments using mammals, insects, fish, and other creatures, without causing them any harm or risk.

The following experiments are suggested:

- Study the influence of pulsed light on a plant's growth.
- Conduct research on the influence of intermittent light on the behavior of insects.
- Study the stress of hens and other birds under the influence of blinking *light-emitting diodes* (LEDs).
- Study the influence of blinking LEDs on people's ability to concentrate or their stress levels.

## How It Works

The basic circuit proposed here consists of a low-frequency oscillator that drives two high-power LEDs. As shown in Figure 3.9.1, the LEDs will blink at alternating times.

The core of this circuit is the 555 *integrated circuit* (IC) wired as an astable multivibrator. The frequency of the light pulses is determined by P1.

The diodes are used to produce a signal with 50 percent of the duty cycle. This is necessary because the LEDs of each channel must be on the same time interval.

The signal at the output of the 555 is applied to two complementary transistors that have as a load high-power LEDs, or small incandescent lamps if the reader prefers.

The *negative positive negative* (NPN) transistor becomes active when the output of the 555 reaches the high logic level, and the *positive negative positive* (PNP) transistor begins when the output is at the low logic level.

As for the values used in the project, the LEDs will blink at frequencies between 0.1 and 10 Hz approximately. Of course, the reader can change C1 to alter this range of frequency according to the

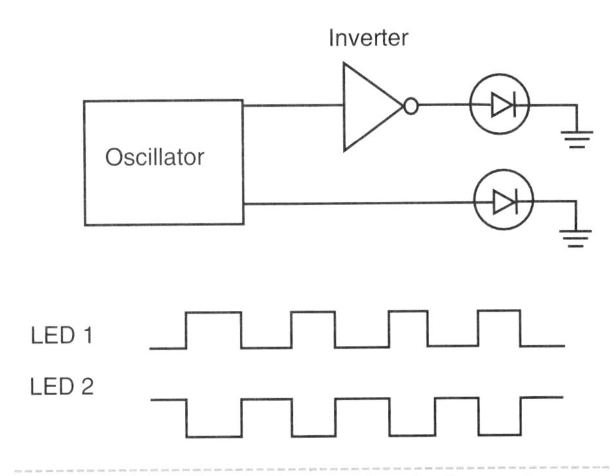


Figure 3.9.1 *Block diagram for the hypnotic LEDs*

experiment in mind, and smaller values for C1 increase the frequency.

The transistors recommended can drive LEDs or lamps with currents up to 500 milliamps. Small 6-volt incandescent lamps can replace the LEDs, depending on the experiment.

The LEDs can be the same or different colors. It is also possible to use only one LED, keeping the other channel off. The circuit is powered from 4 AA cells or a power supply using 6 v x 500 milliamps.

## How to Build

Figure 3.9.2. shows the schematic diagram for the hypnotic LEDs.

The circuit can be mounted using a small *printed circuit board* (PCB), such as the one shown in Figure 3.9.3.

When mounting, the reader must take care when positioning the polarized components such as the IC, the electrolytic capacitors, the LEDs, and the transistors. If the circuit is used to control light sources larger than 100 milliamps, a heatsink will be needed. The circuit should fit inside a small plastic box, with the only external controls being the switch S1 and the potentiometer.

The light sources can be used in different ways. Figure 3.9.4 shows how lights can be placed around a plant specimen to study the effect of variable light. In the same figure, we show how the effects of variable light sources can alter the behavior of an ant colony. Ants are creatures that are very easy to find and easy to work with in many experiments. Many science books show how to construct an ant farm where you can place these live creatures for your experiments..

An interesting exercise is to place the LEDs on a pair of glasses, as shown in Figure 3.9.5, to perform experiments in stress, meditation, and hypnotism.

Of course, when experimenting with people, be sure to have an adult present since injuries can occur. Also, people with certain health concerns, such as epilepsy, should not take part in these experiments.

## Testing and Using

Testing is very simple. Place the batteries into the holder and turn on the S1 switch. The LEDs should alternately blink. Act on P1 to change the frequency.

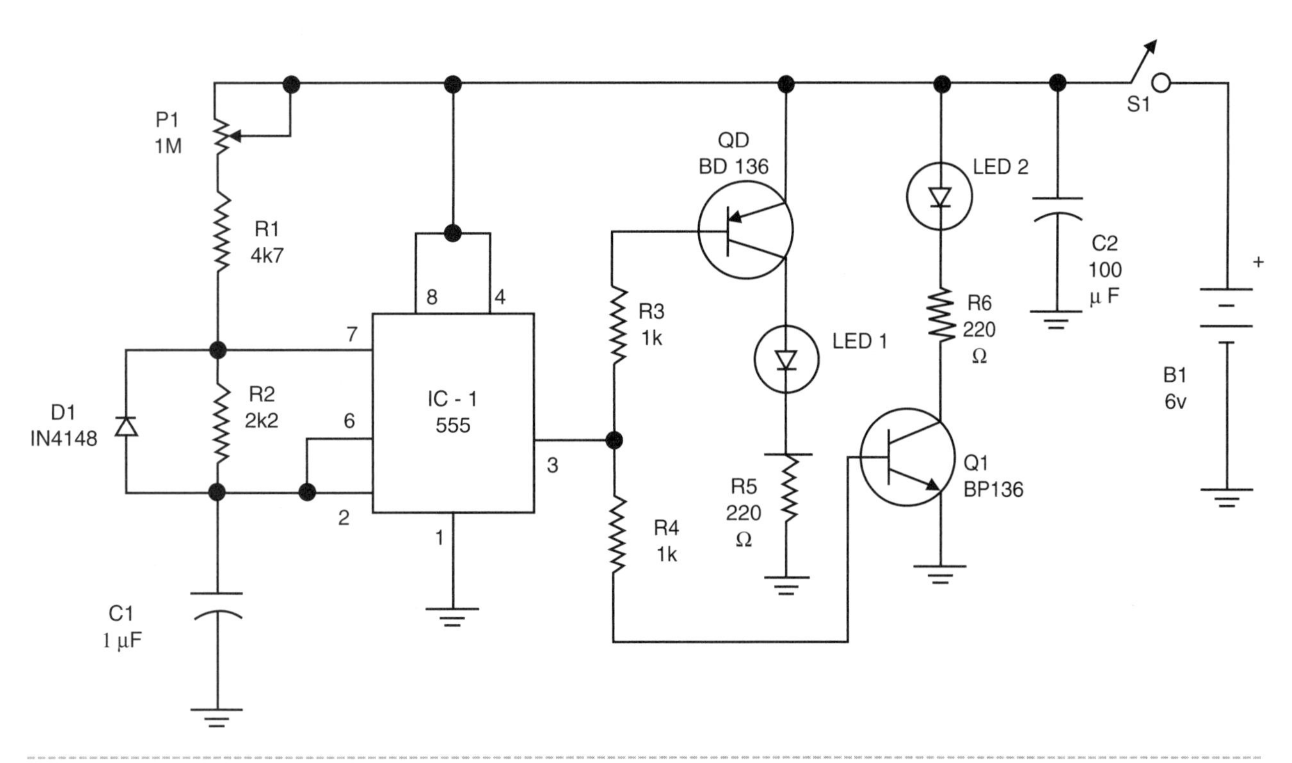


Figure 3.9.2 *Schematic diagram for the hypnotic LEDs*

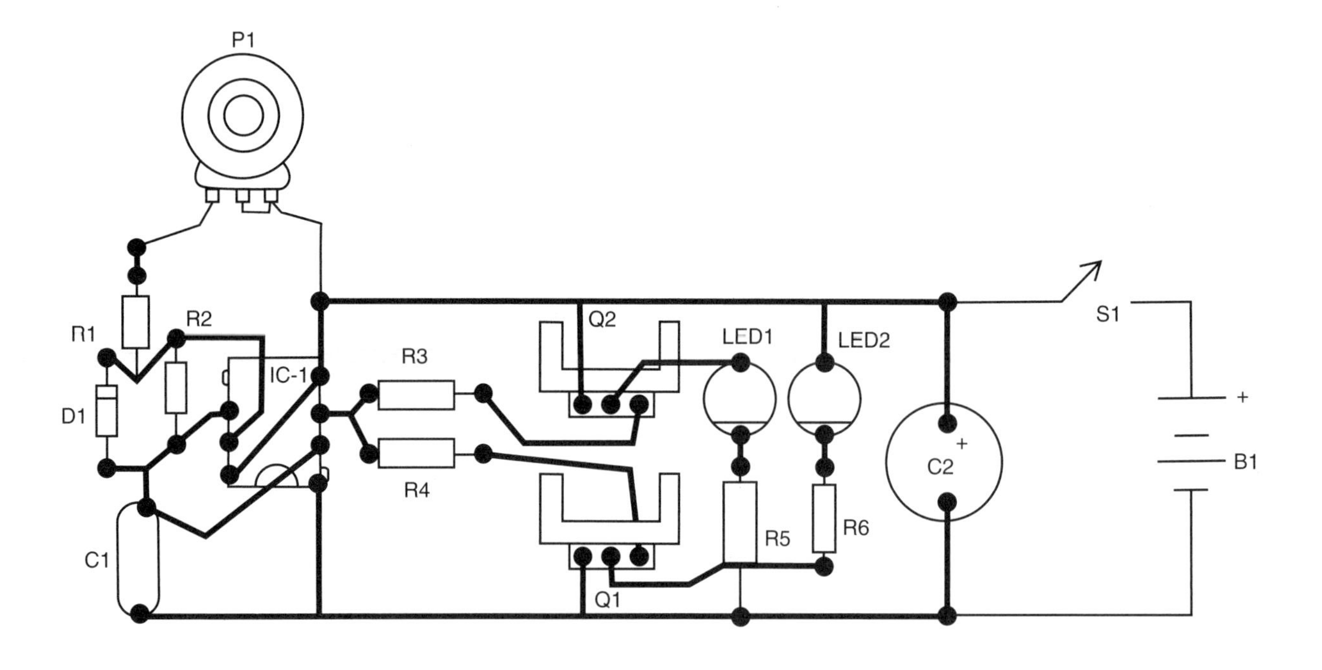


Figure 3.9.3 *PCB for the hypnotic LEDs*

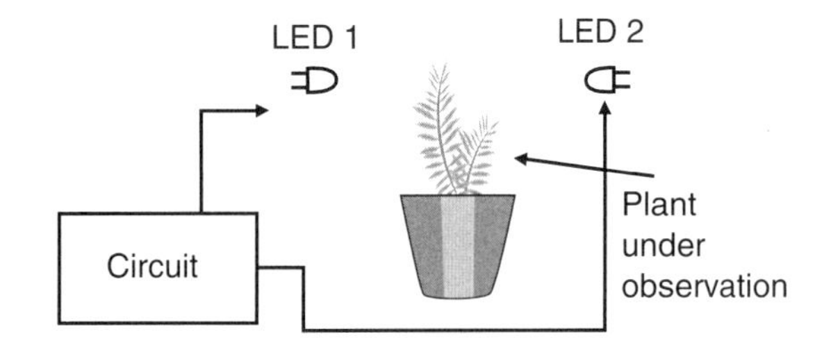


Figure 3.9.4 *Experimenting with plants and insects*

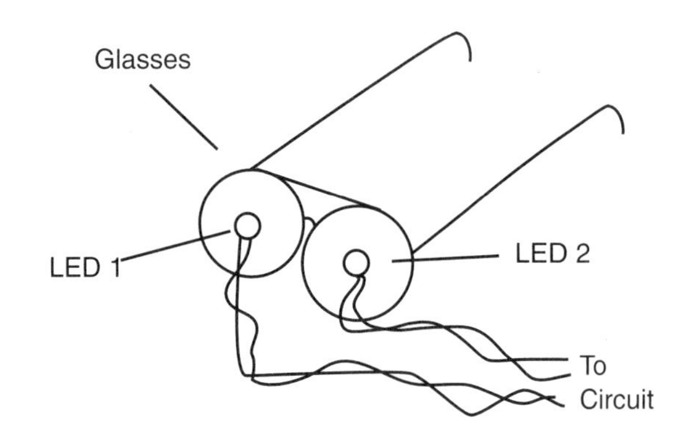


Figure 3.9.5 *Hypnotic glasses*

When everything is ready, place the LEDs around the plant at a distance specified by the experiment and be sure to avoid external light sources. Figure 3.9.6 shows how to set up the experiment.

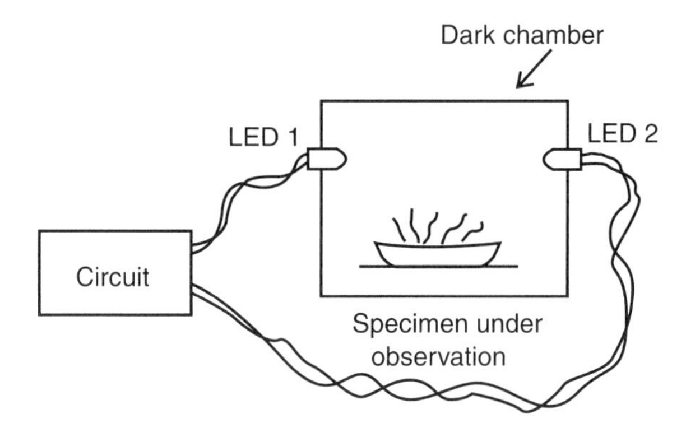


Figure 3.9.6 *An experiment with plant growth and/or circadian rhythms*

## Parts List

IC-1: 555 integrated circuit timer

Q1: BD135 medium-power NPN silicon transistor

Q2: BD136 medium-power PNP silicon transistor

D1: 1N4148 general-purpose silicon diodes

LED1, LED2: Common high-power LEDs (see text)

R1: 4,7 kΩ x 1/8 W resistor, yellow, violet, red

- R2: 2.2 kΩ x 1/8 W resistor, red, red, red
- R3, R4: 1 kΩ x 1/8 W resistors, brown, black, red
- R5, R6: 220 Ω x 1/8 W resistors, red, red, brown
- P1: 1 MΩ linear or log potentiometer
- C1: 1 μF x 12 V electrolytic capacitor
- C2: 100 Ω x 12 V electrolytic capacitor
- B1: A 6 V source or four AA cells and holder
- S1: On/off switch
- Other: PCB, plastic box, wires, solder, etc.

## Additional Circuits and Ideas

A light source can be made to blink or flash independently in many different ways without consuming its power. This section provides some ideas on using some different circuit configurations.

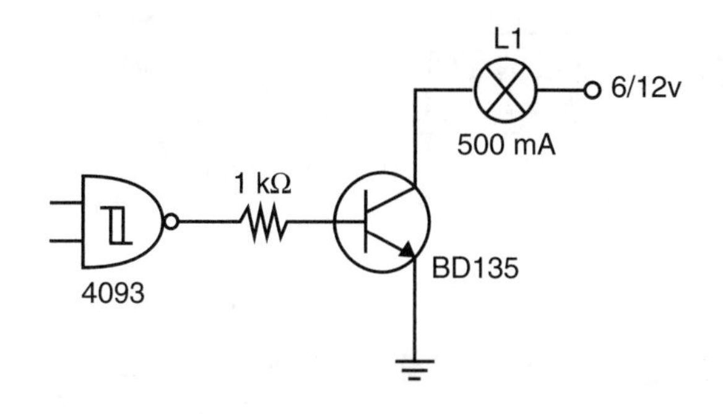


Figure 3.9.8 *Driving more powerful loads*

## Using the 4093 IC

Figure 3.9.7 shows a circuit that can be used to drive two LEDs that will produce alternate flashes of light.

The frequency is controlled by P1, and the circuit can be powered from four AA cells or even a small 9-volt battery. This circuit directly drives the LEDs without the need for a power stage because the *complementary metal oxide semiconductor* (CMOS) IC can sink, or source, the necessary current for this task.

If the reader wants to drive more powerful loads, such as small lamps with currents up to 500 milliamps and voltages that range between 6 and 12 volts, a driver stage such as the one shown in Figure 3.9.8 is recommended.

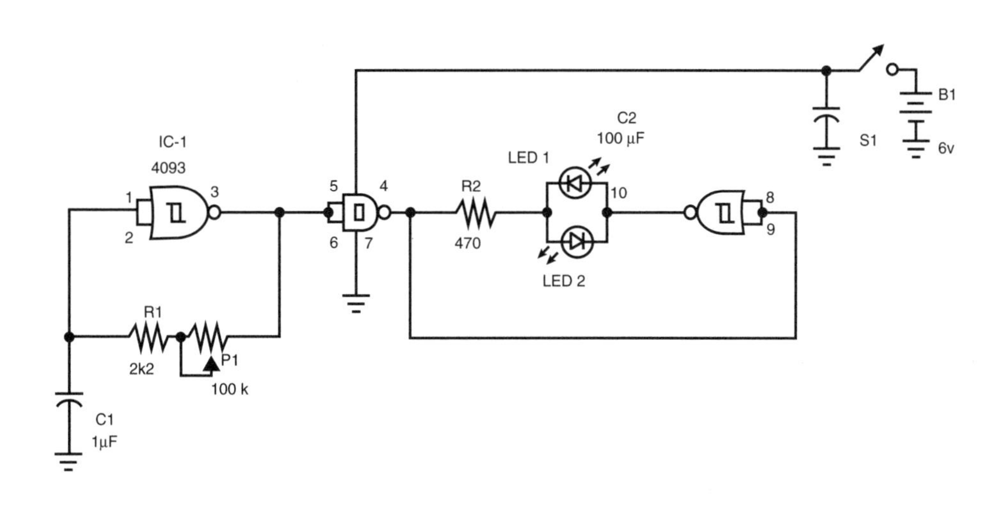


Figure 3.9.7 *Using the 4093 IC*

When powerful loads are used and the circuit operates for long intervals, it is recommended that you use a power supply that replaces the cells. Figure 3.9.9 shows a simple power supply recommended for this circuit.

This circuit can also be used to provide energy for the basic project using the 555 IC. The transformer has a primary according to the AC power line and a secondary rated to 6 v x 250 to 500 milliamps.

## Parts List

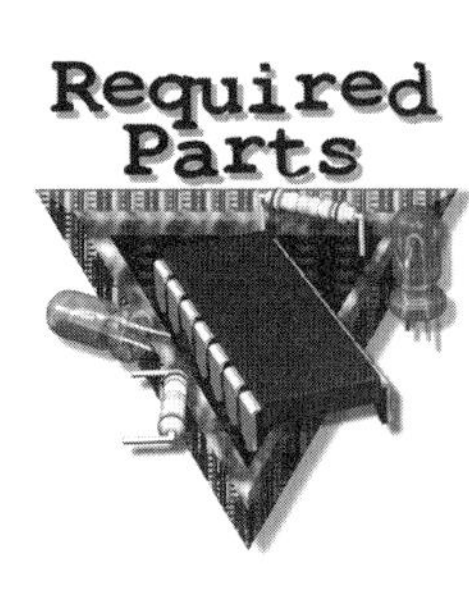


- IC: 1 4093 CMOS integrated circuit
- LED1, LED2: Common LEDs (green and red)
- P1: 100 kΩ potentiometer, lin or log
- R1: 2.2 kΩ x 1/8 W resistor, red, red, red
- R2: 470 Ω x 1/8 W resistor, yellow, violet, brown
- C1: 1 μF x 16 V electrolytic capacitor
- B1: A 6 V source of four AA cells and holder
- S1: On/off switch
- Other:
- PCB, wires, solder, etc.

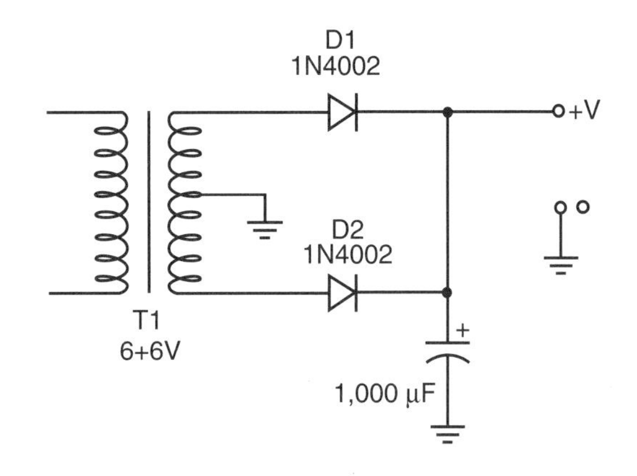


Figure 3.9.9 *A power supply for the circuit*

## Changing the Duty Cycle

The circuits described until now have a duty cycle of 50 percent. This means that the time the LEDs are on is the same for each one.

If experiments with only one LED are made, the duty cycle can be changed, and in this case the LED will produce short light pulses or even long pulses with short intervals between them. This can be made using the circuit shown in Figure 3.9.10.

This circuit can be used to drive common LEDs, high-power LEDs, or low-voltage lamps with currents up to 500 milliamps.

The transistor must be mounted on a heatsink if the driven current is 100 milliamps or higher. It is recommended that an external power supply is used if the light sources drain more than 100 milliamps. For low-power LEDs, a general-purpose PNP type could replace the transistor, such as the BC558 or others.

R3 depends on the current across the LED. Table 3.9.1 gives the values for the resistor according to the LED.

**Table 3.9.1**

Resistor values

| LED or lamp | R3 |
|---|---|
| Common low-power LED | 1 k × 1/8 W |
| Medium-power LED | 470 Ω × 1/2 W |
| High-power LED, jumbo | 47 Ω to 220 Ω x 1/2 W |

## 6/12 V up to 500 Milliamps Lamp No Resistor Needed

## Parts List

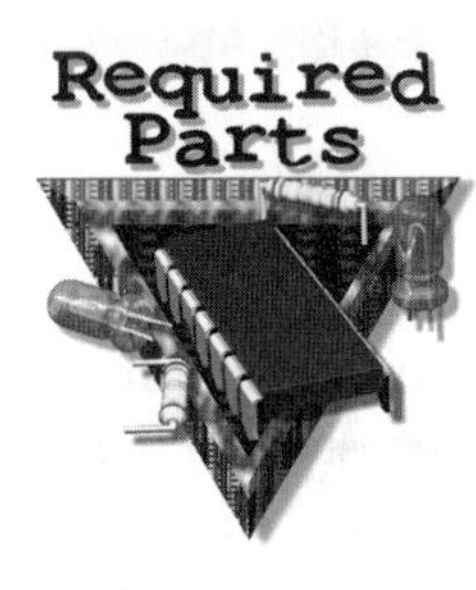


IC-1: 555 integrated circuit timer

Q1: BD136 medium-power NPN silicon transistor

LED: Common or high-power LEDs or lamp

D1, D2: 1N4148 - general-purpose silicon diodes

P1, P2: 100 kΩ potentiometer lin or log

R1, R2: 2.2 kΩ x 1/8 W resistors, red, red, red

R3: See text

C1: 1 μF x 16 V electrolytic capacitor

C2: 100 μF x 12 V electrolytic capacitor

S1: On/off switch

B1: 6 V to 12 V power supply

Other:

PCB, wires, plastic box, etc.

## Experimenting with Infrared Sources

Infrared LEDs are easy to find and cheap, so it would not be a problem to try conducting some experiments using an infrared light, which would replace the original LEDs.

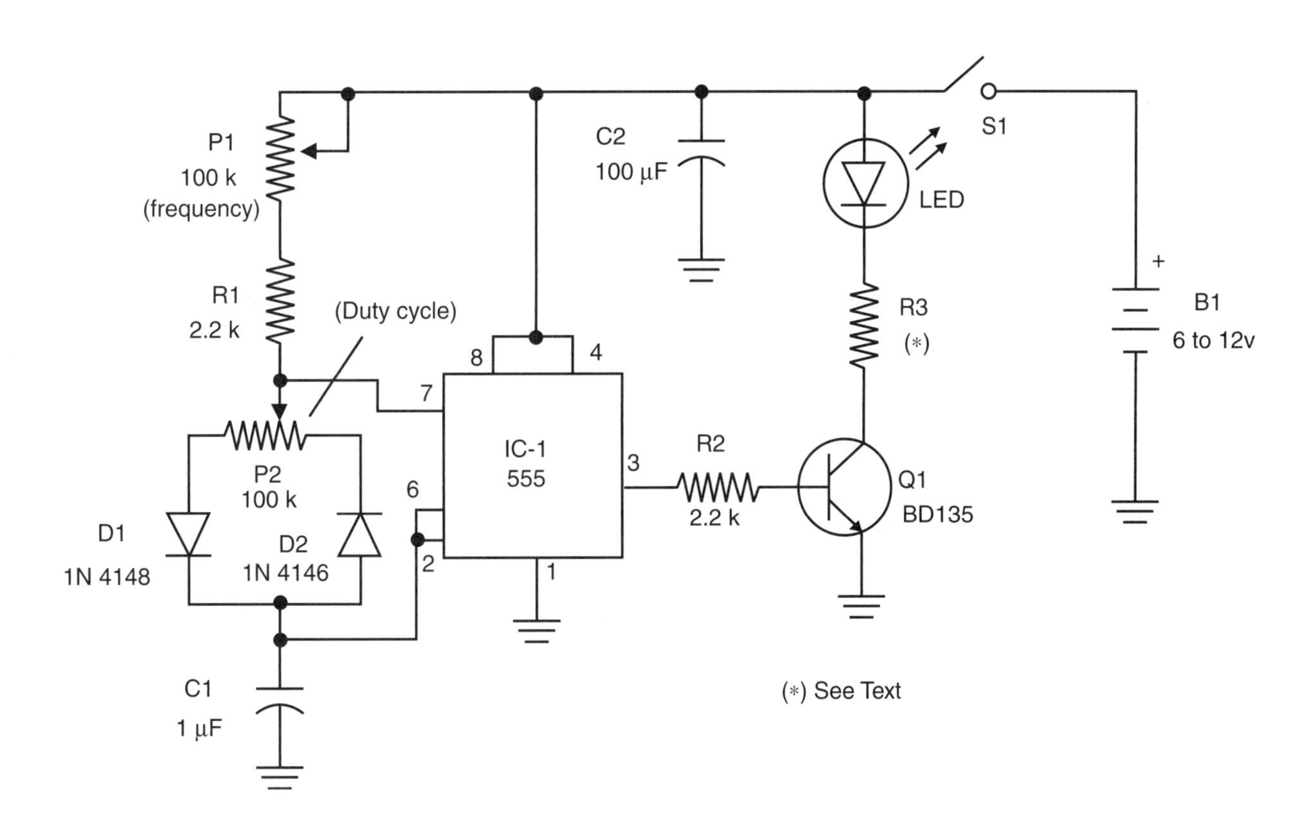


Figure 3.9.10 *Changing the duty cycle*

# Project 10—Insect Repellent

Some insects are affected by audio tones. In a certain species, female insects can produce tones that will repel other females, whereas in other cases certain sounds will cause a disturbance for all the insects.

The basic idea explored in this project is to build an audio oscillator that generates sounds capable of repelling insects. Of course, experiments can be made to see how the same sounds can affect other living beings, such as fish in an aquarium.

Using an audio oscillator, it is possible to scare insects away from certain areas such as a bedroom or, when camping, a tent. This project serves as an interesting application of bionics where an electronic circuit imitates a living being interacting with the insects. The circuit can also be used to study the habits of insects and how external influences (such as sounds) can alter them.

The basic version of the project consists of a simple audio oscillator driving a small piezoelectric transducer. Other versions using high-power circuits will be given too (such versions can be bought in camping stores). The circuit can be used either in scientific works or in practical applications.

Because the current drain will be very low, the circuit can be powered from AA cells, a 9-volt battery, or even a power supply plugged into the AC power line. The low current drain will allow the battery life to last a number of weeks.

An adjusting potentiometer is also included that acts on the frequency, which ranges from 200 to 2,000 Hz typically. However, depending on the application frequency, the range can be altered by replacing C1. Values between 2,200 pF and 0.1 $\mu$F can be tested. It is also possible to replace the transducer, making the circuit oscillate in the supersound range.

## Bionics Experiments and Applications

The tone produced is strong enough to allow this circuit to be used in experiments involving bionics and even some practical applications such as those shown in the following sections.

### Fish Attractor

A circuit can be created to produce a sound that imitates an insect in the water trying to escape, which attracts fish. This effect is used in commercial fish attractors, and it is only necessary to install the circuit inside a bottle, as shown in Figure 3.10.1. The weighty metal piece is necessary to make the bottle dive into the water.

The reader must find the ideal frequency to attract the desired fish, and experiments made in an aquarium can be performed to study the habits of the fish.

### Repelling Other Insects

Experiments can be made to see if the circuit can be used to scare insects such as ants and cockroaches. The reader will have to change the capacitor and the resistor in the frequency circuit to find the ideal tone for the application.

### Effects on Plants

Experiments to verify if the tones have an effect on plant growth can be performed as well. Simply place the circuit near the plant under observation, as shown in Figure 3.10.2.

Because the current drain of the circuit is very low, common AA cells can be used to power the circuit over many weeks. Of course, for even longer experiments or applications, the circuit can be powered from a 3-volt AC/DC converter. Any type of power supply with currents in the range of 50 to 250 milliamps is suitable for this application.

When wiring the power supply to the circuit, be sure to observe the polarity of the connector. Don't use transformerless power supplies for safety reasons.

Figure 3.10.1 *Using the circuit as a fish attractor*

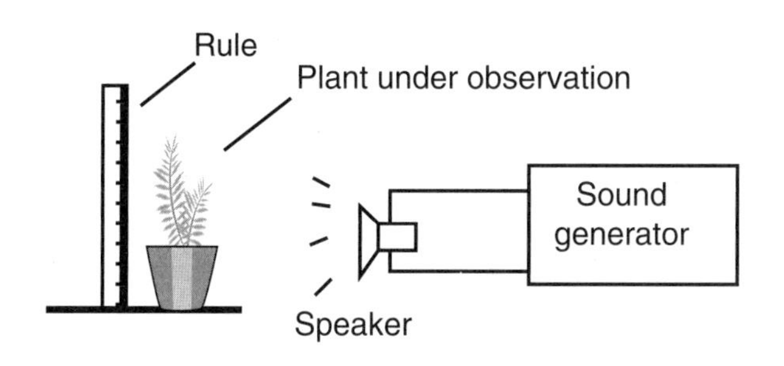


Figure 3.10.2 *Place circuit near plant*

In the next section, we will see how a power stage can be added to increase the sound volume.

## How the Circuit Works

The circuit is made from a 4093 *complementary metal oxide semiconductor* (CMOS) *integrated circuit* (IC), which consists of four Schmitt NAND gates. One of the four gates is wired as an audio oscillator, driving the three other gates that work as a digital buffer or amplifier.

The frequency is determined by C1 and can be adjusted to a large range of frequencies by P1. C1 can be changed if the frequency range covered by the circuit is not large enough for the application the reader has in mind.

The signal from the output of the circuit is square and can be used to directly drive a high-impedance piezoelectric buzzer. Do not use loudspeakers or a low-impedance transducer directly plugged into the output of the circuit. Using this kind of transducer will be discussed later.

## How to Build

Figure 3.10.3 shows the schematic form of the insect repellent.

This circuit is mounted on a small *printed circuit board* (PCB), which is shown in Figure 3.10.4.

All the components can be housed into a small plastic box. Holes should be drilled into the box to allow the release of the sound produced by the transducer. The transducer is a piezoelectric type, such as the ones used in buzzers, but you can also experiment with transducers from ceramic mikes or phones.

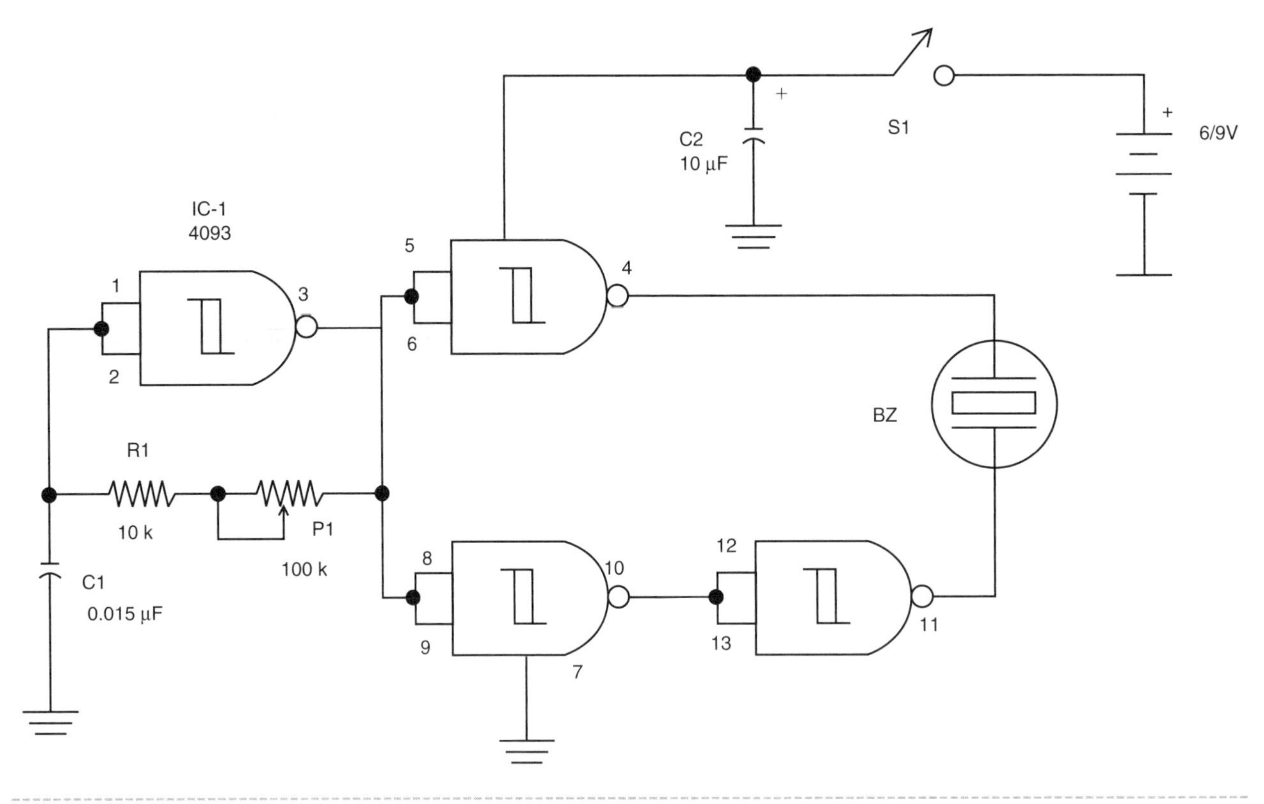


Figure 3.10.3 *Schematic diagram for the insect repellent*

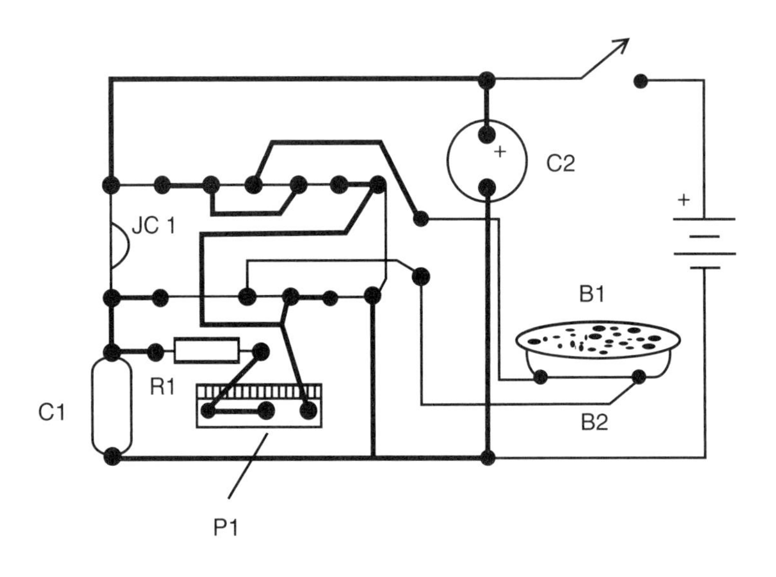


Figure 3.10.4 *PCB for the insect repellent*

## Testing and Using

Turn on the power switch and adjust P1 to the desired audio tone. The ideal tone can be obtained by experimenting, and in a general sense, the perfect tone will be the one that is closest to the sound produced by the actual insect. Experiments shows that certain female insects do not tolerate the presence of other females, so reproducing their sound will probably lead to repelling them.

## Parts List: Insect Repellent

- IC1: 4093 CMOS integrated circuit
- BZ: Piezoelectric transducer (see text)
- R1: 10,000 Ω (10 kΩ), 1/4-watt, 5 percent resistor, brown, black, orange
- P1: 100,000 Ω (100 kΩ) potentiometer
- C1: 0.015 μF ceramic or metal film capacitor
- C2: 10 μF/12 WVDC electrolytic capacitor
- S1: SPST toggle or slide switch
- B1: A 6 or 9 V source or four AA cells or 9 V battery
- Other parts: PCB, cell holder or battery connector, wires, plastic box, solder, etc.

## Additional Circuits and Ideas

Using a CMOS IC is not the only way we have for producing continuous audio tones in experiments and bionics projects. Many other configurations can be used for the same task, some of them driven by powerful transducers such as loudspeakers or tweeters. In the next section, we will discuss some of these configurations.

## Using Transistors

Figure 3.10.5 shows a circuit that uses two transistors to produce a tone in the range determined by C1 and C2, and adjusted by P1. It consists of an astable multivibrator driving a piezoelectric transducer or a small loudspeaker.

Another circuit using two complementary transistors is shown in Figure 3.10.6. This circuit can drive a loudspeaker from 5 to 10 cm (4 or 8 ohm) when powered from 3 to 12 volts.

For voltages up to 6 volts, Q2 must be replaced by a BD136 (or a TIP32) mounted on a small heatsink. In this case, the output power will be up for some watts, ideal for experiments or applications involving large ranges.

The frequency depends on C1 and can be adjusted by P2. Replacing C1 by capacitors in the range of 0.47 to 2.2 μF, the circuit will produce very low-frequency pulses, such as a metronome. Experiment to see the effect of this kind of audio stimulation on insects and animals.

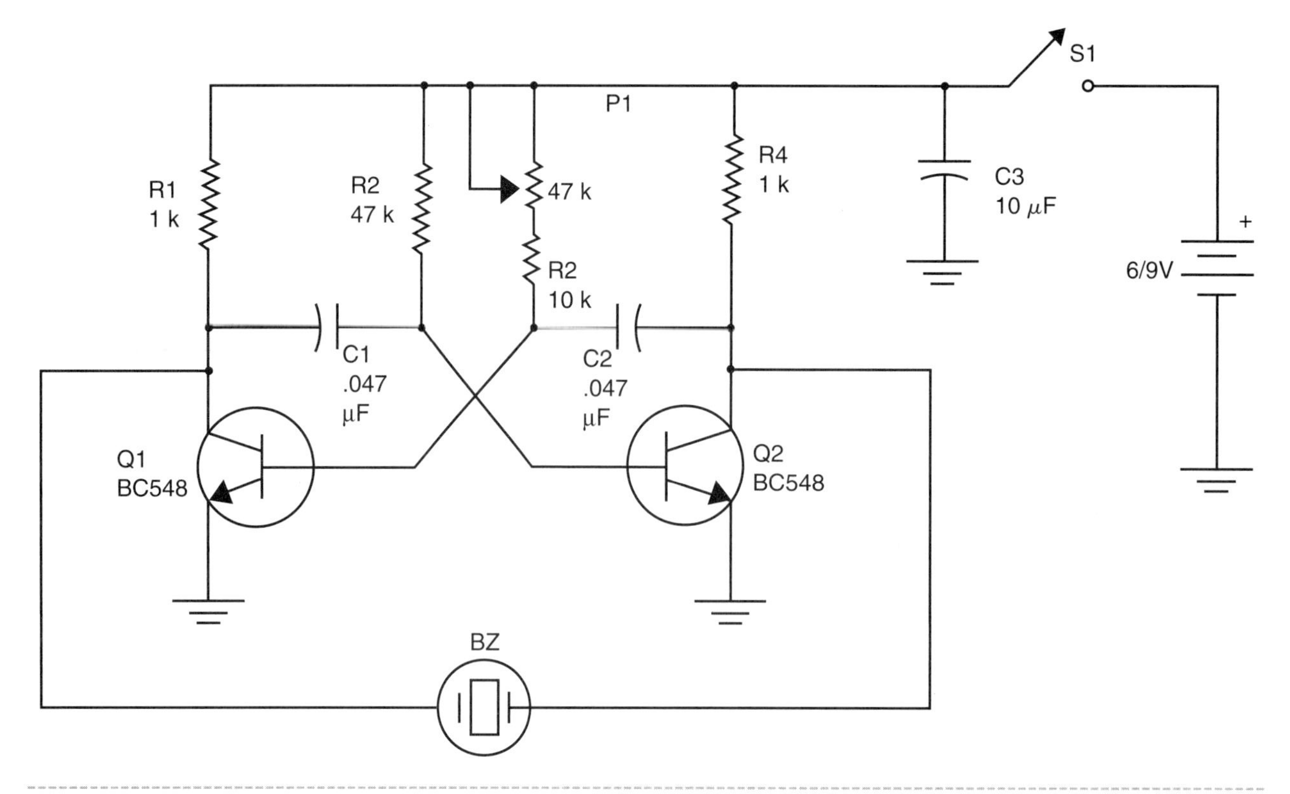


Figure 3.10.5 *Circuit using transistors*

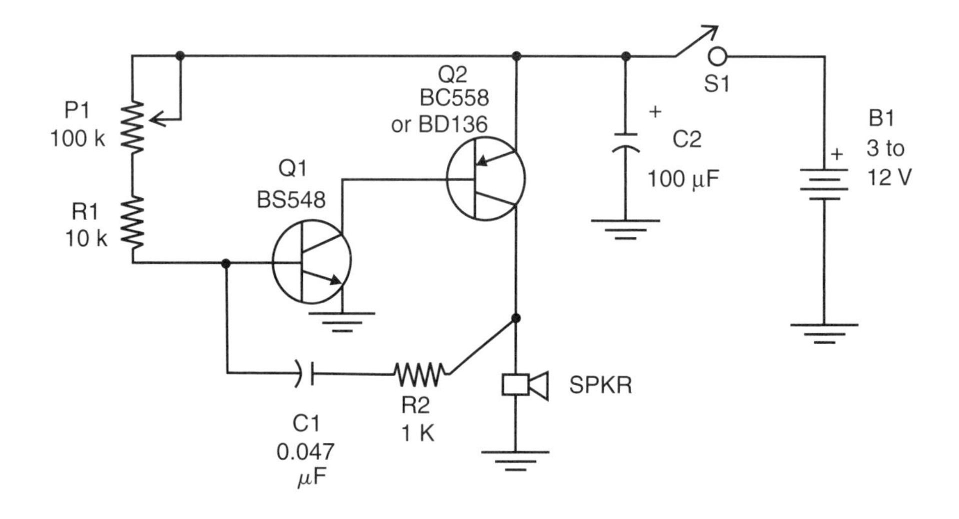


Figure 3.10.6 *Circuit using complementary transistors*

## Parts List

Q1: BC548 or equivalent general-purpose NPN silicon transistor

Q2: BC558 or BD136 general-purpose or medium-power PNP silicon transistor

R1: 10 kΩ 1/8 W resistor, brown, black, orange

R2: 1 kΩ x 1/8 W resistor, brown, black, red

P1: 100 kΩ lin or log potentiometer

C1: 0.047 μF ceramic or polyester capacitor

C2: 100 μF x 6 V electrolytic capacitor

SPKR: 5 to 10 cm 4 or 8 Ω small loudspeaker

B1: 3 to 12 V batteries or cells

Other: PCB or terminal strip, plastic box, wires, solder, etc.

## Using the 555 IC

Figure 3.10.7 shows a tone generator using the 555 IC. This version can drive a piezoelectric transducer and, adding a transistor, a small loudspeaker.

The basic frequency is adjusted by the potentiometer P1 and depends on C1. With the components shown in the circuit, the frequency can be adjusted from 100 Hz to 1 kHz, representing the normal range of sounds that repel or affect most small insects.

The reader is free to change the values of C1, generating sounds of any frequency, including supersounds. Of course, when generating supersounds, an adequate transducer must be used, because small piezoelectric transducers have the reproduction range limited to 10 kHz.

A suitable transducer can be found in piezoelectric tweeters. In this case, the small transformer found inside the loudspeaker must be removed and the transducer directly coupled to the circuit.

**Figure 3.10.7** *Circuit using the 555 IC*

Another idea would be to use a transistor stage to allow the circuit to operate with low impedance loads. This circuit is shown in Figure 3.10.8. This circuit can be used with common loudspeakers if operating in the audible range or it can use piezoelectric tweeters if operating in the supersound range.

## Parts List

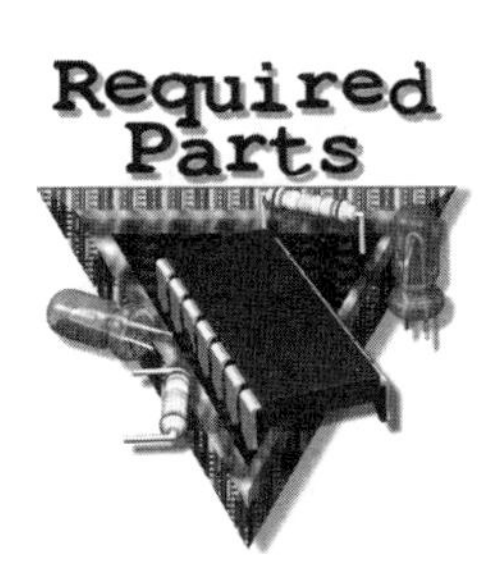


IC-1: 555 integrated circuit timer

P1: 100 kΩ lin or log potentiometer

R1, R2: 2.2 kΩ x 1/8 W resistor, red, red, red

C1: 0.022 μF polyester or ceramic capacitor

S1: On/off switch

B1: 6 or 9 V cells or battery

X1: Piezoelectric transducer

For the transistor output stage:

Q1: TIP31 silicon NPN power transistor

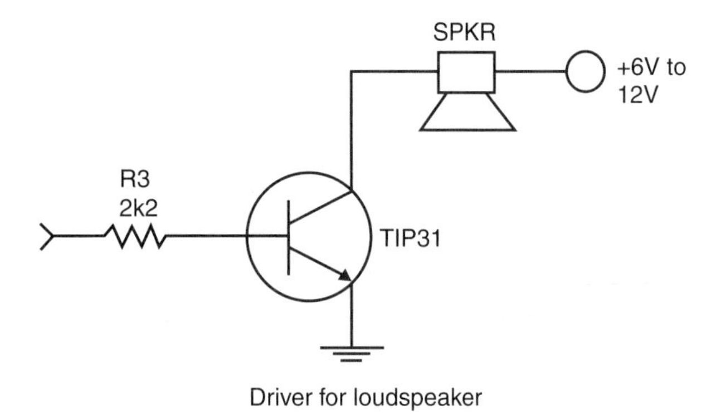


Figure 3.10.8 *Using a transistor drive stage*

R3: 2.2 kΩ x 1/8 W resistor, red, red, red

SPKR: 4 or 8 ohms loudspeaker or tweeter

Other parts: PCB or solderless board, battery connector or cell holder, plastic box, wires, solder, heatsink for the transistor (if used), etc.

# Experiments Using the Circuits

Try conducting some experiments involving living beings to test the influence of continuous tones on their behavior. Such experiments can be conduced with insects, such as ants or flies to discover how a tone can affect their behavior, or fish, as the transducer can be used under water. The potentiometer in the basic version can be replaced by a *light-dependent resistor* (LDR) or a *negative temperature coefficient* (NTC) register to make the circuit sensitive to light or heat.

# Project 11—Bionic Trap



Many insects and other animals are attracted to the light of certain wavelengths or colors. This explains why many kinds of insects beat themselves against fluorescent or even incandescent lamps at night. By choosing the correct color of a lamp, it is possible to attract a certain type of insect and make them fall into a trap.

The bionic trap described here uses a fluorescent lamp and an additional resource for capturing or killing the insects. In the basic version, the insects are collected in a bag, but we will also describe a high-voltage killer later in Project 3.15. It is up to the bionics evil genius to alter the circuit to capture the insect or animals according to the kind of light or radiation suitable for the species.

## The Project

In the basic version, the circuit consists of a high-voltage inverter used to drive a small fluorescent lamp from cells or batteries. The fluorescent lamp is chosen according to the type of insects the evil genius intends to capture. Visible light lamps or *ultraviolet* (UV) lamps with a power of up to 4 watts can be used in this circuit, such as the ones shown in Figure 3.11.1.

The use of an inverter is important if the reader intends to capture insects in the forest or other places where an AC power line outlet would not be accessible. The evil genius can use this device to collect specimens during the night in many different places.

Of course, using cells, the life of the supply is not very long, only 1 or 2 hours, so it is recommended that rechargeable cells be used for this purpose. For a larger autonomy, requiring many hours of supply for the circuit, larger batteries should be used.

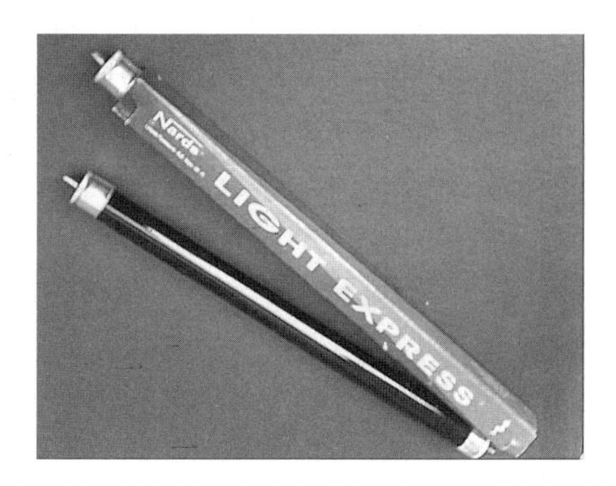


Figure 3.11.1 *Small fluorescent lamps driven by the circuit described here*

## Experiments in Bionics

Many types of insects can be used in bionic experiments or other interesting applications, but the reader must first capture them. This project explains how to build a trap. This trap serves many uses, especially if the reader lives in an area where insects are a problem.

The main applications for the bionic trap are as follows:

- Capture insects for experiments or applications in bionics.
- Capture insects that are disturbing to people.
- Collect insects for studies in biology.
- Kill insects that are harmful to plants.
- Power a fluorescent lamp, such as an emergency lamp.
- Connect the high-voltage output to an electric fence.

## How It Works

To convert a low DC voltage, such as the ones supplied by cells and batteries, into a high AC voltage, the recommended configuration uses a transformer. But to drive a transformer it is necessary to change the pure DC current into a pulsed or alternating current because transformers can't operate with pure DC currents. Consequently, an electronic circuit is necessary to allow the transformer to interface with the DC power supply, as shown in Figure 3.11.2.

The circuit consists of a Hartley oscillator that produces an alternating signal that is applied to the low-voltage coil of a small transformer.

In the secondary of this transformer, a high voltage appears, with peaks reaching more than 200 volts, enough to drive any fluorescent lamp, even the ones weak enough to no longer function anymore when connected to the AC power line.

C2 and C3 determine the frequency of the oscillator. Depending on the transformer, the evil genius must change the values of these components to find the ones that result in the best performance. The reader is also free to change the resistor R1 to find the best performance matching with the characteristics of the transformer.

In the basic version, four cells form the power supply. For the best autonomy, it is recommended that C or D cells or *nickel cadmium* (NiCd) rechargeable cells be used. The circuit will drain between 100 and 300 milliamps, according to the characteristics of the transformer used in the project and the lamp. For larger autonomy, use a 6-volt battery.

Any small fluorescent lamp up to 7 watts can be used. The reader can use white lamps or UV lamps, according to the application. But if the reader connects powerful lamps, even a 40-watt lamp, it will glow but with reduced brightness because the circuit can't source more than a few watts to the load.

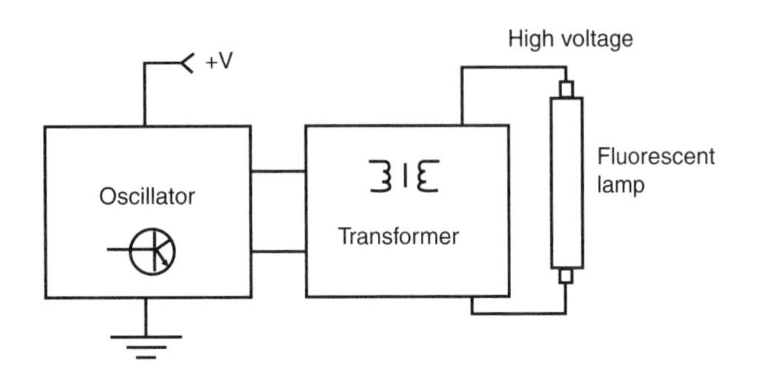


Figure 3.11.2 *Block diagram for the biologic trap*

## How to Build

Figure 3.11.3 shows the complete diagram of the bionic trap inverter.

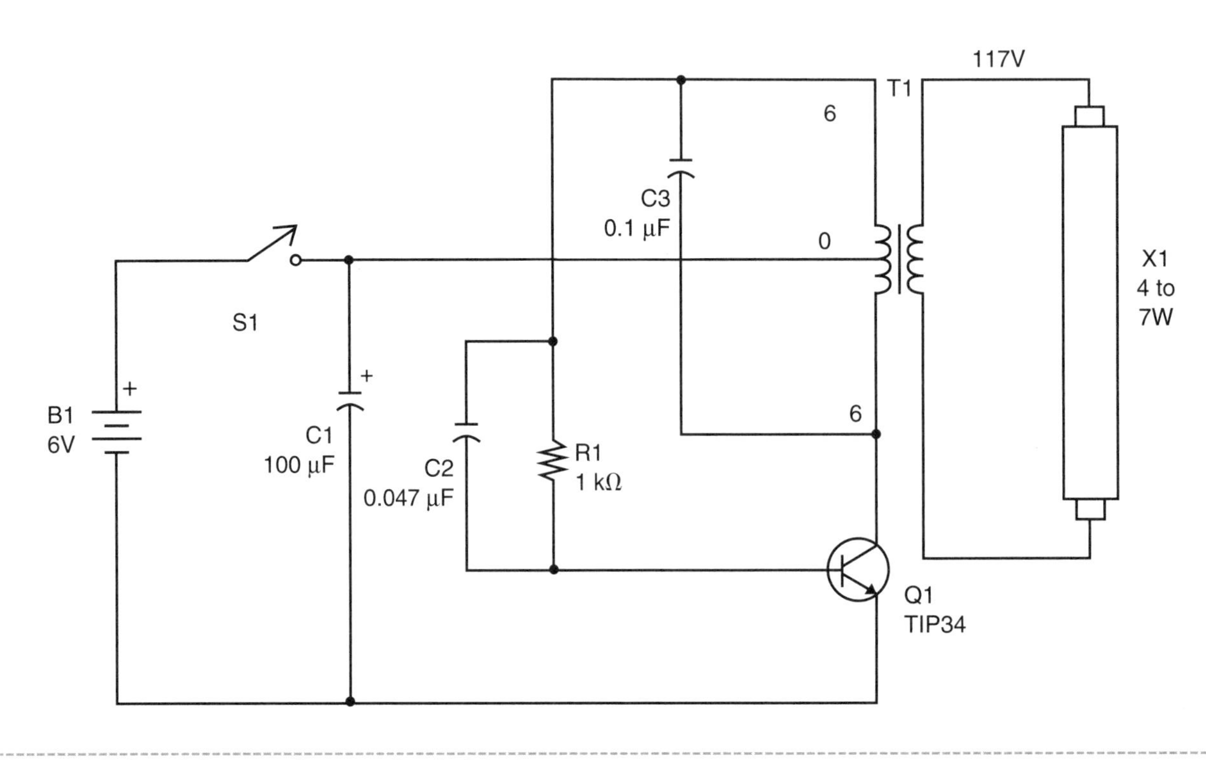


Figure 3.11.3 *Schematic diagram for the bionic trap*

The circuit is very simple, so it can be mounted using a terminal strip as a chassis, as shown in Figure 3.11.4.

The transistor also needs a small heatsink. Take care not to invert the connections for this component when mounting the circuit.

All the components should fit into a small plastic box, and to connect the lamp you can use a long wire, as shown in Figure 3.11.5.

Any type of transformer having a secondary with a *center tap* (CT) and voltages that range from 4.5 to 7.5 can be used. Currents range from 200 to 500 milliamps.

The primary wound can be rated to 110/117/120/127 /220 volts or 240 volts. The higher the voltage, the easier the lamp will be driven.

When connecting the lamp, take care with the isolation of the wires. The high voltage present in this part of the circuit is enough to cause a shock.

C2 and C3 are ceramic or polyester capacitors. The transistor is a TIP31 (with any suffix: A, B, or C) and equivalents such as the BD135 or even the TIP41 can be used. Using a TIP41 or a 2N3055, the circuit can be powered from a 12-volt supply and larger lamps can be driven.

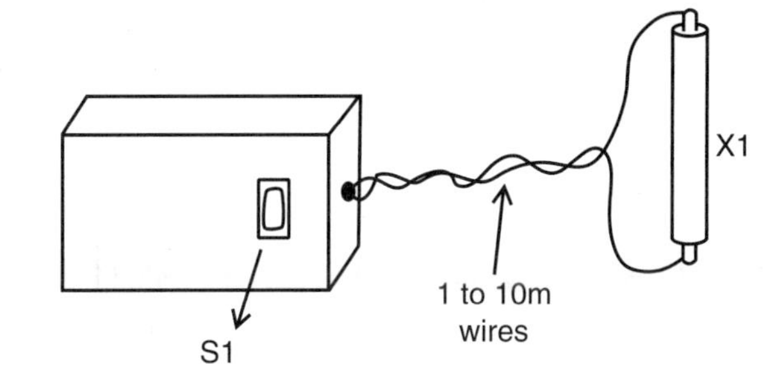


Figure 3.11.5 *The circuit fits into a small plastic box.*

The trap can be made using several methods. One of them is shown in Figure 3.11.6 and consists of a bag placed under a lamp.

The insects attracted by the lamp beat against it and eventually fall into the bag where they are collected. A resource for preventing the insects' escape from the bag can be added, as shown in Figure 3.11.7.

If the evil genius wants to kill the insects, Project 3.15, "Insect Killer," will show you how to do that.



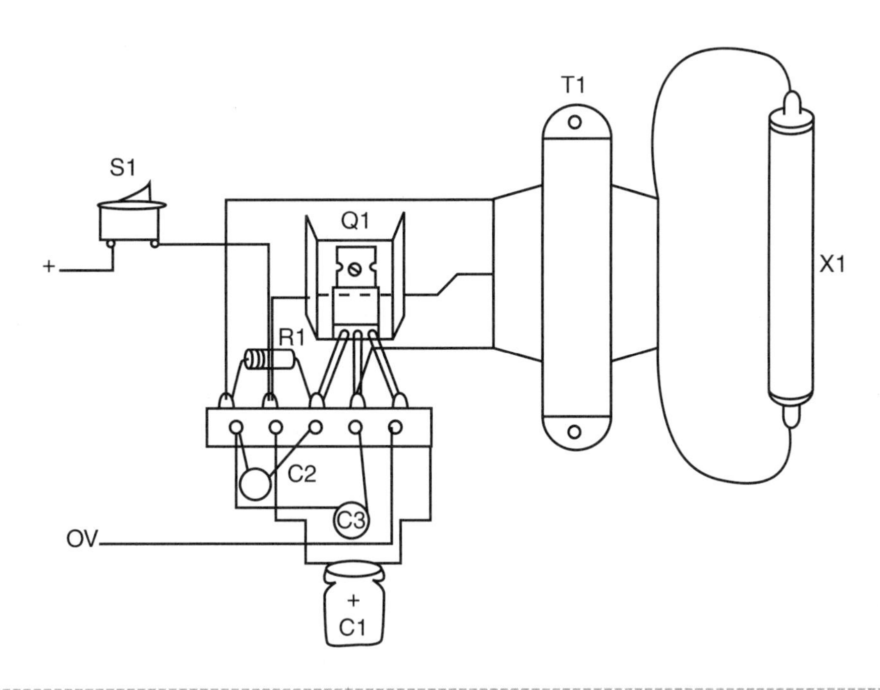


Figure 3.11.4 *Circuit mounting using a terminal strip as a chassis*

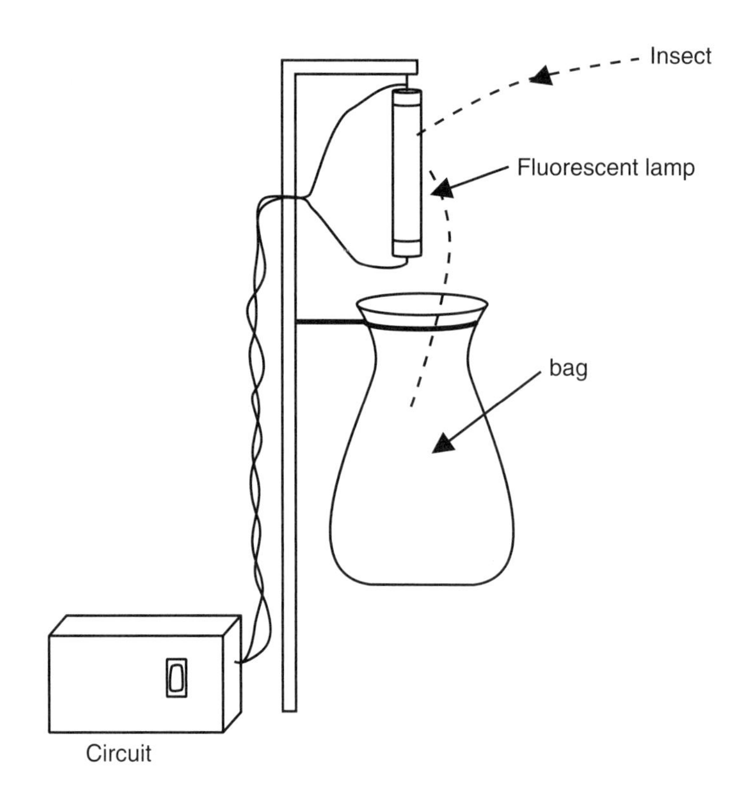


Figure 3.11.6 *The trap using a plastic bag*

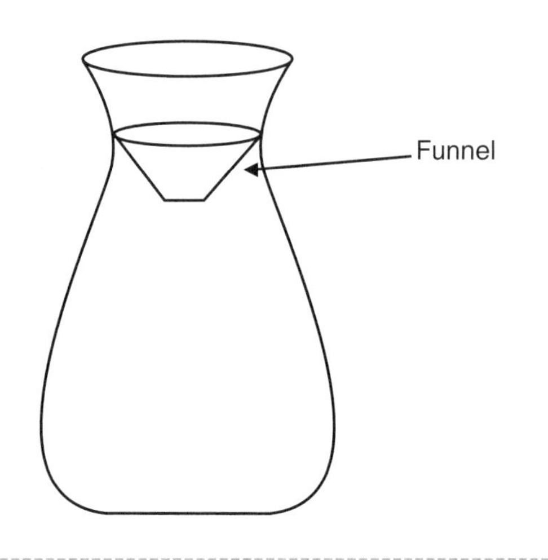


Figure 3.11.7 *Preventing the insects from escaping from the trap*

## Testing and Using

Power the circuit on, closing S1. The fluorescent lamp will glow immediately. (See Figures 3.11.8 and 3.11.9.)

If you lean down and place your ear near the transformer, you should be able to hear the sound of the oscillator. If no sound is heard and the lamp doesn't glow, turn off the circuit and take a good look at your mounting. Something is wrong.

If the transistor heats and the lamp doesn't glow, turn the circuit off and check your mounting. Take care with the positions of the polarized components, because if they are inverted, the circuit won't function.

Once the circuit is working, the reader can try changes in R1, C1, and C2 to get a better performance. A practical idea is to place a 10 kilo-ohms trimpot in a series with R1, adjusting the output power.

When the circuit runs as expected, the evil genius can install it in an area where insects will be collected. Now is the time to experiment with white or UV lamps.

## Parts List

Q1: TIP31 NPN silicon power transistor

R1: 1 kΩ x 1/2 W resistor, brown, black, red

C1: 100 μF x 12 V electrolytic capacitor

C2: 0.047 μF ceramic or polyester capacitor

C3: 0.1 μF ceramic or polyester capacitor

S1: On/off switch

T1: Transformer (see text)

X1: Fluorescent lamp up to 7 W (see text)

B1: 6 V cells or battery (see text)

Other: Terminal strip, plastic box, heatsink for the transistor, connectors to the lamp, wires, solder, etc.

## Additional Circuits and Ideas

A simple inverter would be ideal for the evil genius who is unfamiliar with assembly techniques. It is easy to create and uses no *printed circuit board* (PCB).

But if the reader has more experience in building electronic projects, many other versions can be suggested, such as one that can be powered directly from the AC power line. Let's take a look at those versions.

## Powering the Circuit from the AC Power Line

The evil genius has two options for powering the bionic trap from the AC power line. One of them is a simple power supply, as shown in Figure 3.11.8.

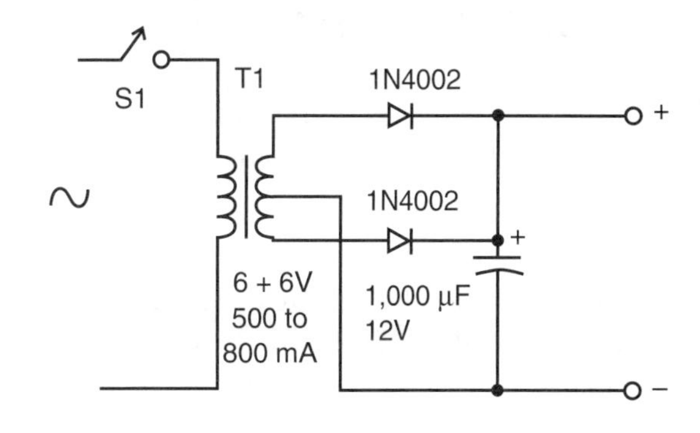


Figure 3.11.8 *A power supply for the circuit*

This circuit uses a small transformer to lower the AC power line voltage to 6 volts and a rectifier to send DC voltage to the inverter. Any transformer producing voltages in the 4.5- to 6-volt range and currents between 500 and 800 milliamps can be used.

The other solution is to directly power the fluorescent lamp from the AC power line using the circuit shown in Figure 3.11.9. It is suggested that you consider buying a ballast rated to the power of the fluorescent lamp. Also, the circuit is not isolated from the power supply line, so be sure to not let any live part of the circuit be exposed, which would cause a dangerous shock.

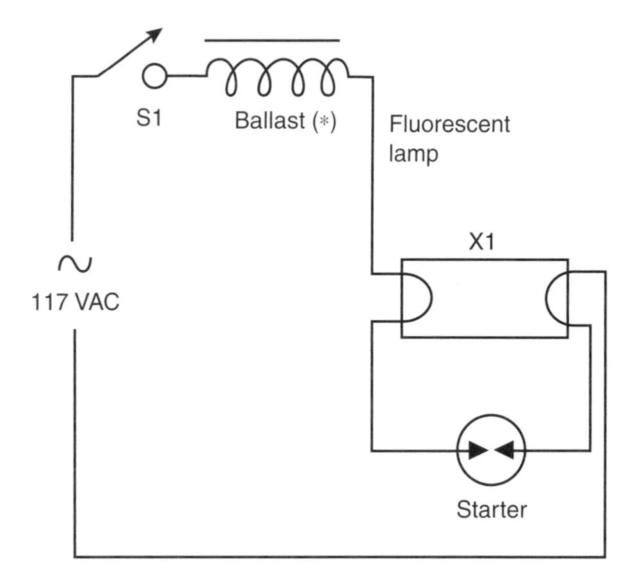


Figure 3.11.9 *Powering the lamp directly from the AC power line*

## A Powerful Inverter

A powerful inverter powered from a 12-volt car battery is shown in Figure 3.11.10. This circuit can drive lamps up to 20 watts and uses two transistors in a push-pull configuration.

The transistor must be mounted on a heatsink because the current drain for the circuit reaches more than 1 amp in some cases. The output power will depend on many factors.

One of the factors that determines the output power is the transformer. Currents that range from 500 milliamps to 1 amp are recommended. As high is the current is, more power can be obtained for the lamp.

You will want to experiment with the values of many of the circuit's components to find the best performance, mainly to match the transformer's characteristics with the circuit. The values for resistors R1 and R2, for instance, can range between 470 and 4.7 kilo-ohms, and C2 and C3 can range between 0.01 and 0.22 $\mu$F. C3 can also be replaced by units between 0.047 and 0.22 $\mu$F.

Additionally, the power source is a 12-volt car or motorcycle battery, and it is important to say that this circuit will power lamps up to 40 watts, but they will not glow at their total power.

## Parts List

- Q1, Q2: TIP42 PNP silicon power transistor
- T1: Transformer (see text)
- R1, R2: 2.2 k$\Omega$ x 1/2 W resistor, red, red, red
- C1: 1,000 $\mu$F x 16 V electrolytic capacitor
- C2, C3: 0.01 $\mu$F ceramic or polyester capacitor
- C4: 0.047 $\mu$F ceramic or polyester capacitor
- S1: On/off switch
- B1: 12 V battery (see text)
- X1: 5 to 20 W fluorescent lamp
- Other: PCB, heatsinks for the transistors, wires, solder, etc.

## Pulsed Light

Experiments with pulsed or stroboscopic lamps can also be used to attract insects. The circuit shown in the stroboscopic lamp section (Project 3.5) is suggested for use in a fluorescent lamp.

## Incandescent Lamps

Although incandescent lamps, due to their spectral characteristics, are not that useful for attracting insects, they can also be used. It is up to the evil genius to discover if white or color lamps are more efficient for this task. Black incandescent lamps are also available, such as the one shown in Figure 3.11.11, but they are very poor UV sources, as suggested by the spectral response.

The advantage of this source is that it can be plugged directly into the AC power line without the need for any other device (only the on/off switch, if desired). Incandescent black light lamps from 60 to 100 watts can be used.

## Other Experiments

The evil genius can probably imagine many other applications linking a light source (visible or UV) with a living being. For instance, it is well known that many types of bees are blind to the color red, and thus they cannot see any source of red light. If you place bees in an area filled only with red light, for them the room will be completely dark. Consequently, they will crash against walls and objects when flying.

An interesting experiment would be to see how insects and many other living beings (fish and plants) react under light sources of different wavelengths (colors). The evil genius could use not only white and UV lamps, but different colored lamps.

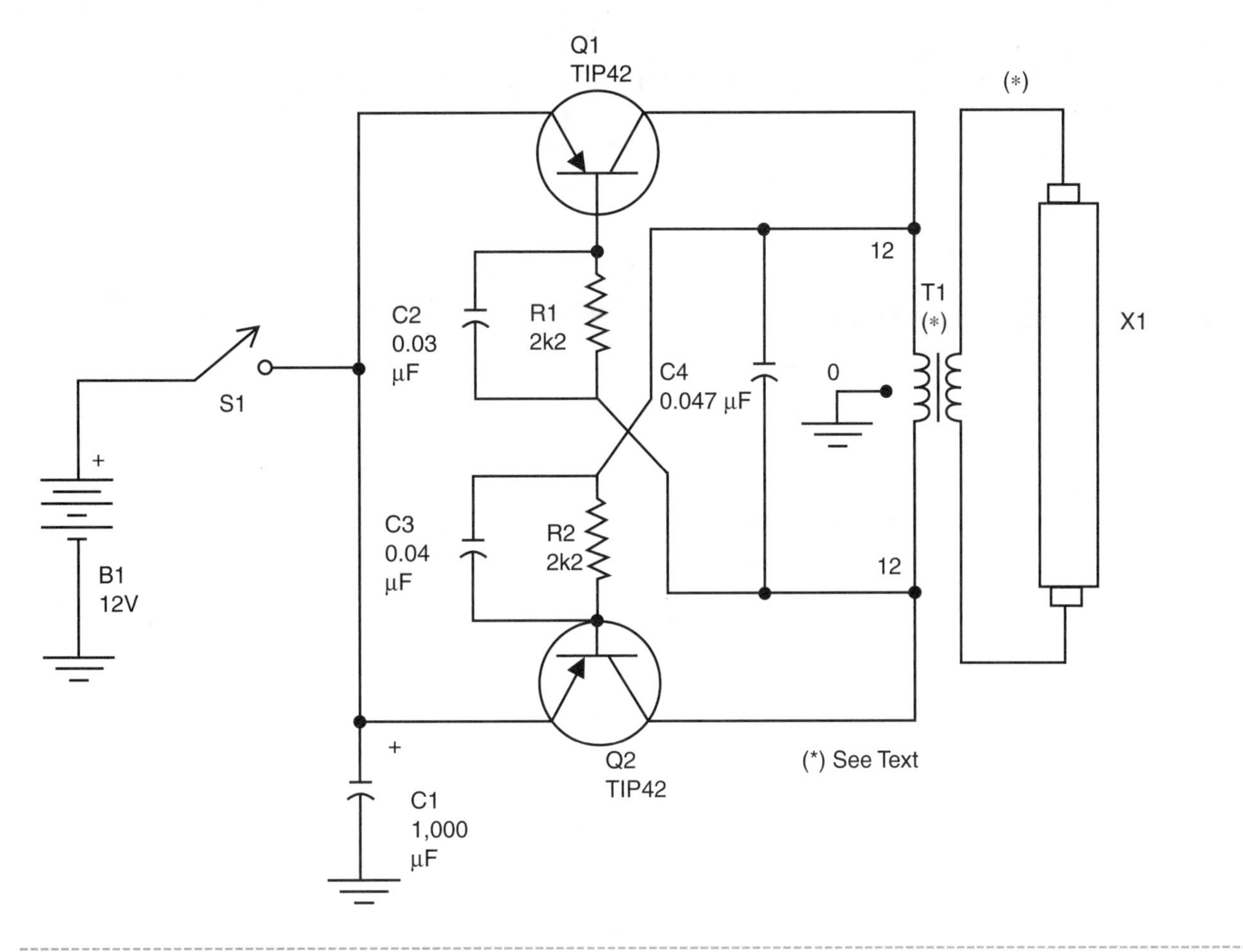


Figure 3.11.10 *A powerful inverter*

Figure 3.11.111 *An incandescent UV Lamp (black light lamp)*

# Project 12—Animal Conditioner

Many animals, such as dogs, cats, and other mammals, can be conditioned to respond to sounds. The interesting point to consider is that many of these animals can hear ultrasounds, which are beyond the hearing range of humans. This means that it is possible to condition a dog, for instance, to answer "inaudible" sounds.

This project will focus on how to mount an audio oscillator to be used in animal conditioning and even an ultrasonic oscillator. The circuits can be powered from cells, meaning that they are small and totally portable.

The evil genius can use the information here to surprise his friends with a remote control that calls his dogs when a button is pressed. The interesting and surprising fact is that no one except the dog will hear anything when the button is pressed because the circuit produces ultrasounds.

Hens on a farm can also be conditioned to the call produced by this circuit, but they are not so sensitive to ultrasounds. The evil genius can thus make interesting experiments to see how the birds respond to his or her "animal conditioner." In a lab setting, the evil genius can also try to discover how animals can distinguish between two sounds with very close frequencies.

The first experiments in animal conditioning were made by the Russian researcher Ivan Pavlov. As shown in Figure 3.12, he had shown that dogs, conditioned by the sound of a bell at meal times, salivated when they heard its sound.

Pavlov's description of how animals (and humans) can be trained to respond in a certain way to a particular stimulus has drawn a tremendous amount of interest ever since he first presented his findings. His work paved the way for a new and objective method of studying animal and human behavior.

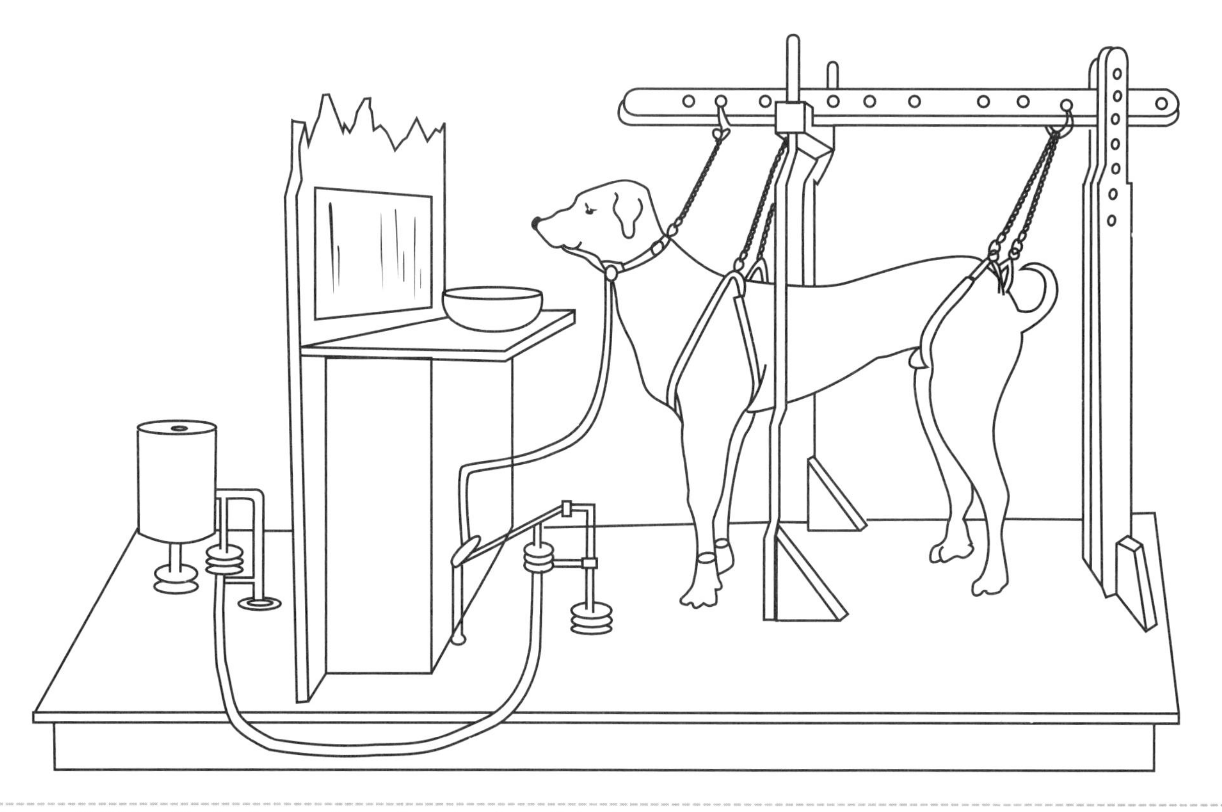

Figure 3.12.1 *Pavlov's experiment in animal conditioning*

## The Project

The basic project consists of two sound or ultrasound oscillators. The first version uses a piezoelectric transducer and can produce sounds only in the audible range. The second version uses a piezoelectric transducer that can produce frequencies of up to 25 kHz, reaching the ultrasonic band. The frequency of the sound can be easily adjusted by trimmer potentiometers in both versions.

The power supply consists of four AA cells. Since the circuit is powered for small time intervals, only when a call is produced, the cell's life will be long.

The current drain depends on the output power, so the economic version is the one that uses the piezoelectric transducer. The other version has a current drain ranging from 80 to 300 milliamps, proportionally reducing the battery life.

## Experiments in Bionics

Many bionics experiments and activities can be performed using the sonic and ultrasonic versions of the conditioner:

- Conditioning your dog to come when the circuit is activated.
- Conditioning other animals using frequencies according to the sensitivity of each specimen.
- Study the ability of different specimens to distinguish sounds of different frequencies or to hear ultrasounds.

## How the Circuit Works

A 555 *integrated circuit* (IC) running in the astable configuration forms the two circuits in the basic version. The frequency is determined by C1, R1, R2, and the resistance adjusted by P1.

The output of the 555 is enough to directly drive a piezoelectric transducer. This type of transducer can reproduce sounds up to 7 or 8 kHz, which will be enough for the applications described in this project.

However, to drive a powerful transducer like a piezoelectric tweeter reaching frequencies of up to 25 kHz, a driver stage is necessary. The version using this stage is also shown in this project.

## How to Build

As explained, the basic project is shown in two versions using piezoelectric transducers or loudspeakers (tweeters).

## Basic Project: Version 1

Figure 3.12.2 shows the schematic diagram for the basic version using a piezoelectric transducer.

The circuit is very simple, and the evil genius can mount it using a *printed circuit board* (PCB) or even a solderless board. The pattern for a PCB is suggested in Figure 3.12.3.

Take care when mounting the polarized components. Any inversion can stop the circuit from functioning.

The transducer is a piezoelectric type used in toys, PCs, and many other applications where a warning sound is produced. The circuit can be housed in a plastic box, such as the one shown in Figure 3.12.4. Remember to cut holes in the box to allow the sound produced by the transducer to be heard.

## Basic Project: Version 2

Figure 3.12.5 shows the schematic diagram for the second version of the animal conditioner.

This version can also be mounted using a PCB or a solderless board. A suggestion for a PCB is shown in Figure 3.12.6.

If you intend to produce sounds only in the audible range, a small loudspeaker (5 to 10 cm) can be used as a transducer. To reach higher frequencies in the ultrasonic range, we recommend using a small piezoelectric tweeter. The transistor must be mounted on a heatsink.

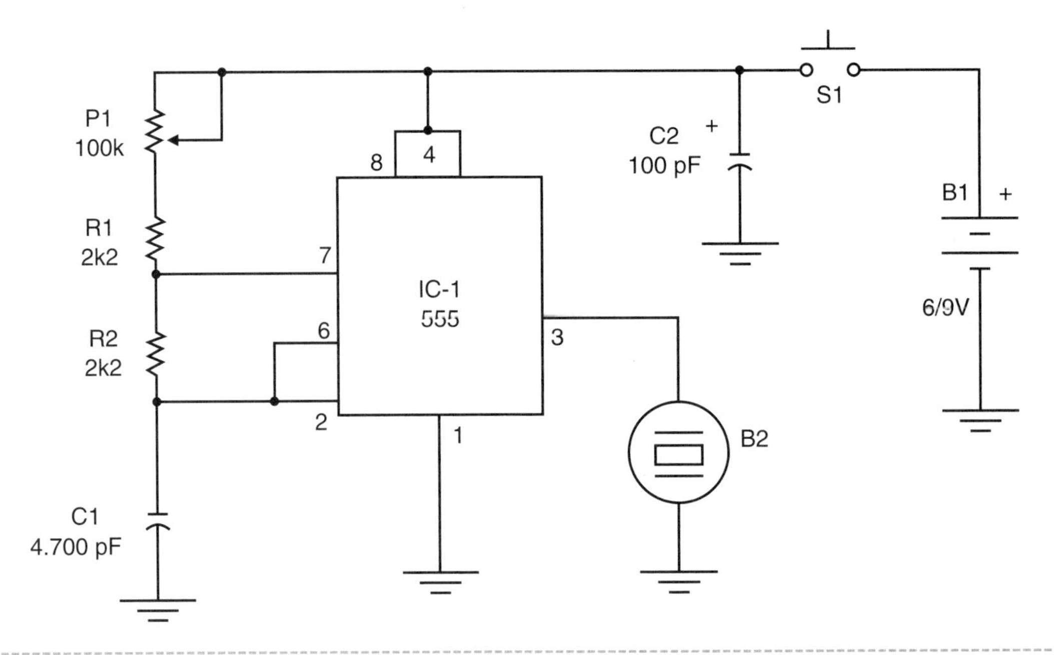


Figure 3.12.2 *Schematic diagram for basic version 1*

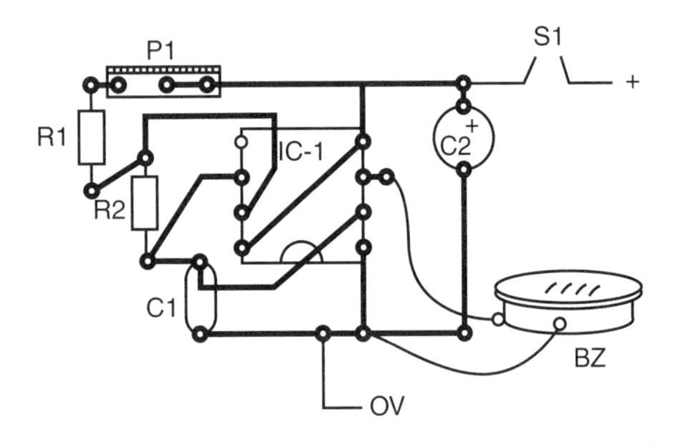


Figure 3.12.3 *PCB for basic version 1*

Figure 3.12.4 *The conditioner fits into a plastic box.*

## Testing and Using

Press S1 to power on the circuit. Adjusting P1, the frequency of the produced sound will change. If no sound is obtained, examine the circuit, looking for mistakes or cold soldering.

Obtaining the sound, adjust P1 to the desired frequency. The ultrasonic sounds are produced when you turn the trimmer potentiometer above the limit of the audible range. This means above the highest treble tone you can hear.

Animal conditioning requires some patience. Press S1 many times near the animal and each time give the animal a reward. After several times, the animal will associate the sound with the reward, running to you each time you press S1.

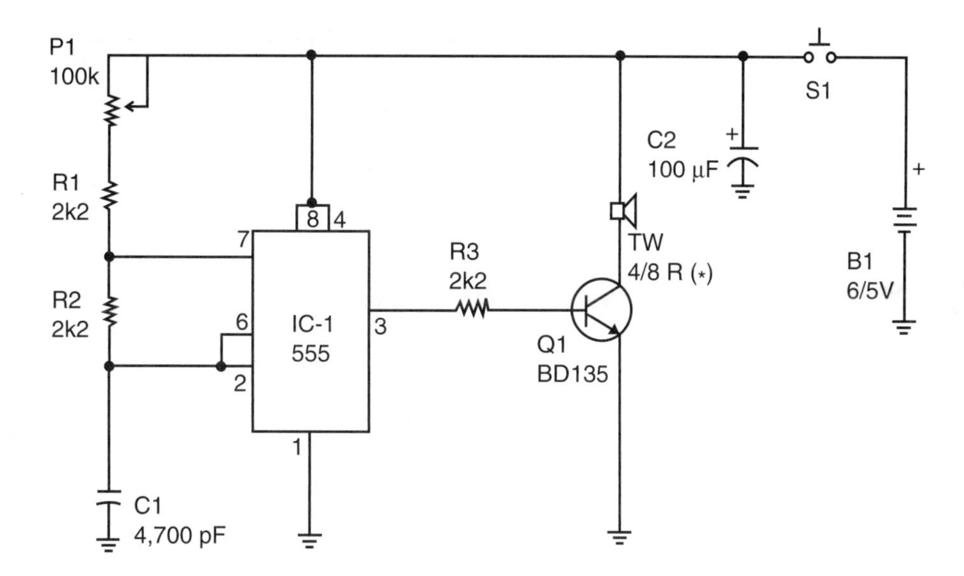


Figure 3.12.5 *Schematics for basic version 2*

## Parts List: Basic Version 1

IC-1: 555 integrated circuit timer

BZ: Piezoelectric transducer

S1: Push button

B1: A 6 to 9 V source or four AA cells or battery

P1: 100 kΩ trimmer potentiometer

R1, R2: 2.2 kΩ x 1/8 W resistor, red, red, red

C1: 4,700 pF ceramic or polyester capacitor

C2: 100 µF x 12 V electrolytic capacitor

Other:

PCB or solderless board, battery connector or cell holder, plastic box, wires, solder, etc.

## Parts List: Basic Version 2

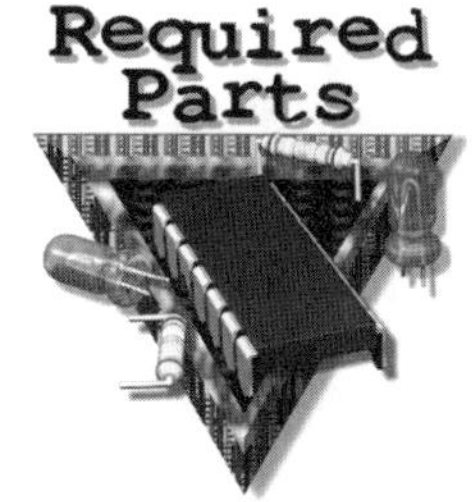


IC-1: 555 integrated circuit timer

Q1: BD135 silicon NPN medium-power transistor

TW1: 4 or 8 Ω piezoelectric tweeter or loudspeaker

S1: Push button

B1: A 6 to 9 V source or four AA cells or battery

P1: 100 kΩ trimmer potentiometer

R1, R2, R3: 2.2 kΩ x 1/8 W resistor, red, red, red

C1: 4,700 pF ceramic or polyester capacitor

C2: 100 µF x 12 V electrolytic capacitor

Other: PCB or solderless board, battery connector or cell holder, small heatsink for the transistor, plastic box, wires, solder, etc.

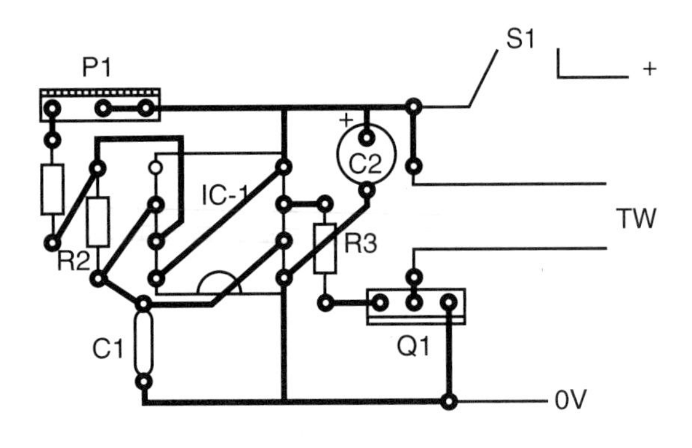


Figure 3.12.6 *PCB for version 2*

## Additional Circuits and Ideas

The basic circuits produce continuous sounds or ultrasounds, which are not the only ones the evil genius can use in animal conditioning. Some circuits that produce modulated sounds or two tones are given in this section, and their usage depends only on the imagination of the reader.

## Modulated Sound Conditioner

The circuit shown in Figure 3.12.7 produces a modulated sound like a siren, reproduced either in a piezoelectric transducer or in a tweeter, as in the basic versions.

Modulation is adjusted by P1 and the tone is adjusted by P2. The circuit can directly drive a piezoelectric transducer or, if the evil genius prefers, a loudspeaker or tweeter, using the same transistor stage as in basic version 2.

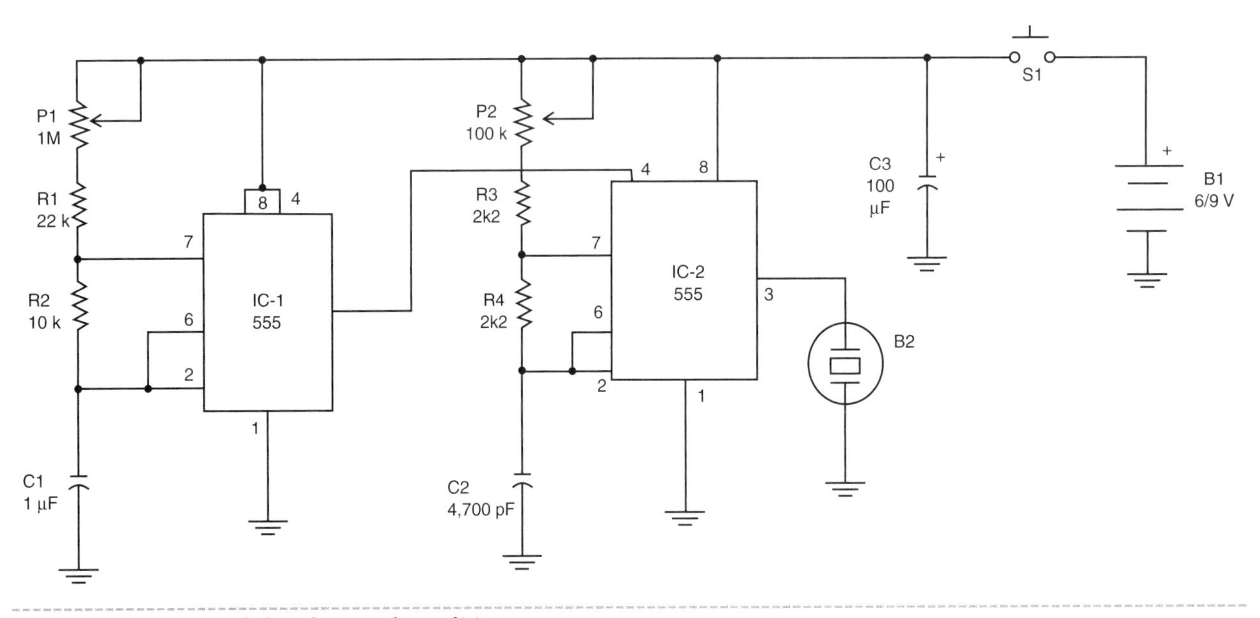


Figure 3.12.7 *Modulated animal conditioner*

## Parts List

```
IC-1, IC2: 555 inte-
  grated circuit timers
P1: 1 MΩ trimmer poten-
  tiometer
P2: 100 kΩ trimmer
  potentiometer
R1: 22 kΩ x 1/8 W resis-
  tor, red, red, orange
R2: 10 kΩ x 1/8 W resis-
  tor, brown, black,
  orange
R3, R4: 2.2 kΩ x 1/8 W
  resistors, red, red,
  red
C1: 1 μF polyester or
  electrolytic capacitor
C2: 4,700 pF ceramic or
  polyester capacitor
C3: 100 μF x 12 V elec-
  trolytic capacitor
BZ: Piezoelectric
  transducer
S1: Push button
B1: 6 or 9 V cells or
  battery
Other: PCB, battery con-
  nector or cell holder,
  plastic box, wires,
  solder, etc.
```

## Two-Tone Conditioner

Another interesting circuit for animal conditioning is shown in Figure 3.12.8. This circuit produces alternating tones at a rate determined by R1. The reader can experiment with the resistor in the range of 1 to 4.7 MΩ.

The tones are determined by R2 and C2. The evil genius is also free to experiment with values that range from 10 to 100 kΩ. Higher values result in lower frequencies or bass tones.

The circuit can be mounted on a small PCB or on a solderless board, and it fits into a plastic box, which should have holes for the transducer. Figure 3.12.9 shows the PCB suggested for mounting this circuit. When mounting, be careful with the positions of the polarized components.

The circuit can also drive a loudspeaker using the output stage shown in version 2 of the basic project. In this case, remove C1 and C2 and replace them with 1 kΩ resistors.

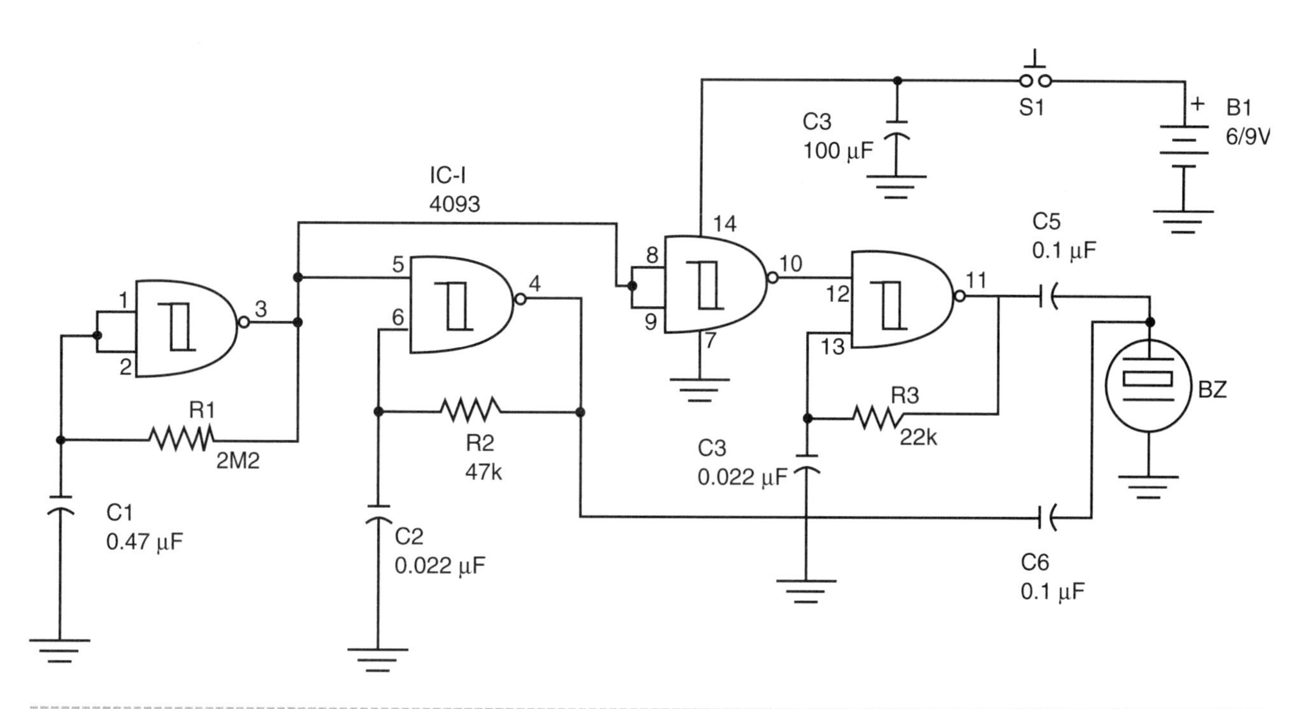


Figure 3.12.8 *Circuit for a two-tone conditioner*

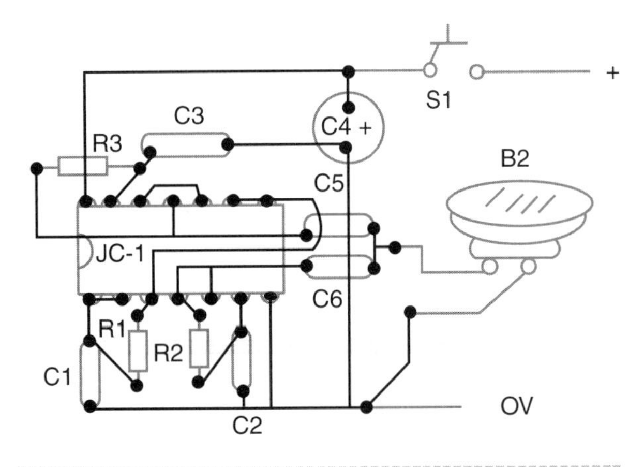


Figure 3.12.9 *PCB for this project*

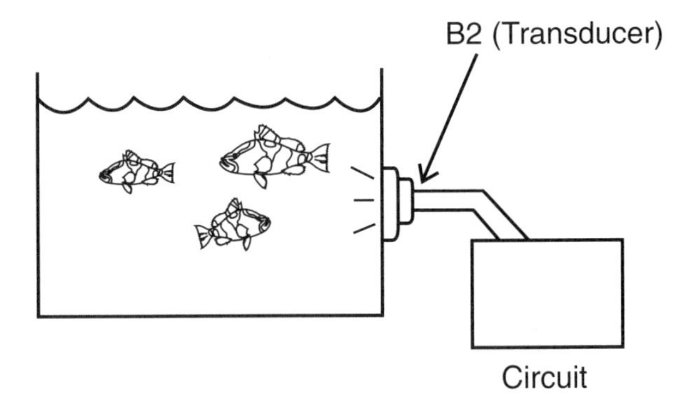


Figure 3.12.10 *Experimenting with fish in an aquarium*

## Parts List

IC-1: 4093 complementary metal oxide semiconductor (CMOS) integrated circuit

BZ: Piezoelectric transducer

R1: 2.2 MΩ x 1/8 W resistor, red, red, green

R2: 47 kΩ x 1/8 W resistor, yellow, violet, orange

R3: 22 kΩ x 1/8 W resistor, red, red, orange

C1: 0.47 μF ceramic or polyester capacitor

C2, C3: 0.022 μF ceramic or polyester capacitor

C4: -100 μF x 12 V electrolytic capacitor

S1: Push button

B1: 6 or 9 V cells or battery

Other: PCB, battery connector or cells holder, wires, plastic box, solder, etc.

## Other Ideas for Experiments

Many living beings can be conditioned by sounds, so try some experiments with fish in an aquarium, as suggested by Figure 3.12.10. The experiments can reveal the sensitivity of some species to sounds of determined frequencies.

# Project 13—White Noise Generator



White noise is defined as a random sound within a prescribed volume and tonal parameters. A more sophisticated way to define it is to say that white noise is a signal with equal power per frequency unit (hertz) over a specified frequency band. Another definition would say that white noise is noise with an autocorrelation function of 0 everywhere but at 0. It is also called Johnson noise. White noise also has a constant frequency spectrum.

What is important to the bionics evil genius is that white noise affects people and animals in specific ways, with the main effect on people being one of relaxation and sleep. That is why natural sources of white noise such as the sea, wind, or rain have a particular influence on humans, causing many to drop off to sleep or reach a state of relaxation.

The main sources of natural background noise are the molecular-level agitation of matter via temperature changes and natural electric discharges in the atmosphere. When we put a shell over our ear to hear the "sea" noise, we are adding an acoustic amplifier to our natural hearing organs. The shell increases the white noise level produced by the thermal agitation of the air molecules to a level above the audible threshold. This is an audible white noise.

The idea in this project is to mount a white noise generator to be used in many interesting applications involving human reactions to this kind of sound. Powered from batteries and coupled to an amplifier, it can be used to make you sleep in noisy places or help you reach a state of relaxation. You can also use it to excite other living beings such as plants, insects, or even fish in an aquarium to see how they react to these sounds.

The basic circuit is very simple but needs an external amplifier. Versions with an added amplifier will be suggested too.

## The Project

Electronics components such as resistors, diodes, transistors, and so on can be used to produce white noise, covering a large band of frequencies. In our projects, we will use silicon transistors because they are cheap and easy-to-find devices.

The basis of our project is a transistor used as a white noise source coupled to a preamplifier. The signals coming from the output of the preamplifier are coupled to an external audio amplifier driving a loudspeaker.

The amount of white noise obtained from the output of the amplifier depends on its power. Amplifiers that range from 0.5 to 50 watts are recommended for most of the applications suggested here.

Coupled to the amplifier, the bionics evil genius can use the white noise generator to help someone sleep; cover the noise produced by cars, tools, and other noisy places; and see how the noise affects living beings such as mammals, insects, and plants.

## Experiments in Bionics

Why white noise has a special effect on people is not well known. Maybe it is because the sound doesn't have a definite pattern on which the brain can concentrate its attention. The dispersive characteristic of the noise makes the brain think about "nothing," helping it reach a sleepy or relaxed state.

Some suggested experiments and projects are as follows:

- Build a "relaxation circuit" to interact with humans.
- Create a noise-canceling circuit using white noise.
- Determine how white noise affects plants and other living beings at your home.

## The Electronic Circuit

The thermal noise in a semiconductor junction (emitter base junction in a transistor) is amplified by the circuit and can be applied to the input of any audio amplifier. A loudspeaker reproduces the signal at the output of the amplifier, which is formed by a *negative positive negative* (NPN) general-purpose transistor with a gain determined basically by R1.

The circuit needs a voltage of up to 9 volts to produce noise at a level high enough to drive the amplifier, so the power supply must be a 9-volt battery or a 12-volt source. Because current drain is very low, a 9-volt battery will have an extended life in this application, providing energy for many weeks.

Any general-purpose NPN transistor can be used as a white noise source, and types such as the BC548, 2N2222, and BC547 can be used. Figure 3.13.1 shows the representation of white noise.

An important point to consider is that the circuit is sensitive to external noises, such as the sound produced by the AC power line. To avoid this noise, the circuit must be connected to the input of the amplifier using a shielded cable. It is not recommended that you use an external power supply connected to the AC power line because it can introduce noise in the circuit.

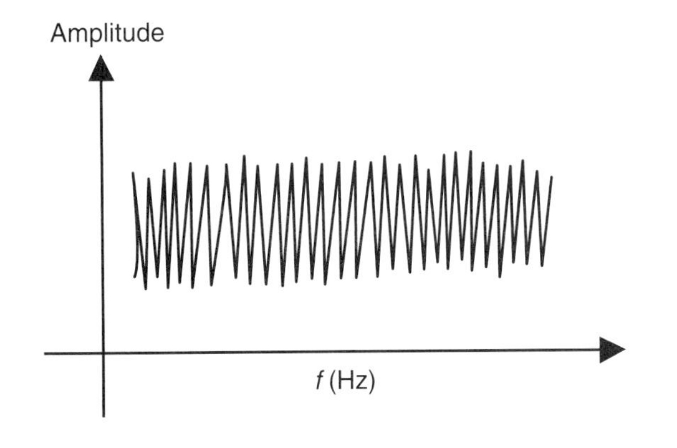


Figure 3.13.1 *White noise*

## How to Build

The complete schematic diagram of the basic version of a simple white noise generator is shown in Figure 3.13.2.

The circuit can be mounted on a small *printed circuit board* (PCB) such as the one shown in Figure 3.13.3. Observe the position of the polarized components, mainly the transistors.

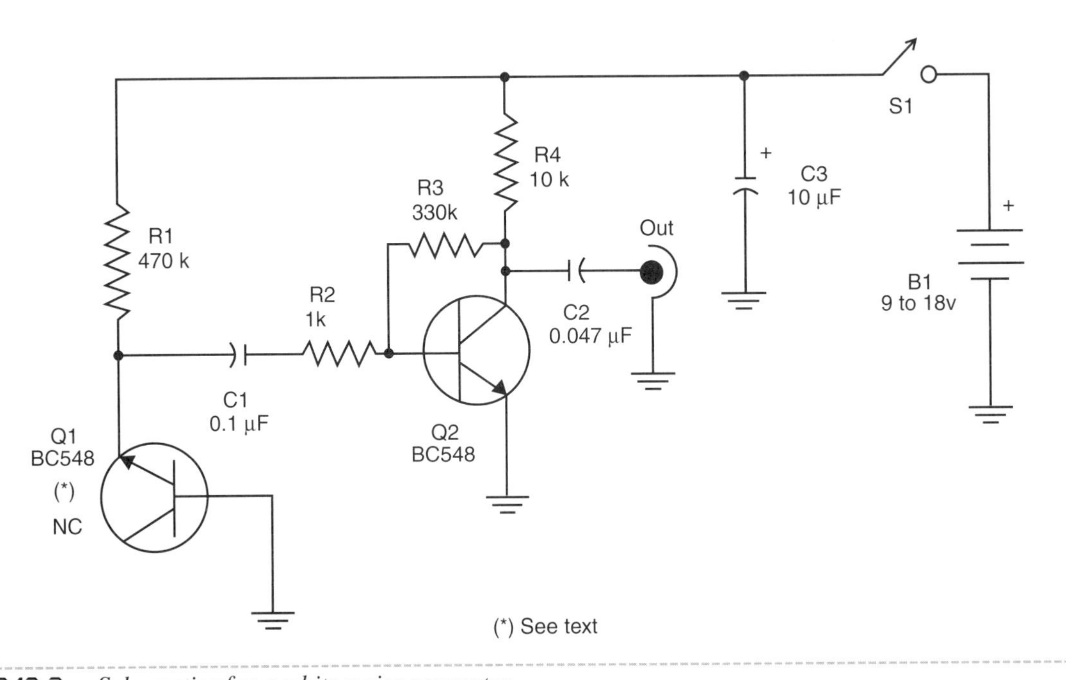


Figure 3.13.2 *Schematics for a white noise generator*

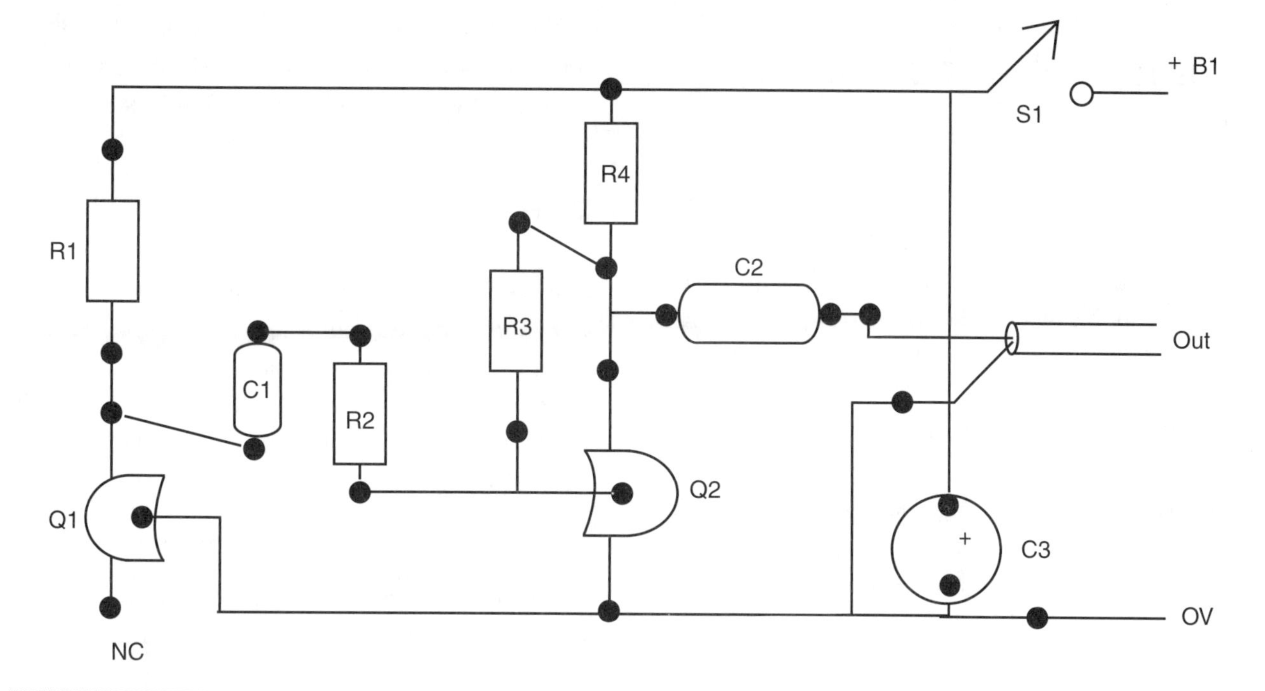


Figure 3.13.3 *PCB for the white noise generator*

## Testing and Using

Plug the output of the circuit to the input of any audio amplifier, as shown in Figure 3.13.4. Then turn on the power supply and adjust the volume control of the power amplifier to get a sound like the one produced by the sea or wind.

When using the circuit for relaxation purposes, adjust the volume to a comfortable level. When using it to cover ambient noise, adjust the volume to the level necessary.

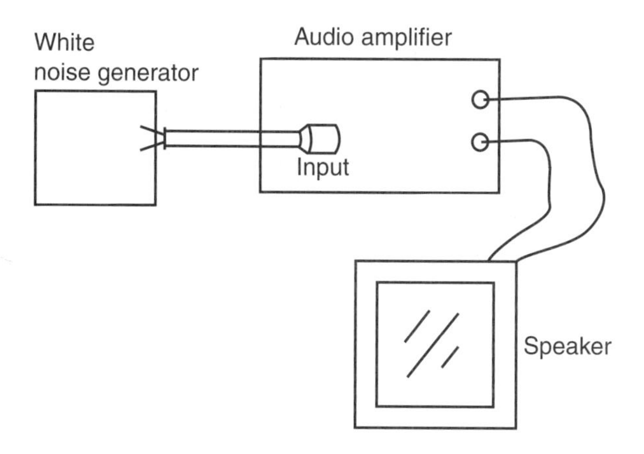


Figure 3.13.4 *Testing the white noise generator*

## Parts List

Q1, Q2: BC548 general-purpose NPN silicon transistors

R1: 470 kΩ x 1/8 W resistor, yellow, violet, yellow

R2: 1 kΩ x 1/8 W resistor, brown, black, red

R3: 330 kΩ x 1/8 W resistor, orange, orange, yellow

R4: 10 kΩ x 1/8 W resistor, brown, black, orange

C1: 0.1 μF ceramic or polyester capacitor

C2: 0.047 μF ceramic or polyester capacitor

C3: 10 μF x 16 V electrolytic capacitor

S1: On/off switch

B1: 9 to 18 V battery

Other: PCB, battery connector, shielded cable, wires, solder, etc.

## Additional Circuits and Ideas

Many experiments that involve humans interacting with electronic circuits can be made using a white noise generator. Some suggestions for a few exercises are given in this section.

### Stochastic Resonance

Stochastic resonance occurs when a modulated signal that is too weak to be heard or otherwise detected under normal conditions becomes detectable. This is due to a resonance phenomenon that is produced between the weak deterministic signal and a signal with no definite frequency called *stochastic noise*. A signal with no definite frequency is called noise and can be produced by random phenomena such as thermal movement of atoms in a material, or wind, or rain.

This phenomenon explains why some people can hear "voices" in the noise produced by the wind or grass excited by the wind. The Native Americans say that they can hear the voices of their ancestors in the noise of the grass, as explained in the movie *The Grass Harp*.

You can make experiments trying to record the voices of the past or other dimensions using a white noise generator as a source of stochastic noise.

### Adding an Amplifier

Figure 3.13.5 shows how to use the LM386 audio amplifier, described in the electric fish experiment, with the white noise generator. Make sure that the white noise generator is powered from a 9- to 12-volt source and the LM386 is powered from another supply producing 6 volts.

### White Noise Generator Using an Operational Amplifier 741

Another source for white noise is the circuit shown in Figure 3.13.6. The advantage of this circuit is that the amplitude of the white noise can be controlled by P1, which sets the gain of the amplifier. This is important for getting the necessary amount of signal to excite the input of the external amplifier.

Another advantage is that the circuit has low-impedance output. This means that it is less sensitive to noise, especially the noise produced by the AC power line.

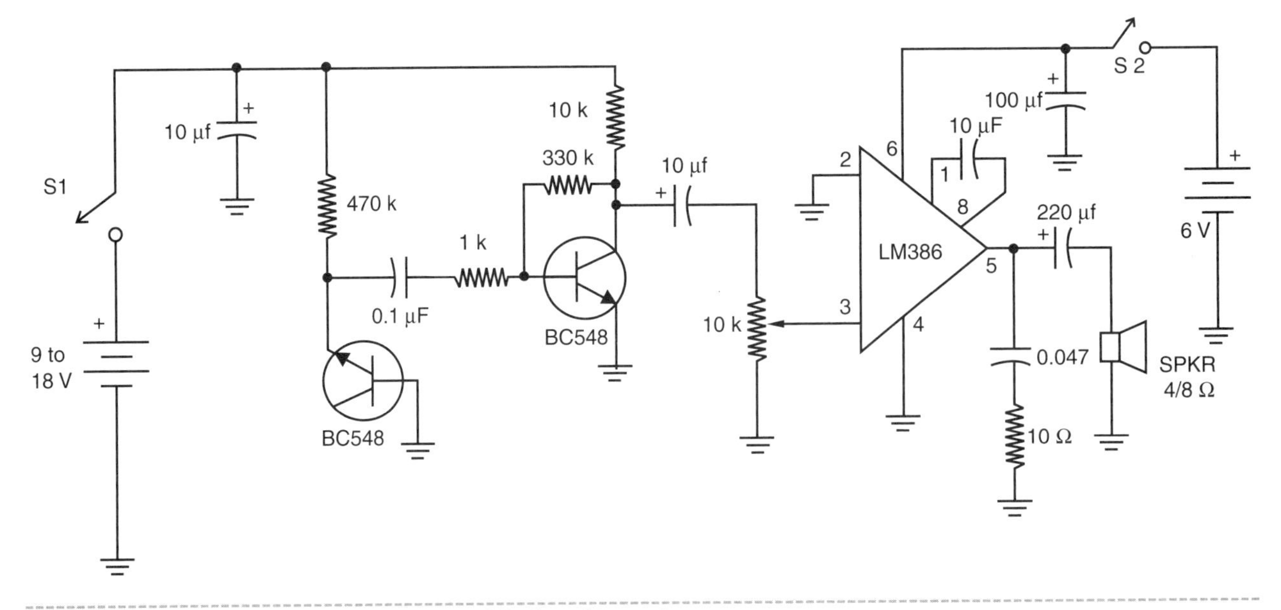


Figure 3.13.5 *Using the LM386 to amplify the white noise*

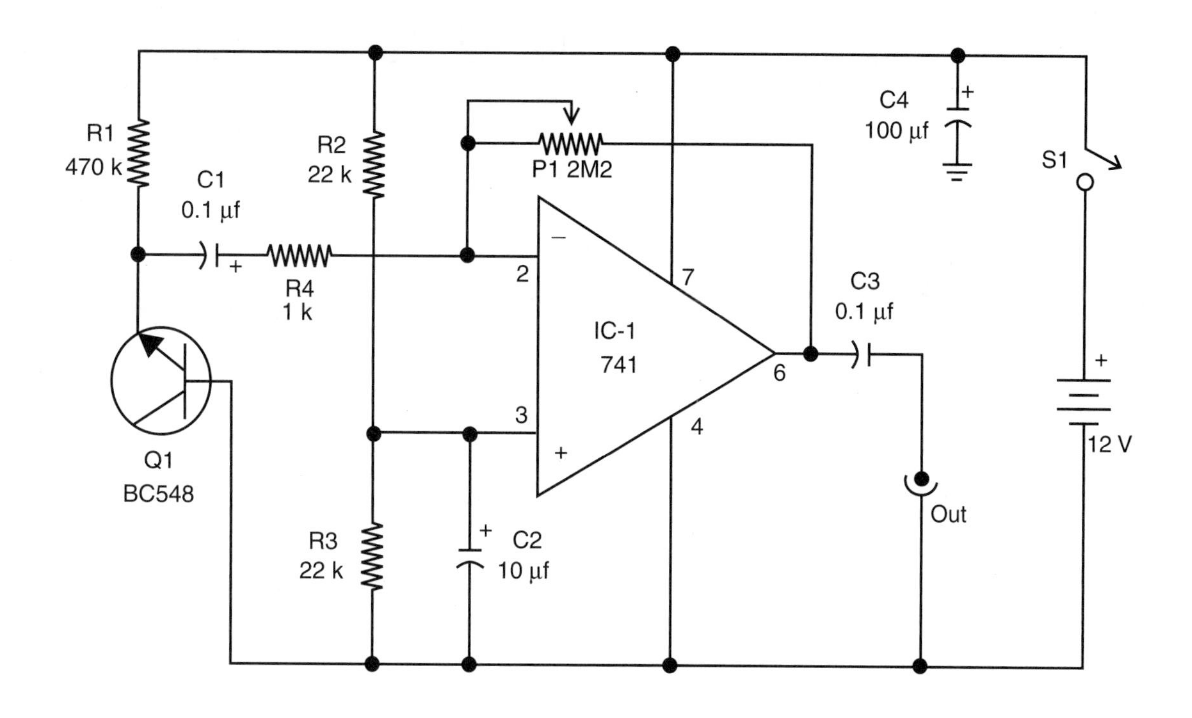


Figure 3.13.6 *Circuit using an operational amplifier*

The circuit is powered from a 9- or 12-volt power supply and the current drain is also very low, extending the battery life.

Any NPN general purpose can be used as a noise source, and the operational amplifier 741 can be replaced by equivalents.

## Parts List

- IC-1: 741 integrated circuit operational amplifier
- Q1: BC548 or equivalent, any general-purpose NPN silicon transistor
- R1: 470 kΩ x 1/8 W resistor, yellow, violet, yellow
- R2, R3: 22 kΩ x 1/8 W resistor, red, red, orange
- P1: 2.2 MΩ potentiometer, lin or log
- C1, C3: 0.1 µF ceramic or polyester capacitor
- C2: 10 µF x 16 V electrolytic capacitor
- C4: 100 µF x 16 V electrolytic capacitor
- S1: On/off switch
- B1: 9 to 18 V battery
- Other:
- PCB, shielded cable, wires, solder, etc.

## Mounting a Noise-Canceling Helmet

If you live in a noisy place, you can install earphones in a helmet and plug them into the output of the amplifier, as shown in Figure 3.13.7.

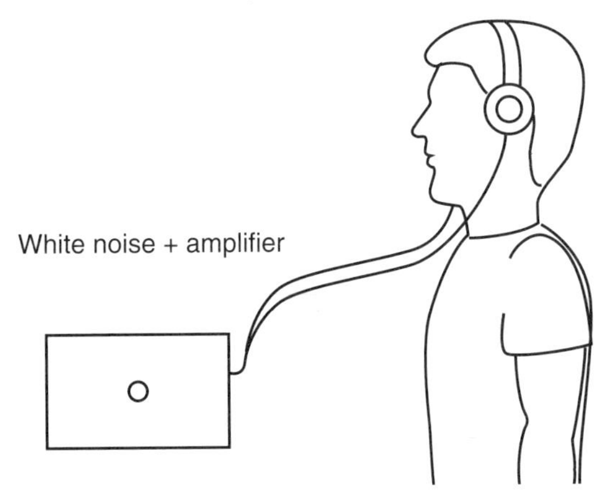


Figure 3.13.7 *A noise-canceling helmet using a white noise generator*

## Other Applications

The influence of white noise on plants and animals is not well known. The evil genius can make some important discoveries using white noise to stimulate plants or observe its effect on insects.

# Project 14—Bionic Ear



What would a bionics evil genius think about being able to hear very weak sounds from far away such as light footsteps on a carpet or even conversations through walls? Steve Austin, the six-million-dollar bionic man, had those same abilities. He could hear the weakest sounds, even ones produced from many miles away. Of course, the fantastic ears of the bionic man can't be reproduced by simple electronic equipment. Our natural ears are the most sensitive sensor that nature can create, and no other sensor, even an electronic one, could be superior to them.

The project described here imitates nature, creating a circuit that can receive weak sounds and reproduce them via headphones. The sensitivity of the project can be upgraded to pick up sounds that even creatures in nature cannot detect. Our project consists of a very sensitive amplifier coupled to a microphone with special resources to pick up weak sounds from distant places or even through walls. For example, one device can pick up a conversation that is far away or on the other side of a wall, in the same way a spy microphone picks up sounds.

Figure 3.14.1 *The ears act as antennas, picking up weak sounds.*

## Bionic Applications

The basic idea involved in this project is to create an ear, linking an electronic amplifier circuit to resources, such as sound lenses and sensors, in order to pick up very weak sounds. The electronic circuit is created by our technology, and the acoustic capabilities for picking up the sounds are based on solutions found in nature.

Such solutions would consist of the shape and characteristics of the ears of many animals that depend on their hearing for their survival. As Figure 3.14.1 shows, the ears of a rabbit operate as acoustic shells, picking up sounds coming from a determined direction and concentrating them into the sensor, or the ear drum.

A bionic ear can be used for many interesting applications such as the following:

- You can use it to hear conversations through walls or in distant places.
- The sensitivity of the ear can help you find leakage in water plumbing.
- Noise sources can be easily located.
- You can explore the nature of hearing or recording sounds produced by many living beings.

## How the Circuit Works

The audio amplifier has no secrets: It is the same one used to amplify the signals generated by the electric fish in Project 3.1. The difference is that the circuit used here receives the signal picked up by a very sensitive electret microphone.

An integrated amplifier, the LM386, has a gain of 200 programmed by an external capacitor plugged in between pins 1 and 8. This circuit can source about 500 milliwatts to an earphone with an impedance of about 8 ohms, which is enough to provide good volume.

The electret microphone has an internal field effect transistor, giving a preamplification to the audio signal. The signal in the output of the microphone passes across a volume (sensitivity) control formed by a potentiometer and is then amplified to the input of the amplifier.

The circuit is powered from four AA cells. Because the current drain is not high, the cells' life will be extended to many hours of continuous use.

The electret microphone will be installed in acoustic resources according to the application. Figure 3.14.2 shows some of the resources recommended for this project.

The acoustic shell or the acoustic apparatus for hearing through walls is very important for concentrating all the possible sound into the microphone. These aspects will be described later in this section.

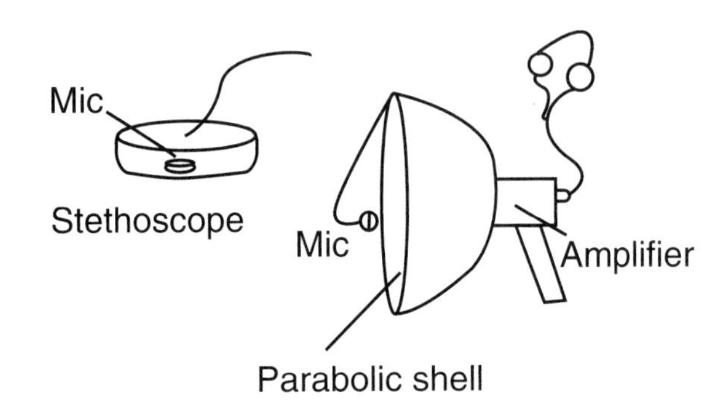


Figure 3.14.2 *Acoustic resources concentrate the sound into the microphone.*

## How to Build

Figure 3.14.3 shows the complete diagram of the electronic circuit for the bionic ear.

The *printed circuit board* (PCB) is the same as that recommended for the electric fish project, because R1 and the microphone can be mounted externally, as shown by Figure 3.14.4.

Of course, the other difference in the project is that, instead of a loudspeaker, the output load is an earphone. The same figure shows how to connect this transducer.

The circuit fits into a small plastic box, and to be easily used a handle is added, as shown in Figure 3.14.5.

The microphone is connected to the circuit by a long, shielded cable to avoid noise, mainly the hum produced by the AC power supply lines.

When mounting, take care with the position of the polarized components such as the *integrated circuit* (IC), electret microphone, power supply, and electrolytic capacitors. Any inversion will prevent the circuit from operating.

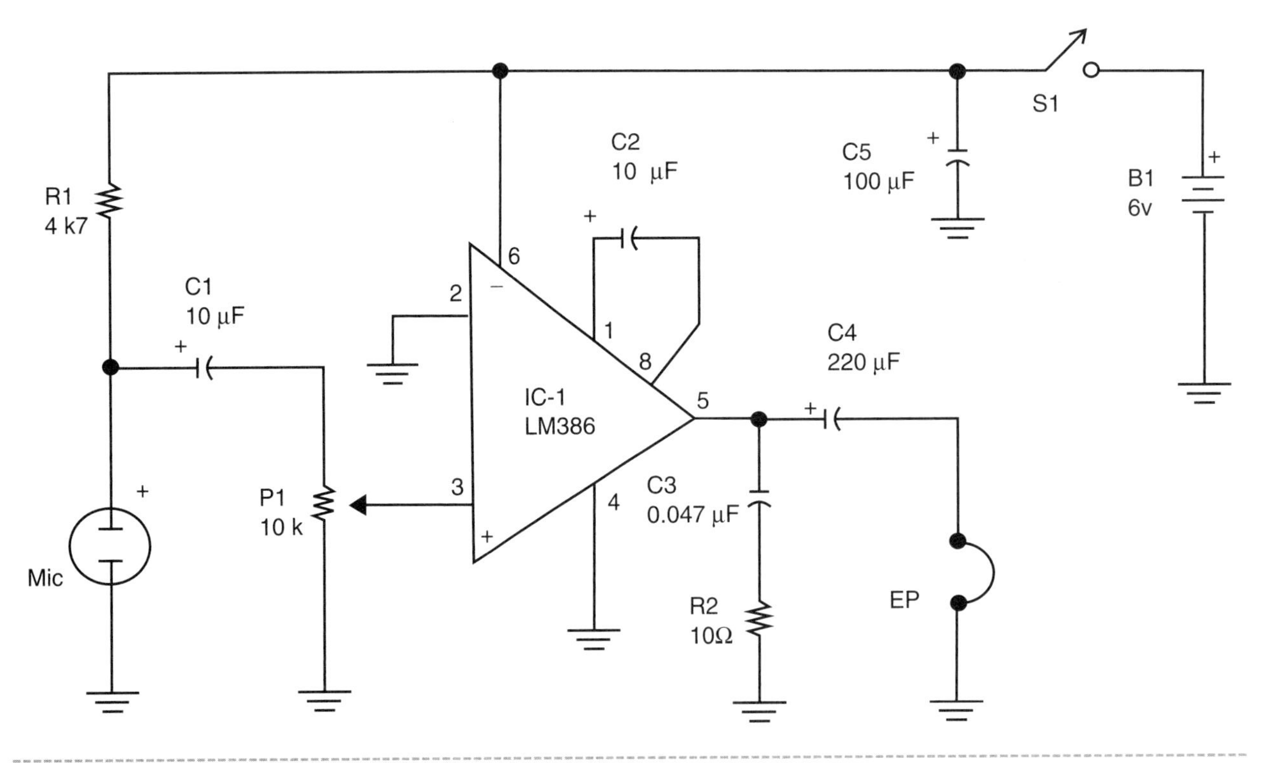


Figure 3.14.3 *Bionic ear*

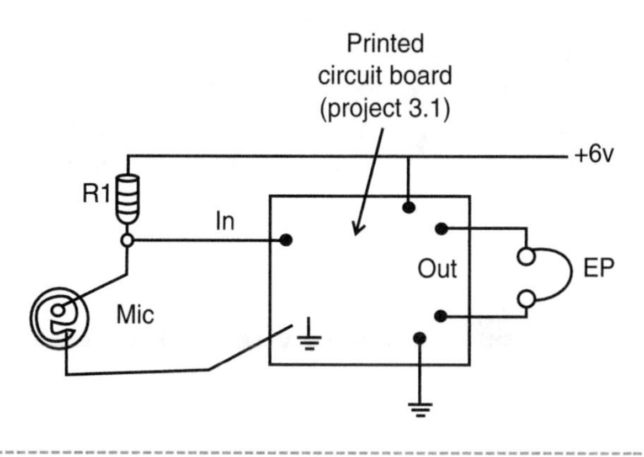


Figure 3.14.4 *The circuit uses the same PCB as Project 3.1.*

Figure 3.14.5 *Caption*

## Acoustic Resources

The simplest acoustic aid is a small plastic shell with the microphone placed at its center, as shown in Figure 3.14.6. Using this, it is possible to pick up sounds from one particular direction.

An exactly parabolic shell is the ideal for picking up weak remote sounds, such as the ones found in nature. The reader can hear the song of birds or pick up a conversation from a long distance, as suggested by Figure 3.14.7.

The larger the parabolic shell is, the more sound can be concentrated into the microphone. For practical purposes, shells with diameters between 40 and 80 centimeters are recommended.

If the evil genius wants, he or she can add an output line to a small tape recorder to the circuit, as shown in Figure 3.14.8.

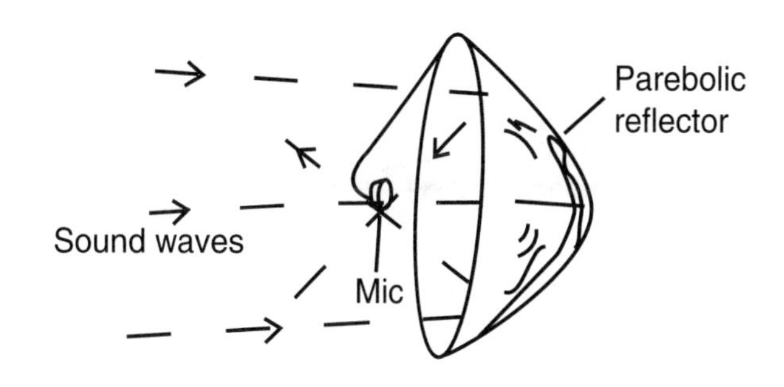


Figure 3.14.6 *A directional microphone*

Figure 3.14.7 *Using a parabolic shell*

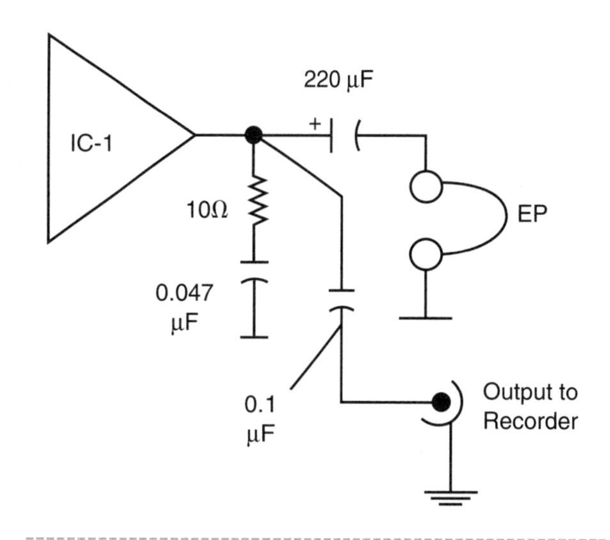


Figure 3.14.8 *Adding a recorder output*

Another acoustic resource would be a stethoscope made with a heavy piece of metal and a plastic sponge, as shown in Figure 3.14.9. Using this unit, the evil genius can hear sounds through walls.

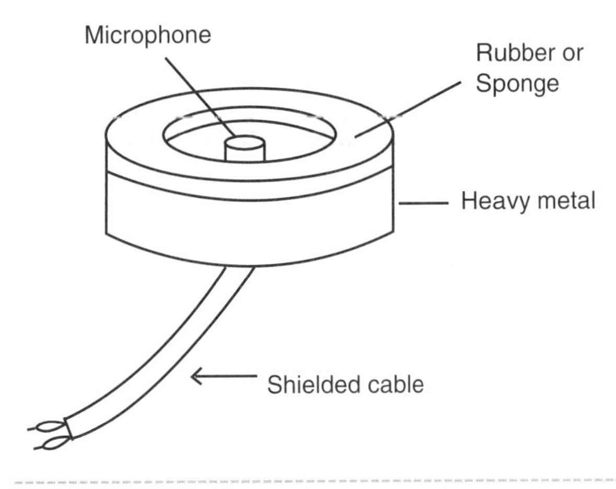


Figure 3.14.9 *The stethoscope probe*

## Testing and Using

Testing is very easy. Place the cells in the cell holder and turn the device on. The ambient sounds will be reproduced loud and clear in the headphones when you open the sensitivity control. When using the device, all you have to do is point the microphone or stethoscope at the sound source.

## Parts List

IC1: LM386 (any suffix) integrated circuit audio amplifier

R1: 4.7 kΩ x 1/8 W resistor, yellow, violet, red

R2: 10 Ω x 1/8 W resistor, brown, black, black

C1, C2: 10 μF x 12 V electrolytic capacitor

C3: 0.047 μF ceramic or polyester capacitor

C4: 220 μF x 12 V electrolytic capacitor

C5: 100 μF x 12 V electrolytic capacitor

P1: 10 kΩ potentiometer log

S1: On/off switch

MIC: Two terminals electret microphone

B1: A 6 V source or four AA cells and holder

EP: Earphone 8 to 32 Ω

Other: PCB, earphone jack, plastic box, acoustic apparatus, wires, solder, etc.

# Additional Circuits and Ideas

In principle, any small audio amplifier powered from cells or a battery can be used to amplify the signals of an electret microphone. The evil genius can also use other types of transducers to pick up signals from nature, and two suggestions are provided here.

## Using the TDA7052

An audio amplifier suitable for applications where electret microphones are used is the TDA7052. The basic circuit for this application is shown in Figure 3.14.10.

The circuit is powered from four AA cells and can produce about 170 milliwatts) to 8-ohms earphones in the basic configuration. The input for the electret microphone is the same, and the sensitivity or volume control is also performed by a 10 kΩ potentiometer. Earphones with an impedance that ranges from 8 to 64 ohms can be plugged into the output of the circuit.

A small PCB for this circuit is shown in Figure 3.14.11. Another way to mount this circuit, for experimental purposes, is to use a solderless board.

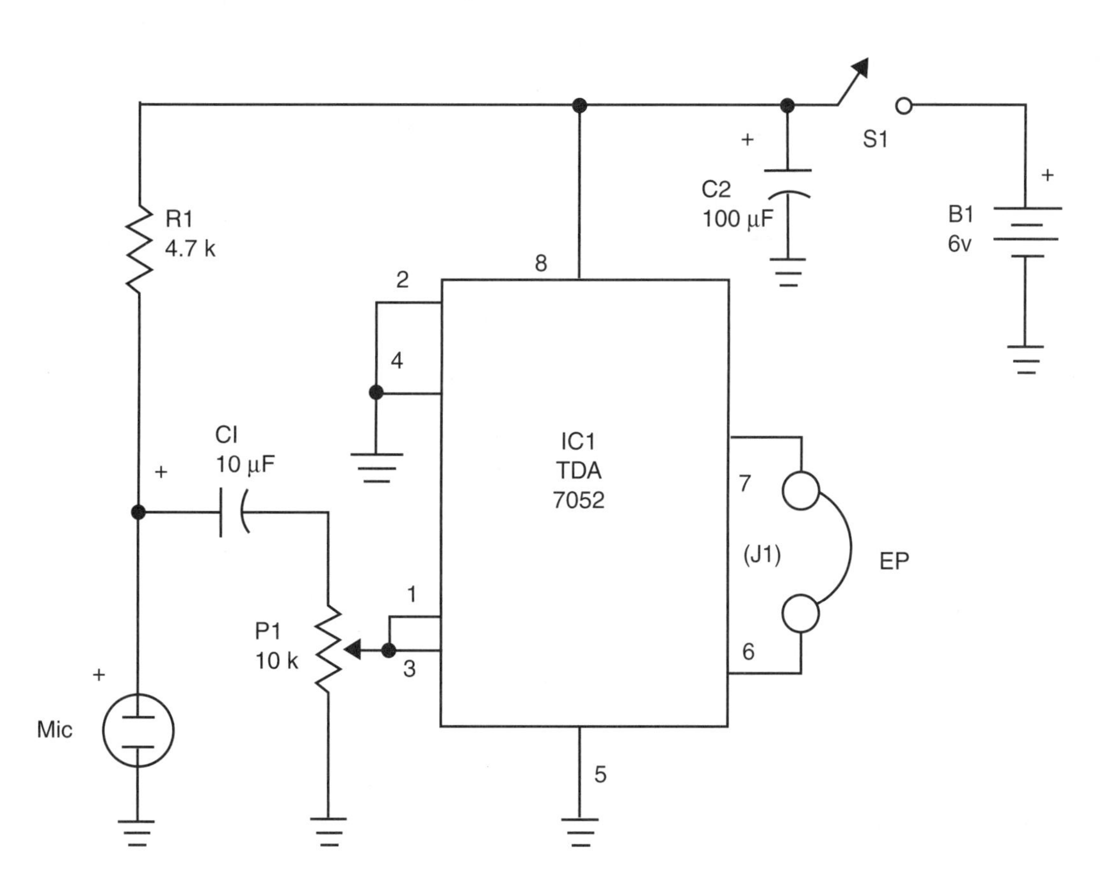


Figure 3.14.10 *Using the TDA7052 IC*

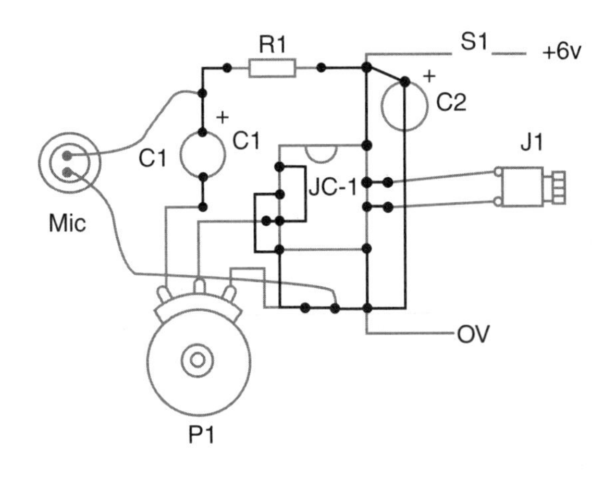


Figure 3.14.11 *PCB for the TDA7052 amplifier*

## Parts List

IC-1: TDA7052 integrated circuit audio amplifier

R1: 4.7 kΩ x 1/8 W resistor, yellow, violet, red

P1: 10 kΩ potentiometer log

C1: 10 μF x 12 V electrolytic capacitor

C2: 100 μF x 12 V electrolytic capacitor

S1: On/off switch

B1: A 6 V source or four AA cells and holder

J1: - Output jack for the earphones

MIC: Two terminal electret microphone

Other:

PCB, plastic box, cell holder, wires, knob for the potentiometer, solder, etc.

## Hearing Magnetic Fields

Although some living beings, such as pigeons, can sense magnetic fields, no discovery has ever been made of a creature that produces magnetic fields for some special use.

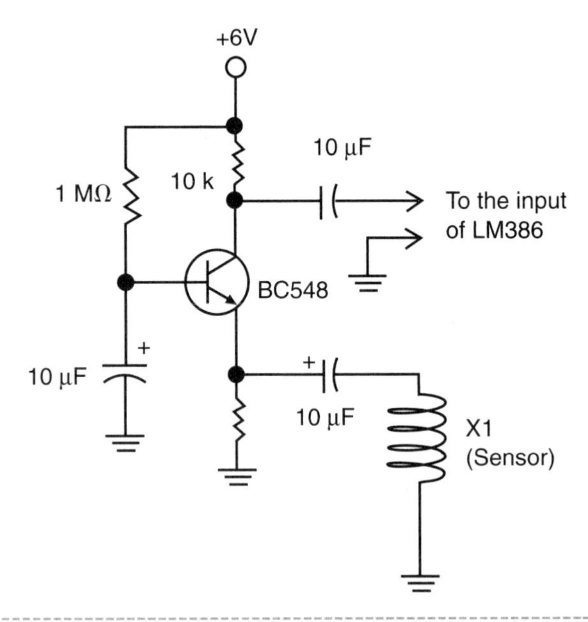


Figure 3.14.12 *Using a magnetic sensor*

If the evil genius wants some adventure trying to discover some "magnetic creature," a magnetic sensor can replace the microphone in our basic project. The assembly process for this is shown in Figure 3.14.12.

This circuit is also useful for detecting sources of magnetic noise such as power lines and electric appliances.

During usage, all that is required is to place the sensor in the suspected area and try to hear the sound of magnetic fields, which will be heard as a hum in the earphones.

The circuit is housed in a small, plastic box. X1, the sensor, consists of 500 to 10,000 turns of enameled wire 28 to 32 AWG in a plastic or cardboard form, as shown in Figure 3.14.13.

The primary winding of any transformer can be used as a sensor; simply remove the metal core. A ferrite rod inside the coil will increase its sensitivity. Just place the sensor near the AC power lines, transformers, or electric appliances, and hear the noise produced by their magnetic fields.

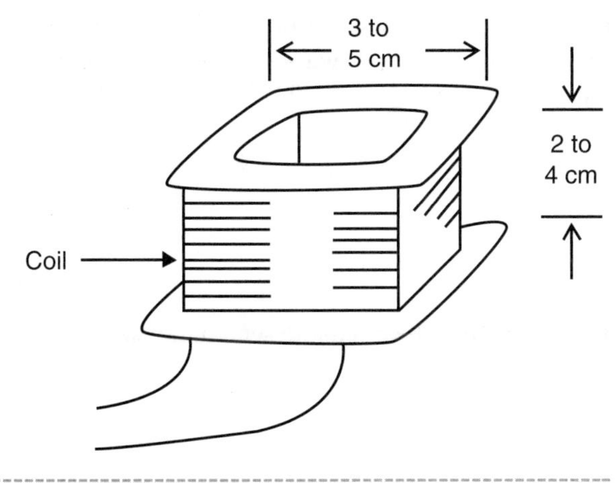


Figure 3.14.13 *The sensor*

# Project 15 — Insect Killer

Project 3.11 described a biologic trap where visible and *ultraviolet* (UV) light were used to attract insects. In that project, the insects were collected in a bag placed under the lamp. The idea suggested here is a trap that kills insects and other very small creatures by a high-voltage electric discharge.

Many pests such as cockroaches, slugs, and caterpillars can be killed by the discharge produced by the high-voltage electrodes. The evil genius can add this circuit to the biologic trap or use it alone, placing it where insects are present.

The circuit is powered from the AC power line, but it is safe because the high-voltage electrodes are isolated. However, even though it is not dangerous, it can cause severe shocks if touched, so the evil genius must take care with its installation, placing it where no one has access to the electrodes. The circuit can also be used as an electric fence, keeping animals confined, as suggested by Figure 3.15.1. The uncovered wire must be isolated in this application.

Finally, we have to say that the power consumption of the circuit is very low. The evil genius can plug the circuit into the supply line without the danger of a large energy bill at the end of the month. The circuit drains less than 5 watts from the power supply line since it operates in a pulsed manner.

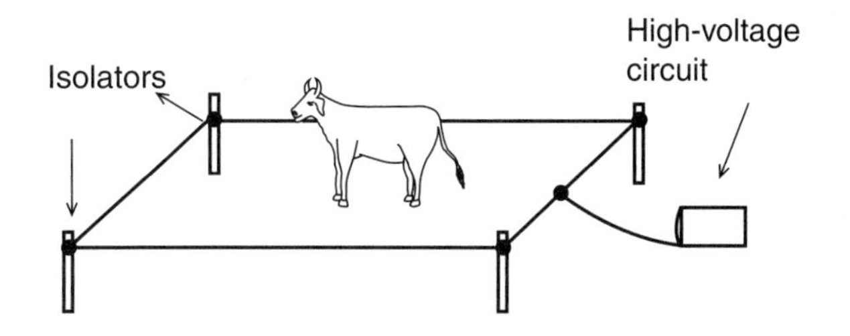


Figure 3.15.1 *The circuit can be used as an electric fence.*

## Bionics Applications

The great advantage of the electric killer is that it uses no chemical substances and therefore is an "ecological" solution for pest elimination. It can be used in many applications:

- Killing pests without the use of chemical substances.
- Capturing insects and killing them using the biologic trap.
- Making experiments with insect conditioning.
- Creating an electric fence to confine animals.

## How the Circuit Works

The circuit consists of a relaxation oscillator using a neon lamp to trigger a *silicon-controlled rectifier* (SCR).

The capacitor C1 charges through R1 and D1 until the voltage across the lamp rises enough to trigger it. At this moment, the neon lamp lights up and the capacitor C2 is discharged through the gate of the SCR. The result is that the SCR conducts the discharge current of C1 that flows across the low-voltage winding of the transformer.

The high-voltage pulse produced at the secondary is applied to the lamp that flashes for a moment. The pulse rate can be controlled by P1, and the power of the flash depends on the capacitor C1.

The transformer can be any type that is used in power supplies, with a primary rated to 117 *volts alternating current* (VAC), a secondary of 9 to 12 volts, and currents that range between 250 and 600 milliamps.

Although the primary of the transformer is rated to 12 + 12 volts (24 volts), the pulses applied to this component can reach 80 volts or more. This means that voltage induced in the primary is not 117 volts but higher with pulses rising to 400 volts or more.

In some cases, the transformer, which is not designed to support this voltage, can present some leakage, with sparks between the turns of the coil. If this happens, find another transformer for the application.

Depending on the components used, to compensate tolerances C2 can be changed, and values that range from 0.1 to 0.47 $\mu$F can be tested.

To operate from the 220/240 VAC power line, change R1 to 1 k$\Omega$, replace D1 with a 1N4007, and use the TIC106D for the SCR. It is not necessary to mount the SCR on a heatsink because it only works for short time intervals without generating large amounts of heat.

## How to Build

Let's start with the complete schematic diagram for the insect killer, shown in Figure 3.15.2.

Since the circuit is very simple and no *integrated circuit* (IC) is used, it can be mounted using a terminal strip as a chassis. Figure 3.15.3 shows the component's placement for this mounting.

This simple way to mount is ideal for beginners who don't have the resources to make a *printed circuit board* (PCB). When mounting, be careful with the positions of the polarized components, such as the diodes, capacitor, and SCR.

## The Trap

The trap is made using a piece of wood and some uncovered wires, as shown by Figure 3.15.4.

The distance between wires can vary from 0.4 to 1 centimeters depending on the size of the insects to be killed. This distance is determined by the fact that the insect must touch two wires at the same time to receive the electric discharge. The wires to the circuit must be isolated, and the maximum length is 3 meters.

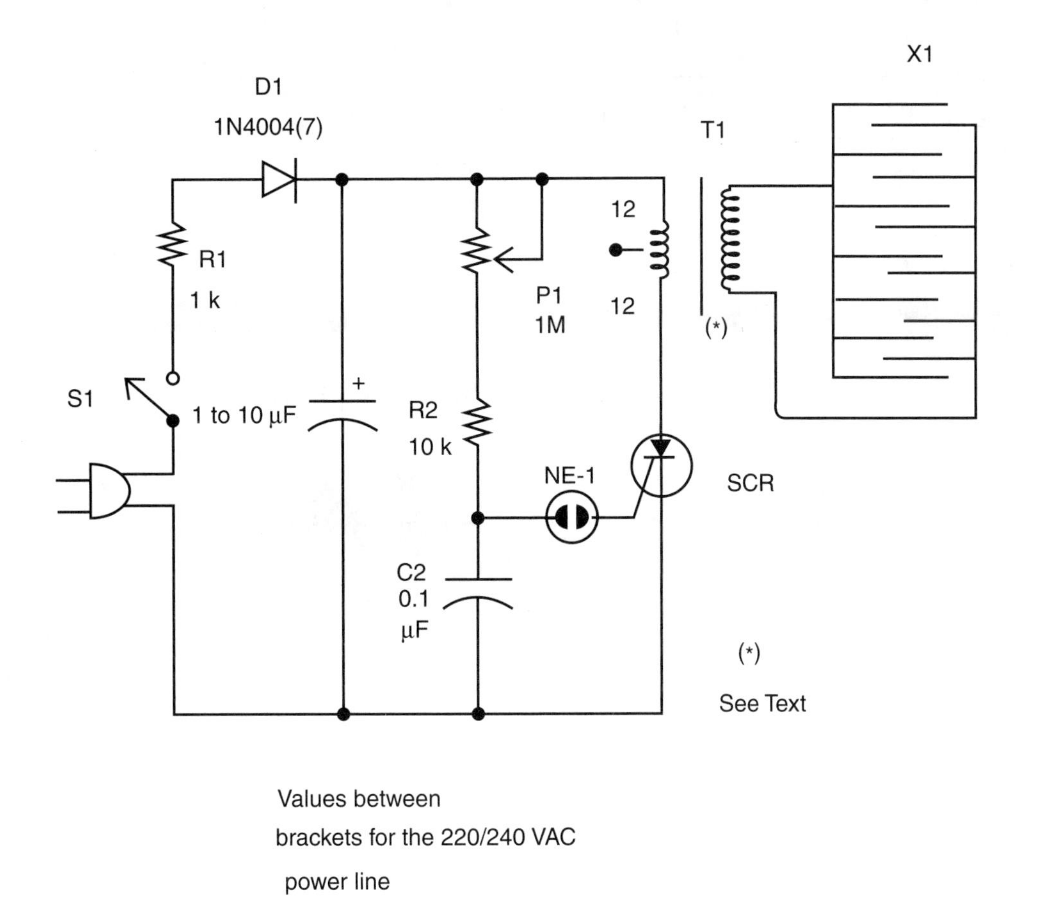


Figure 3.15.2 *Insect killer schematic diagram*

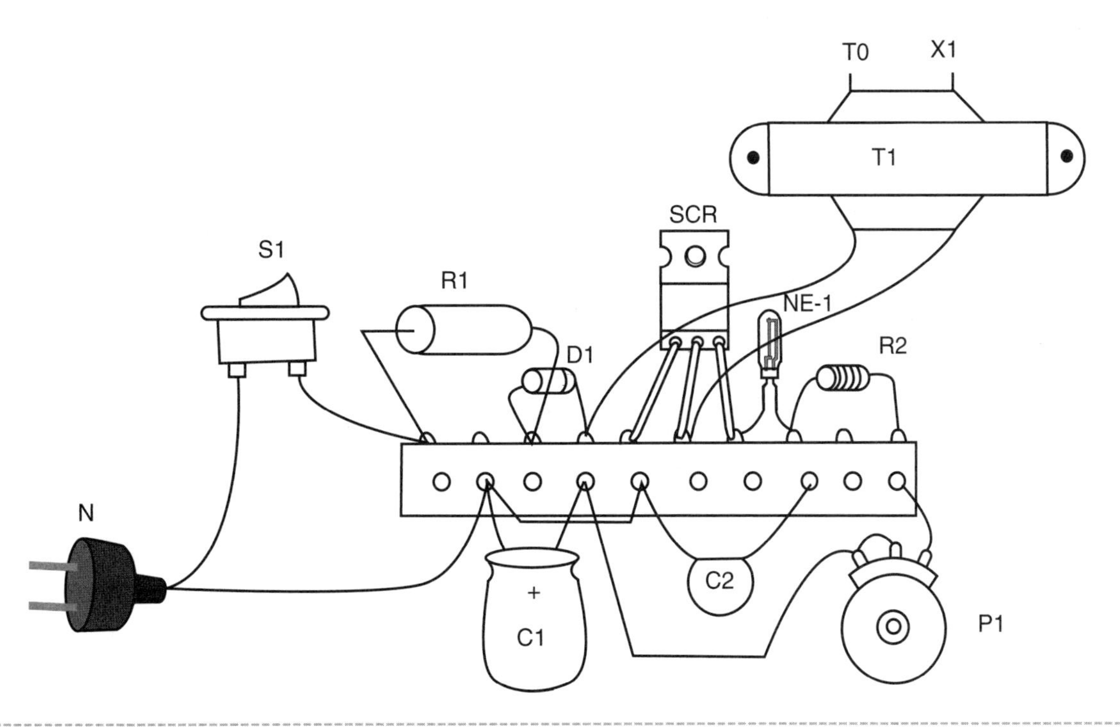


Figure 3.15.3 *Component placement on a terminal strip*

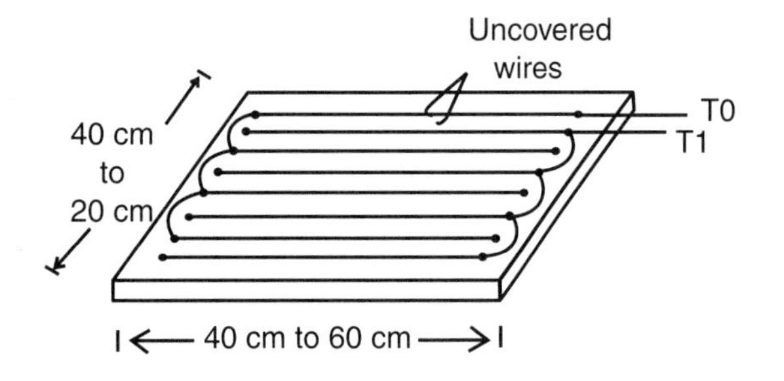


Figure 3.15.4 *The trap*

## Testing and Using

Testing is easy. Plug the circuit into the AC power line and connect the output of the circuit to a fluorescent lamp, as shown in Figure 3.15.5.

Adjusting P1, the lamp will flash, indicating the generation of high-voltage pulses. If you are a courageous evil genius, plug the trap into the circuit and verify the discharge, touching your fingers to the wires. Make adjustments until you are able to produce the brightest flashes in the fluorescent lamp.

Now you can use the insect killer. Use some substance (sugar for cockroaches, for instance) and place the killer in a place were you judge the insects to be.

## Important

The circuit doesn't have problems with shorts. Even if an insect dies making a bridge between wires, this won't cause any problems for the circuit.

## Parts List

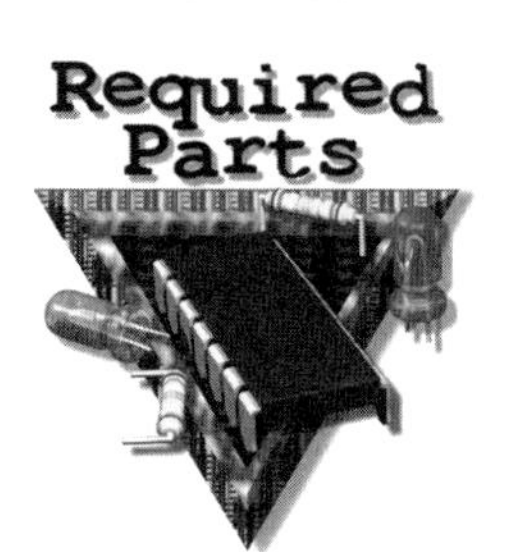


- SCR: TIC106B (117 VAC power line) or TIC106D (220/240)
- D1: 1N4004 (117 VAC) or 1N4007 (220/240 VAC) silicon rectifier diode
- NE-1: NE-2H or equivalent neon lamp
- R1: -470 Ω x 10 W (117 VAC) or 1 kΩ x 10 W (220/240 VAC) wire-wound resistor
- R2: 10 kΩ x 1/8 W resistor, brown, black, orange
- R3: 47 kΩ x 1/8 W resistor, yellow, violet, orange
- P1: 1 MΩ lin or log potentiometer

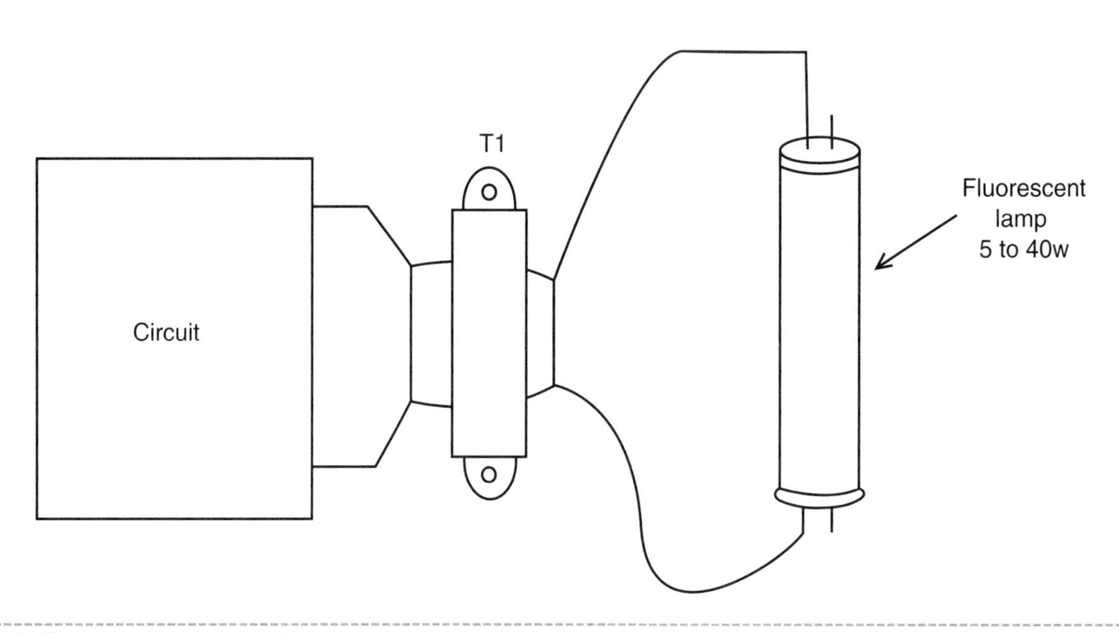


Figure 3.15.5 *Testing with a fluorescent lamp*

C1: 4.7 to 22 $\mu$F x 200 V (117 VAC) or 400 V (220/240 V) electrolytic capacitor

C2: 0.1 $\mu$F x 100 V or more polyester capacitor

T1: Transformer (see text)

X1: Electrodes (see text)

Other: PCB or terminal strip, power cord, plastic box, wires, etc.

## Additional Circuits and Ideas

High voltages can be generated by many different electronic configurations. Some of them are discussed in the next section.

## Very High-Voltage Killer

Very high voltages can be generated using a horizontal transformer or a fly-back transformer, such as the one shown in Figure 3.15.6 in the basic circuit.

This transformer will replace T1 in the basic circuit and can generate voltage pulses reaching up to 10,000 volts, certainly enough to kill any insect, regardless of its size! For this purpose, the evil genius must make the primary of the transformer, placing 20 to 30 turns of common wire at the core of the transformer, as shown in the figure.

Another change in the project is to place the wires with a certain amount of space between them. Distances between 0.5 and 1 centimeters will be necessary to prevent sparks from being produced between them. Figure 3.15.7 shows how to connect the flyback output to the electrodes.

Figure 3.15.6 *Using a flyback transformer*

High voltages can also be obtained from spark coils, such as the ones used in cars or motorcycles. In this case, however, the evil genius must take care when mounting and using the circuit, because the primary and secondary are not isolated. This means that the wires in the trap will not be isolated from the power supply line, and any accidental touch can cause a severe shock.

## Using a Laser

A very interesting idea to be developed is shown in Figure 3.15.8. The basic idea is a scanner that detects the passing of a fly both by an image sensor or another way.

When the fly is detected, a laser is triggered, striking the insect. Of course, simple LED lasers would not be powerful enough to knock down a fly, so *helium-neon* (HeNe) lasers or even more powerful ones must be used (with much care, of course).

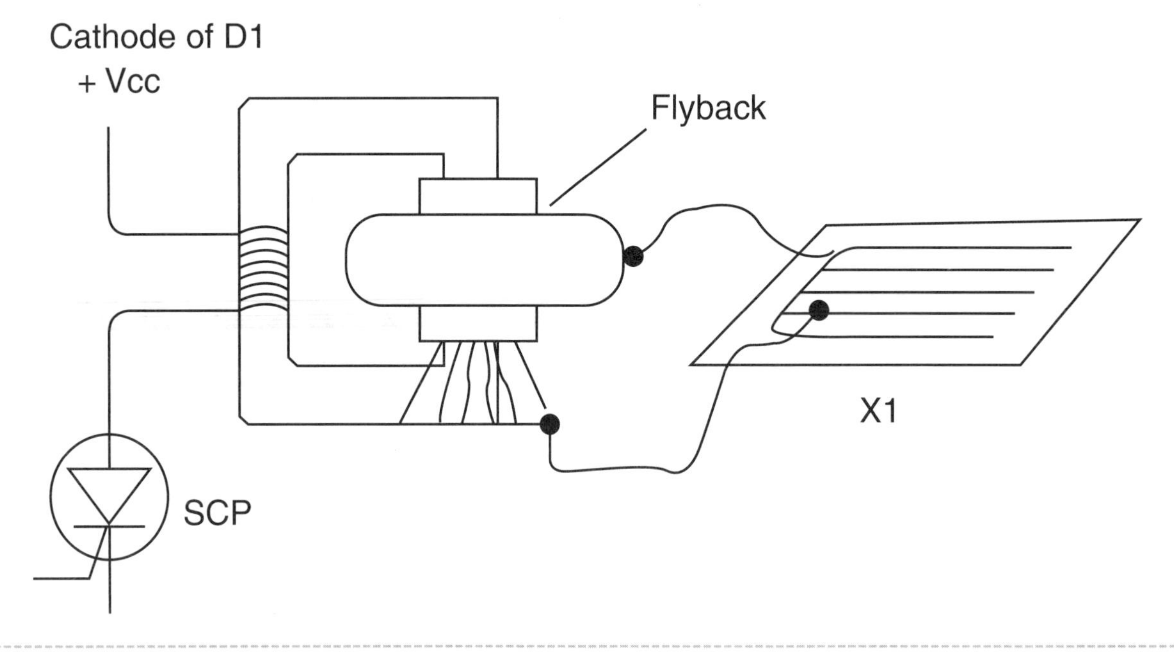


Figure 3.15.7 *Connecting the circuit to the trap*

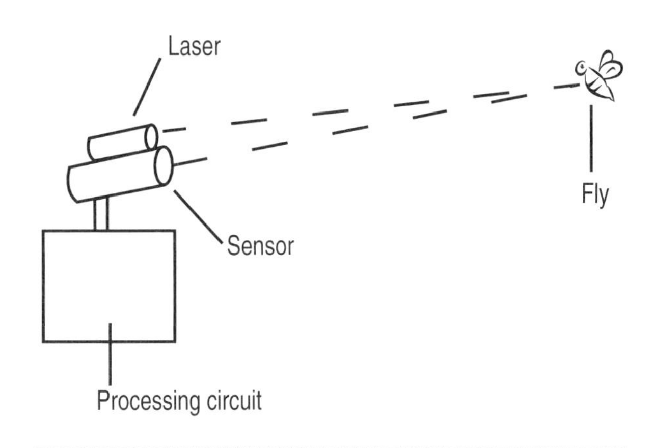


Figure 3.15.8 *Fly killer using a laser*

# Project 16—Bionic Tactile Organ

One of the objectives of bionics is the creation of artificial human organs based on the performance of natural organs. The five senses are included in this case, and some of them are explored in this book with simple projects that the evil genius, without a large amount of experience in electronics, can undertake.

Now it is time to explore the tactile sense with a simple project that would be ideal for a demonstration or that could be used as a starting point for a more complex sensory project with wider applications. The idea is that the sense of touch is transmitted from an artificial hand to transducers installed in a glove or even on a flat surface where a person's hand is placed, as shown in Figure 3.16.1.

The first question to be asked is, what kind of electric magnitude should we choose to transfer to a person's fingers as a tactile sensation: vibration, electric shock, mechanical pressure, or temperature?

The idea explored in this project is vibration, because it is an electric magnitude that's easy to work with using simple circuits. Thus, our project will transmit the sensation of vibration using small transducers.

The idea of a remote, sensing tactile organ is very interesting and can be used for a variety of different purposes. For instance, a bionic tactile organ can be installed in an artificial hand, as shown in Figure 3.16.2. The project is safe because it is powered by cells and is easy to mount in the basic version.

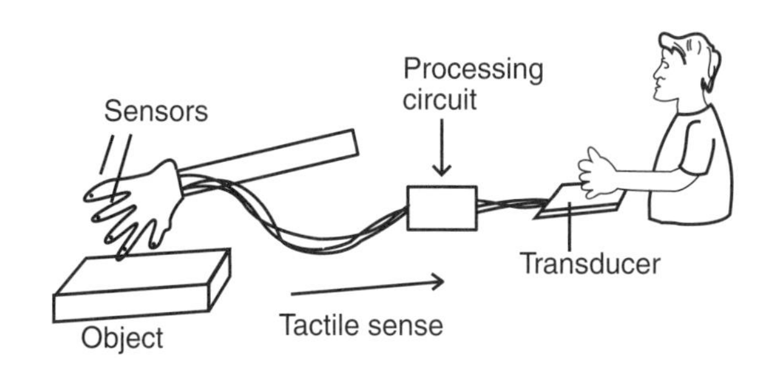


Figure 3.16.1 *The basic idea of an artificial tactile organ*

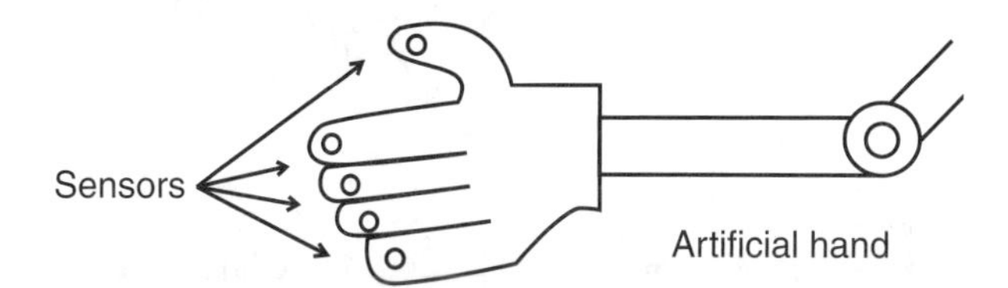


Figure 3.16.2 *An artificial hand with an artificial tactile organ—a bionics solution*

## The Applications in Bionics

An artificial tactile organ can be used with an artificial hand or arm and in many other applications. The evil genius can create a robot able to send tactile information to the operator. For instance, this robot will be able to handle objects with precision, sensing the pressure necessary to pick up and place an egg without breaking it.

- Create a bionic hand with tactile sense.
- Design a remote sensing device for general use.
- Demonstrate how senses can be converted into electric signals.

## How the Circuit Works

The circuit is formed by many oscillators (one for each finger) with a frequency determined by the amount of pressure on an electrode. The simplest electrode is made with conductive foam, which is also used to protect *integrated circuits* (ICs) from an *electrostatic discharge* (ESD). Figure 3.16.3 shows how this foam is used.

When the foam is pressed, its ohmic resistance falls, changing the frequency of the oscillator. The basic frequency is determined by C1, the resistance of X1 (the sensor), and R1. The circuit is calculated to

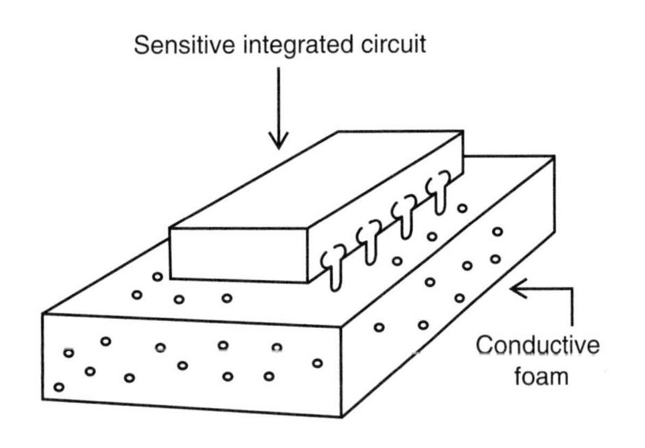


Figure 3.16.3 *Sponge used to protect ICs from ESD*

produce a central frequency in the audible range, something around 1 kHz.

In this project, we will use a 555 IC for each oscillator, but other ICs can be configured for the same task, as we will suggest in the "Additional Circuits and Ideas" section.

The output of this circuit is powerful enough to drive a transducer, which will be the interface between the sensors and the fingers of the operator. Small piezoelectric transducers can be used for this task, and they can be installed inside a glove or, if the reader prefers, on a surface where the fingers can be placed.

The circuit is powered from common AA cells. The current drain is very low and four cells can power all five circuits necessary for each finger of a hand. Details of the construction are given in the following section.

## How to Build

Figure 3.16.4 shows the schematic diagram of one oscillator. The evil genius will mount five circuits like this to transmit the tactile sense to each finger.

The circuit can be mounted on a solderless board or a *printed circuit board* (PCB), such as the one shown in Figure 3.16.5.

P1 adjusts the frequency according to the characteristics of the sensor. C1 can be tested using a large range of values between 220 pF and 1 nF. The ideal value is one that matches the resistance of the sensor, providing an audible tone.

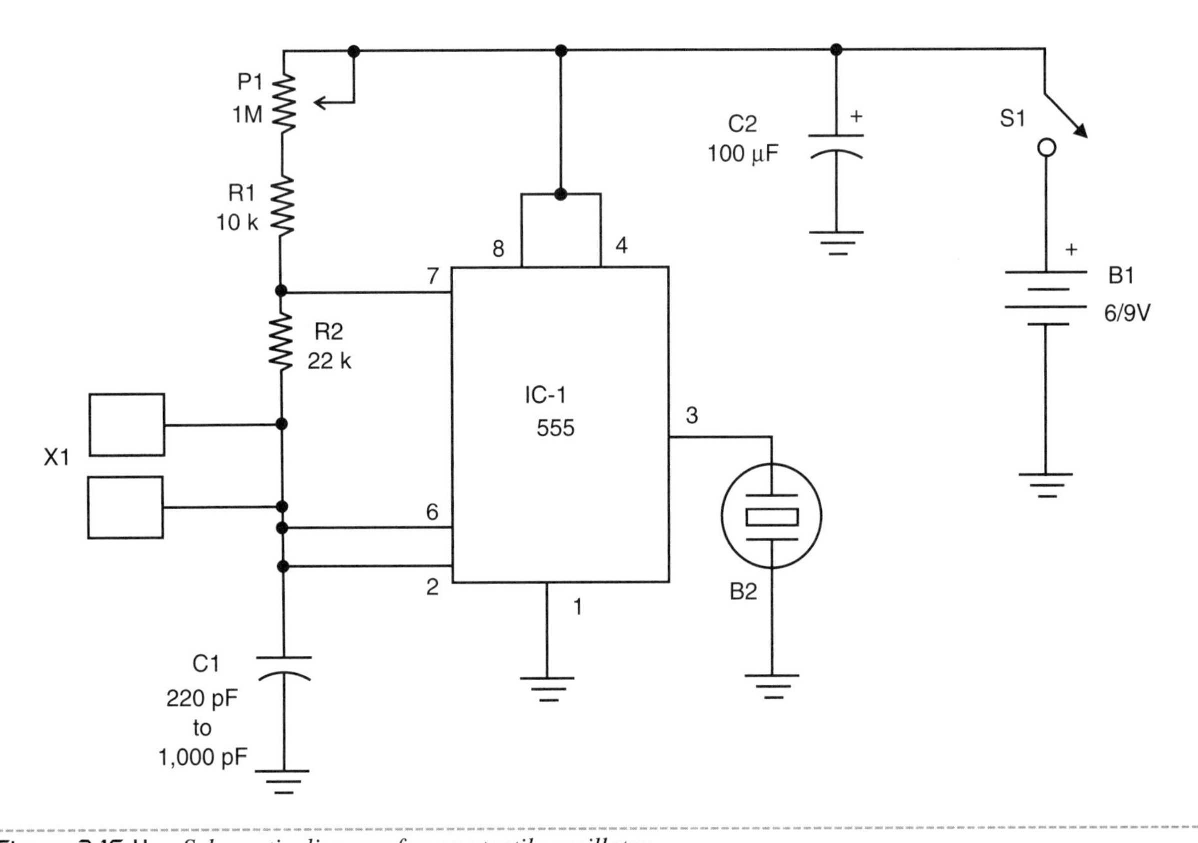


Figure 3.16.4 *Schematic diagram for one tactile oscillator*

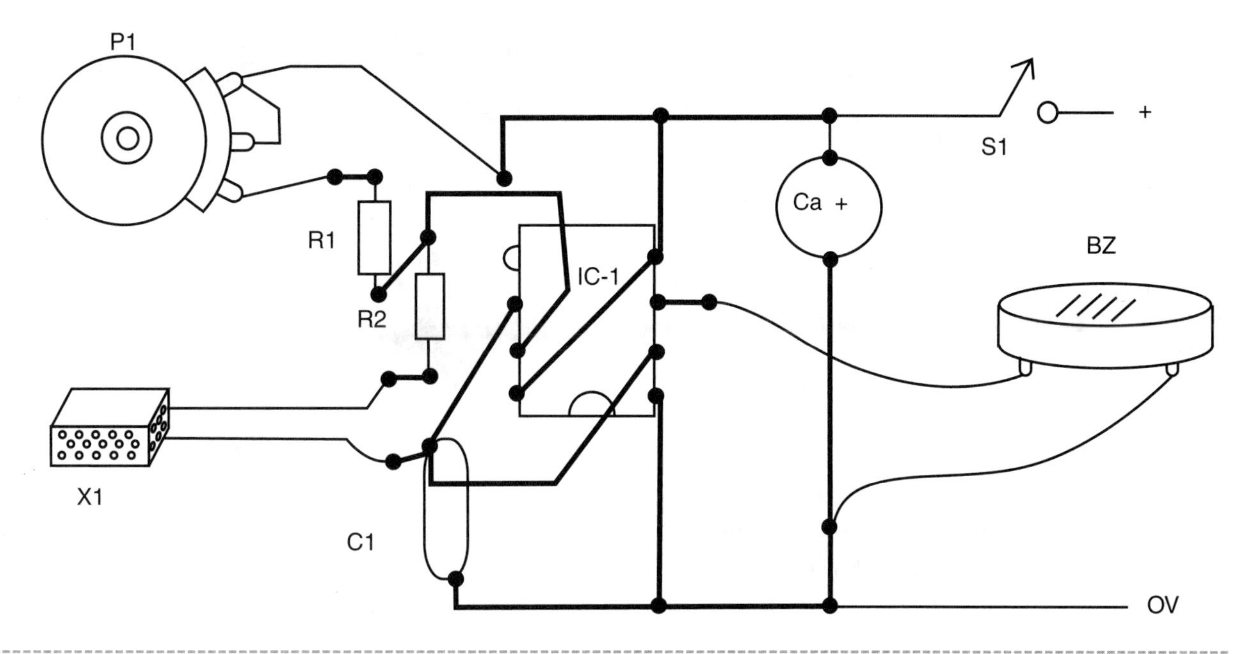


Figure 3.16.5 *PCB for the bionic tactile organ*

P2 adjusts the level of excitation or sensitivity in the sensor. Low levels are better because they fall in a range where our tactile sensors are most sensitive.

The circuit can be housed in a small plastic box. If many oscillators are used, cables should be used to connect them to the remote hand. All the oscillators can be powered from the same power supply.

Do not use power supplies from the AC line because they are dangerous in an application where hands or body parts have contact with devices in the circuit.

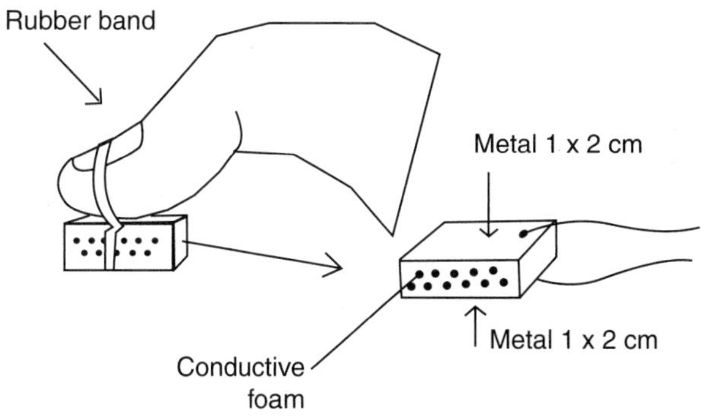

Figure 3.16.6 *Details of the sensors*

## The Sensors

Figure 3.16.6 shows how the sensors can be mounted on an experimental bionic finger or hand. Small pieces of conductive foam are cut and placed between the electrodes, which are formed by flexible metal plates or grids. Find a foam soft enough to be easily compressed under pressure.

The transducer can be mounted inside a glove, as shown in Figure 3.16.7.

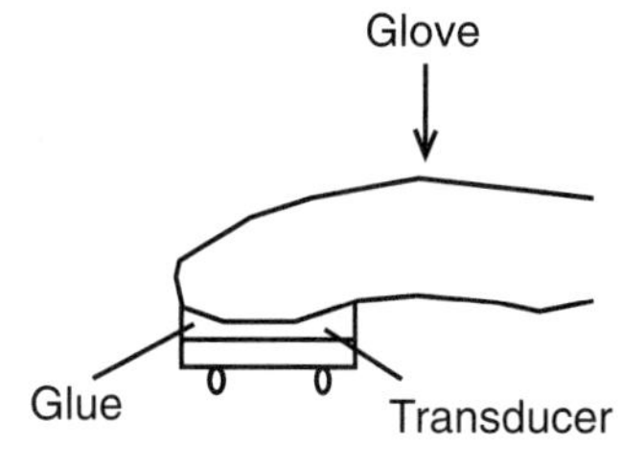


Figure 3.16.7 *Placing the transducer inside a glove*

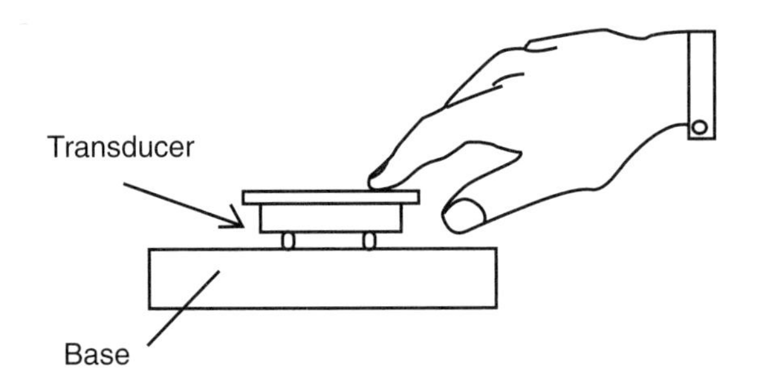


Figure 3.16.8 *Another solution for the transducer's mounting*

Another solution is to mount the transducer on a flat surface, such as a piece of wood, where the subject can place his or her hand, as shown in Figure 3.16.8. One must be able to touch the vibrating membrane or even the piezoelectric crystal.

## Testing and Using

Power the circuit on and place your finger on the transducer. Adjust P1 until you sense a light vibration. Then, pressing the sensor, the vibration will change, giving more tactile sensation or vibration. The amount of vibration is proportional to the pressure in the transducer.

When using the glove device, you will touch the objects with the sensors and try to determine shape and consistency from the vibrations in your fingers. Of course, how well you determine those characteristics will depend on the number of sensors used in the hand.

## Parts List Material for One Finger

IC-1: 555 integrated circuit timer

S1: On/off switch

B1: A 6 V source or four AA cells

BZ: Piezoelectric transducer

X1: Pressure sensor (see text)

P1: 1 MΩ potentiometer

R1: 10 kΩ x 1/8 W resistor brown, black, orange

R2: 22 kΩ x 1/8 W resistor red, red, orange

C1: 220 pF to 1,000 pF ceramic capacitor

C2: 100 μF x 12 V electrolytic capacitor

Other:

PCB, cell holder, plastic box, wires, solder, etc.

## Additional Circuits and Ideas

A bionic tactile device has many functions and is limited only by the imagination of the evil genius. You can work with different types of sensors and use

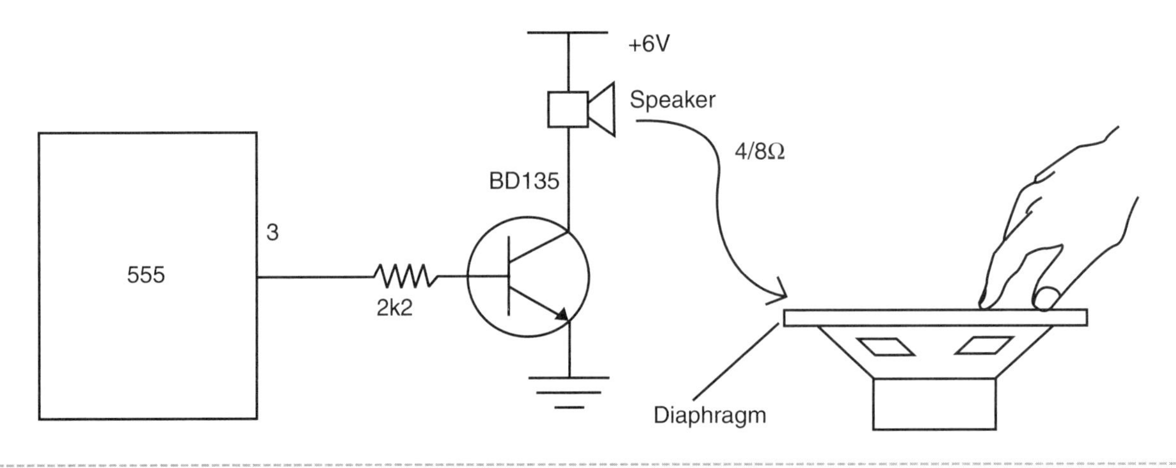


Figure 3.16.9 *Using a loudspeaker as a transducer*

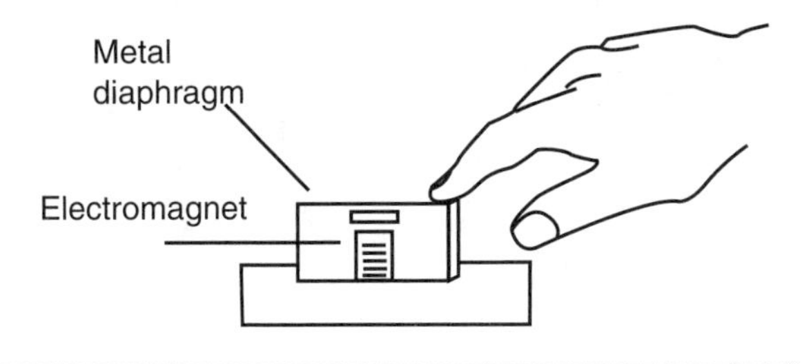


Figure 3.16.10 *Using electromagnets*

different techniques for transmitting the sensations to a person's fingers or a receiver. Some suggestions for upgrading or creating new solutions are given in the following section.

## Using Other Transducers

The first idea to explore is the use of a small loudspeaker as a transducer in the output stage. As shown in Figure 3.16.9, a person touches the loudspeaker, which will need drivers that use transistors. The loudspeaker produces a mechanical vibration that can be sensed by the fingers.

Small electromagnets with a metal diaphragm can also be used, as shown in Figure 3.16.10. The electromagnets are formed by 100 to 500 turns of 30 to 32 AWG wire around a small screw.

## Circuit Using the 4093

Figure 3.16.11 shows how a 4093 IC can be used as a pressure-controlled oscillator suitable for this application. The advantage of this configuration is that you can use only one IC to drive four transducers from the signals picked up by three sensors.

## Working with Electric Stimulation

Another interesting project using a tactile transducer is shown in Figure 3.16.12. The circuit applies electric

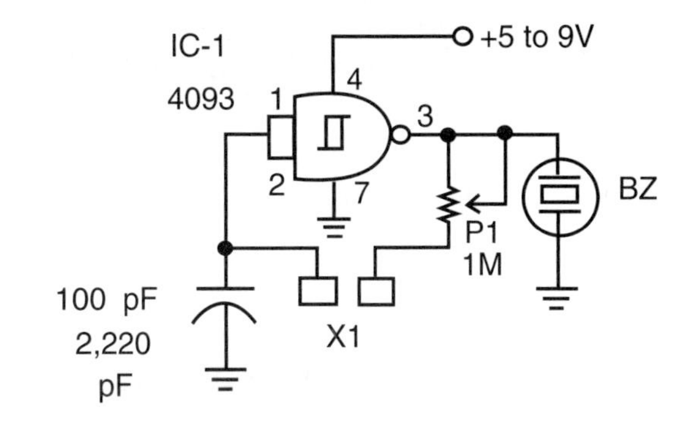


Figure 3.16.11 *Using the 4093 IC*

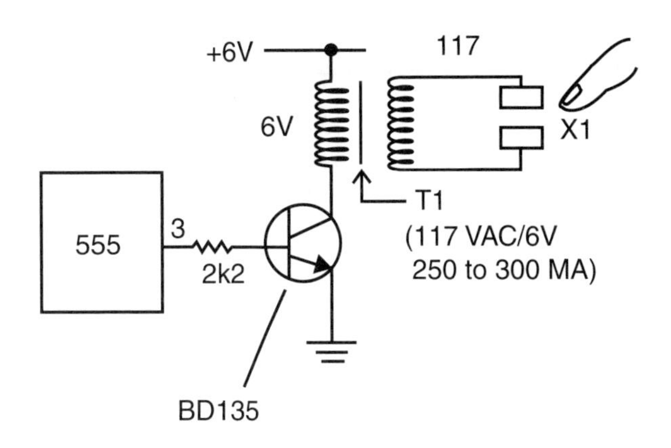


Figure 3.16.12 *Using electric stimulation*

shocks to one's fingers from the stimulus picked up by the sensors.

P1 is adjusted to put the circuit in the threshold of stimulation; that is, where the vibrations begin to be sensed. Then any change in pressure will lower the frequency, increasing the voltage applied to the sensor. The central frequency of the oscillator must be found to provide the best sensation of pressure in the receiver.

# Project 17—Lie Detector

Under stress, the human body undergoes organic transformations that include electrical changes or electrical characteristics. These changes can be detected by external devices such as a lie detector. The basic idea of a lie detector is that, under stress, the skin's electric resistance changes, revealing if he or she is lying.

Of course, detecting when someone is lying is not so simple. Interpreting the changes in skin resistance and preparing the subject for an interrogation require special skills from a professional who works with a lie detector.

Thus, the project described here is not infallible and serves as an object of curiosity to demonstrate how such a machine operates. The evil genius can use it with his or her friends more for fun than as a serious device that can detect a lie under any condition. The changes in skin resistance can also be used as a feedback circuit. Real lie detectors are not easy to use. The specialists are trained not only to observe the signal indicators but to observe how the person reacts and small details not seen by untrained persons. These specialists study for years to become successful operators of lie detectors.

Finally, the project can be used to detect changes in the external resistance of other living beings, such as plants or fish in an aquarium, or in areas where changes are related to biological activities. Coupling an electronic device with a living being is what bionics is all about, as this book will convey.

## Projects in Bionics

Mounting the lie detector, the reader can perform the following experiments and applications:

- Show how an experimental lie detector functions.
- Detect changes in a living being's body resistance under certain conditions, performing experiments with stress, circadian rhythms, and so on.
- Use the circuit for visual feedback.
- Detect changes in environments where living beings are present.
- Detect biological activity in some experiments. (Any biological activity produces electric signals. Those signals can be observed using this circuit as a biologic amplifier.)

## How the Circuit Works

The basic circuit consists of a Wheatstone Bridge where one of the arms measures the resistance of the subject under interrogation. The null or zero indicator, which appears when the bridge is balanced, is a common analog galvanometer. Figure 3.17.1 shows the basic circuit of a Wheatstone Bridge.

Simply place the electrodes in the hands or on any point of the subject's skin and then balance the bridge, using a potentiometer for this task. Any change in the resistance of the subject will be sensed by the bridge, which goes out of balance, and is indicated by the galvanometer.

The circuit is powered from four AA cells or a 9-volt battery. Because the current drain is very low, these supplies will have an extended life.

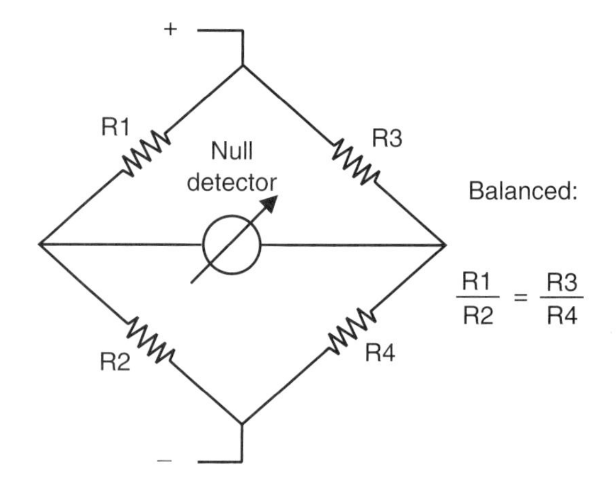


Figure 3.17.1 *The Wheatstone Bridge*

It is also important to make sure that the current flowing through the electrodes is extremely low and does not cause any sensation for the subject or pose any danger to cause a shock. The presence of the transistor is just to act as an amplifier for the very low currents that flow across the skin of the subject.

This basic circuit can be changed and many upgrades can be added to improve its performance. Making these changes will be covered in the "Additional Circuits and Ideas" section.

## Building the Lie Detector

Figure 3.17.2 shows the complete diagram of the lie detector in its basic version using an analog galvanometer as a null detector.

Since few components are used, the ideal for the evil genius who is learning about electronics is to use a terminal strip as a chassis, as shown in Figure 3.17.3. Of course, the more experienced evil genius can use a solderless board or a *printed circuit board* (PCB) to mount the circuit.

The galvanometer is a unit with zero at the center of the scale, as shown in Figure 3.17.4, with full-scale currents ranging between 50 $\mu$A and 1 mA.

But if the reader doesn't find this type of galvanometer at a store or even after looking through old, abandoned equipment, a common galvanometer with zero in the extreme of the scale can be used. The only difference will be the adjustments when operating the device.

The electrodes can be constructed in a couple of different ways. One of them is to use two small metallic plates upon which the subject can place his or her fingers. The other consists of larger metallic electrodes to be kept in the hands. Figure 3.17.5 shows these two options.

The circuit can be housed in a small plastic box, as shown by Figure 3.17.6. When mounting, be careful with the position of the polarized components, such as the transistor, galvanometer, and B1.

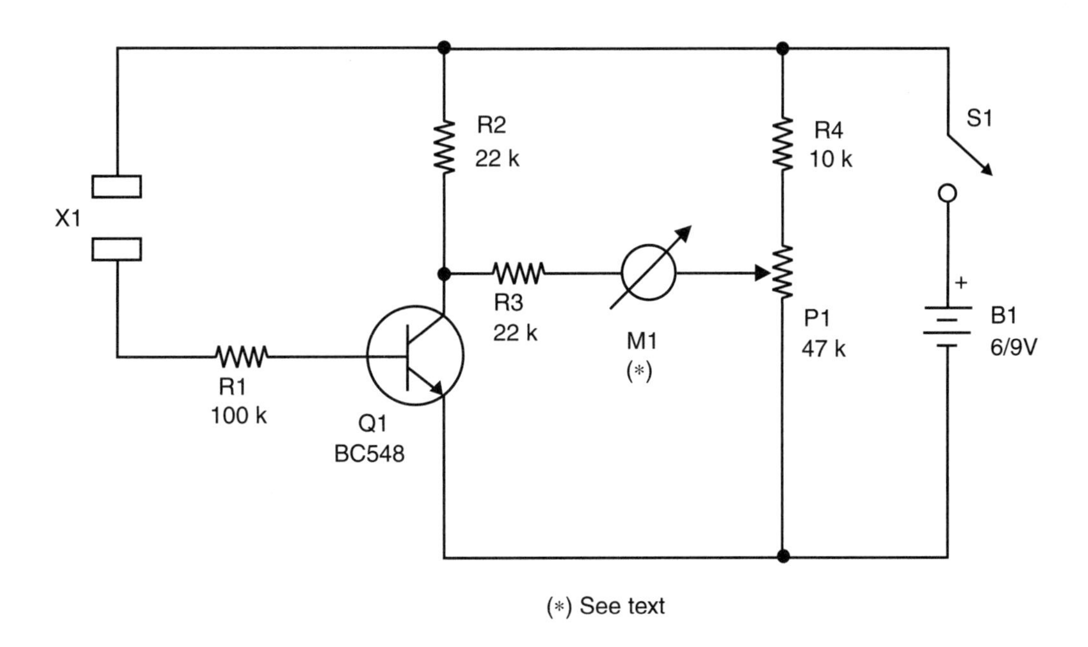


Figure 3.17.2 *The schematics for the lie detector*

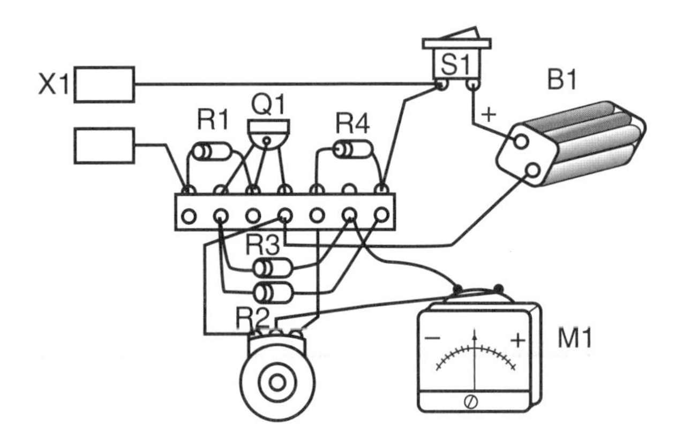


Figure 3.17.3 *Components mounted on a terminal*

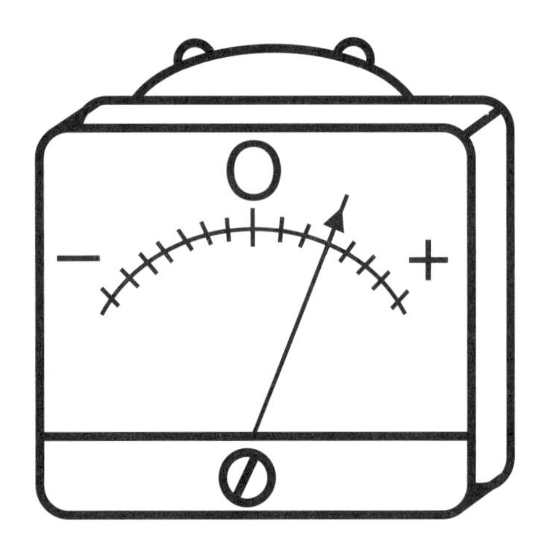


Figure 3.17.4 *The recommended galvanometer*

## Testing and Using

To test the device, turn on the circuit, close S1, and place your fingers on the electrodes. Keep constant pressure on the electrodes.

Then adjust P1 to have an indication of the middle of the scale in M1. This is valid even you are using a galvanometer with zero at one extreme. Observe that the galvanometer will indicate any change in the pressure of your fingers on the electrodes.

In an interrogation, the subject would need to keep constant pressure on the electrodes and not let the indicator move. If any movement occurs, it is because the person is lying. During a test, you would only have to ask the questions and keep an eye on the indicator.

When performing experiments with plants or other living beings, the procedures are the same: Start from the point where the bridge is balanced. Any change in the balance will be indicated by the galvanometer.

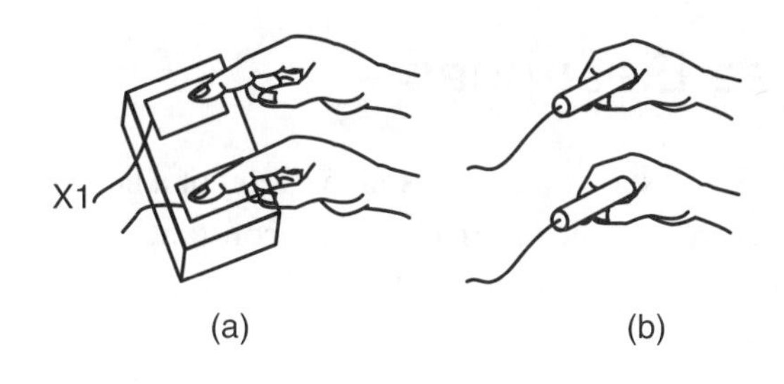


Figure 3.17.5 *The electrodes*

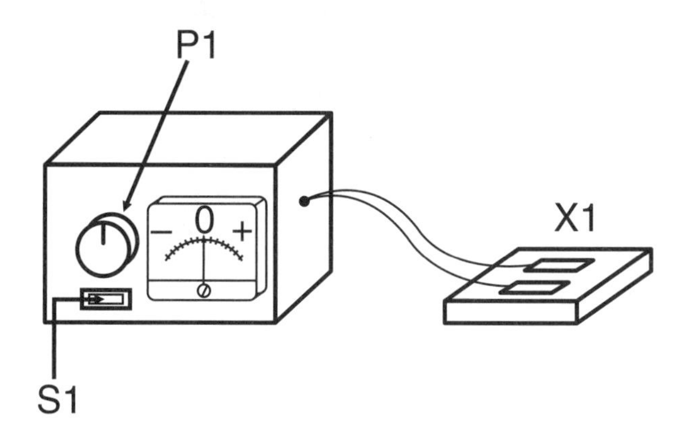


Figure 3.17.6 *Housing the project*

## Parts List

Q1: BC548 general-purpose NPN silicon transistor

M1: Galvanometer (see text)

X1: Electrodes (see text)

R1: 100 kΩ x 1/8 W resistor, brown, black, yellow

R2, R3: 22 kΩ x 1/8 W resistors, red, red, orange

R4: 10 kΩ x 1/8 W resistor, brown, black, orange

P1: -47 kΩ potentiometer, lin or log

S1: On/off switch

B1: 6 or 9 V AA cells or battery

Other: Terminal strip or PCB, plastic box, cell holder or battery connector, wires, solder, etc.

## Additional Circuits and Ideas

Many improvements can be used to increase the performance of this circuit or change it. Some of them are described in this section.

## Darlington Configuration

The transistor used in the basic project is a common bipolar type with gains that range from 125 to 800. The sensitivity of the circuit can be increased with the use of two transistors wired in a Darlington configuration, as shown in Figure 3.17.7.

When wired as a Darlington pair, the gain of the stage will be the product of the individual transistors' gains. For instance, if each transistor has a gain of 400, the total gain of the stage will be 160,000 ($400 \times 400$). This circuit can detect very small changes in skin resistance due to the high gain of the Darlington stage.

The parts list is the same, but the transistor is replaced by two BC548s.

## Three Electrodes

Figure 3.17.8 shows an interesting version of a lie detector using three electrodes. Since constant pressure must be exerted on three different points, it is more difficult to maintain the balance of the circuit and easier to detect any change during the interrogation.

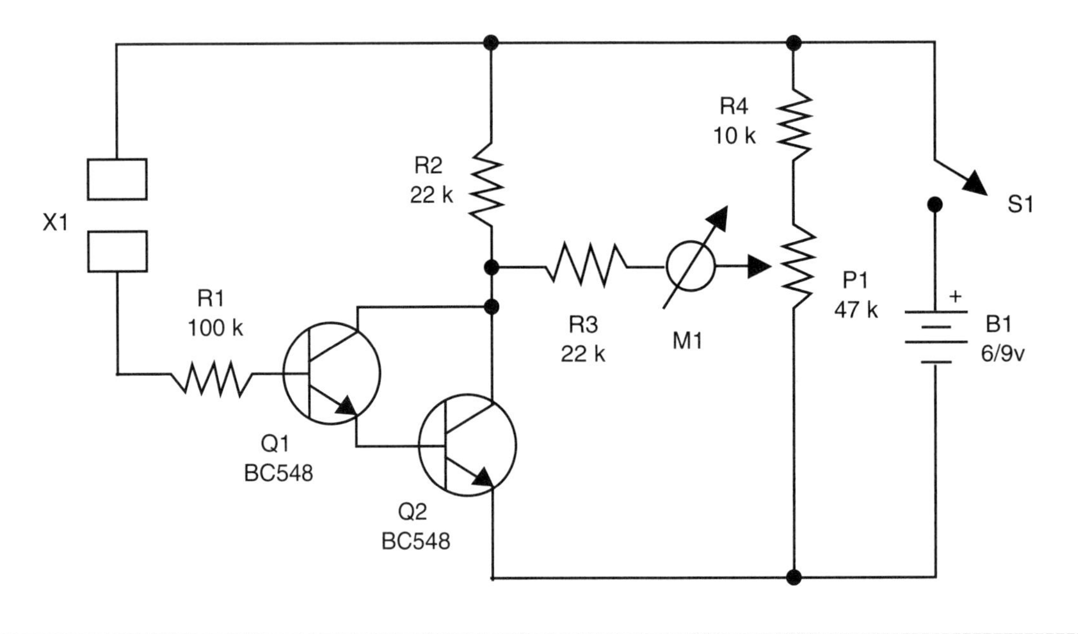


Figure 3.17.7 *Using a Darlington pair to increase the gain*

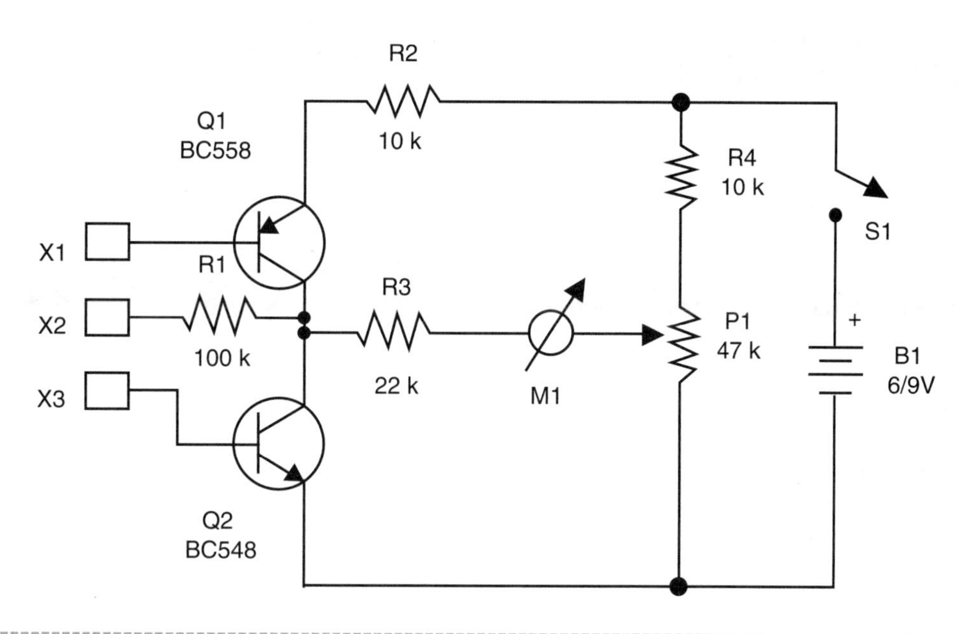


Figure 3.17.8 *A three-electrode lie detector*

## Parts List

Q1: BC558 general-purpose PNP silicon transistor

Q2: BC548 general-purpose NPN silicon transistor

M1: Galvanometer (see basic project)

X1, X2, X3: Electrodes (see basic project)

R1: 100 kΩ x 1/8 W resistor, brown, black, yellow

R2, R4: 10 kΩ x 1/8 W resistor, brown, black, orange

R3: 22 kΩ x 1/8 W resistor, red, red, orange

P1: 47 kΩ potentiometer, lin or log

M1: Galvanometer, as in the basic project

S1: On/off switch

B1: 6 or 9 V cells or battery

Other: Terminal strip or PCB, cell holder or battery connector, plastic box, wires, etc.

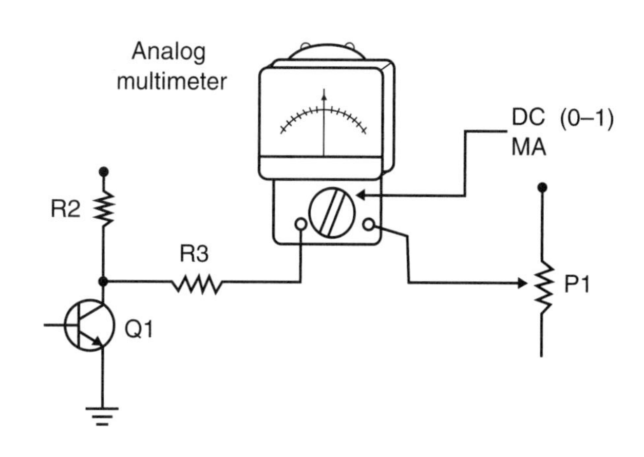


Figure 3.17.9 *Using a multimeter as a replacement for M1*

## Using a Multimeter

Digital multimeters are not suitable for this application because the numbers in the display change so fast that we can't see what they are saying. In this project, what is important is not a fixed number but the ability to detect changes. Thus, an analog multimeter is the best replacement for the galvanometer and is wired as shown in Figure 3.17.9.

The scale of the multimeter is the lowest for currents (normally some microamps) and the measurements for adjustments are the same as in the basic project.

Any multimeter with sensitivity starting at 1,000 ohms/volt can be used. As we will see in the next projects, the multimeter can be used in many experiments involving the interface of living beings and electronics.

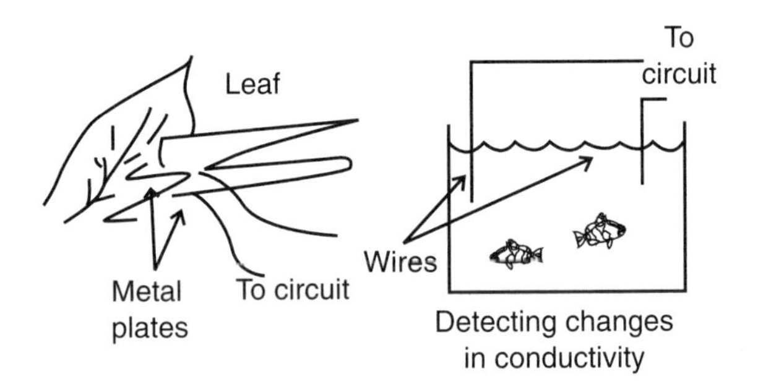


Figure 3.17.10 *Using the lie detector with other living systems*

## Other Experiments

The evil genius can also connect the lie detector to plants and many other creatures not to detect lies, but to detect changes in biologic activities. Figure 3.17.10 shows how to connect electrodes to a couple different specimens.

The important point to observe in the connections is that the circuit detects changes in the resistance the same way that the bio-amplifier does (as in Project 3.6).

# Project 18—Bionic Smell Generator

Smell is one of the chemical senses. The other is taste. They are so called because they sense chemicals. Smells are chemicals because the senses are stimulated by chemical substances present in the air.

Using these senses, we sample our environment for information. These senses are very important in nature because many living beings depend on them to find food, for sexual purposes, and even defense. Living beings are continuously testing the quality of the air they breathe and they use this sense to be informed of other information, such as the proximity of food or another individual.

Our sensory systems can detect only chemicals having certain properties. For instance, odor molecules must be small enough to be volatile (or dispersed in the air) so that they can vaporize, reaching the nose and then dissolving in the mucus. Since volatile molecules can travel through the air, smell, unlike taste, can signal long distances.

Our olfactory sense also serves a recognition function. All individuals have their own unique smell (some more pleasant than others!) and can recognize and be recognized by the smell. Some estimates suggest that we can distinguish around 10,000 different smells. Also, it has been proposed that smell can influence mood, memory, emotions, mate choice, the immune system, and the endocrine system (hormones).

Experiments have been made in bionics to help anosmics (people who have lost some or all of their sense of smell) recover this sense. This process entails connecting their olfactory nerves to bionic sensors and has led to good results.

Although smells are complex, it was discovered that they are formed by only three chemical substances that are mixed in some proportion. The combination, as in the case of basic colors, results in all the smells we can sense.

Based on this fact, a company called TriSenx (www.trisenx.com) launched a device that plugged into a computer and that could reproduce any smell based on those basic substances, as suggested by Figure 3.18.1.

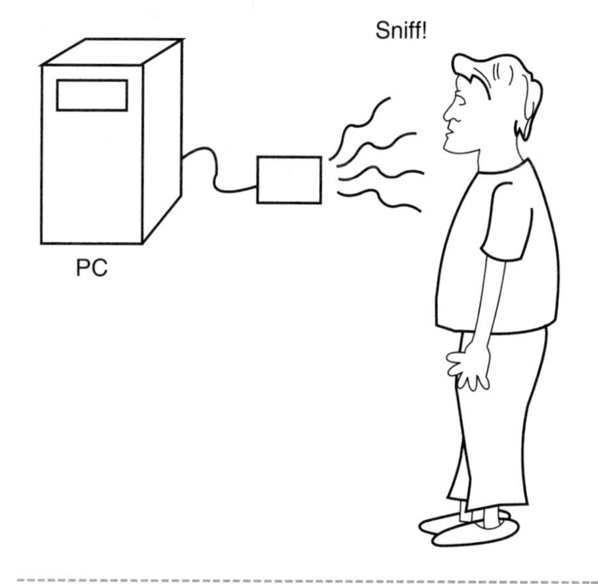


Figure 3.18.1 *Sending smells by the Internet*

What we are going to describe in this project is a simple electronic device that can produce smells based on chemicals. A circuit produces an artificial smell to interact with the smell sense of living beings, providing the bionic characteristic of the project.

Imitating nature, we can generate smells for our comfort, to scare animals and or other humans, or to add realism to a presentation. The smell chosen depends on the application needed. We believe that the evil genius will make a good choice, not as suggested by Figure 3.18.2.

The circuit is timed and powered from the AC power line. No dangerous substances are used, although you can use them (see Figure 3.18.2), and the application depends only on the imagination of the evil genius. The main applications are provided in the following section.

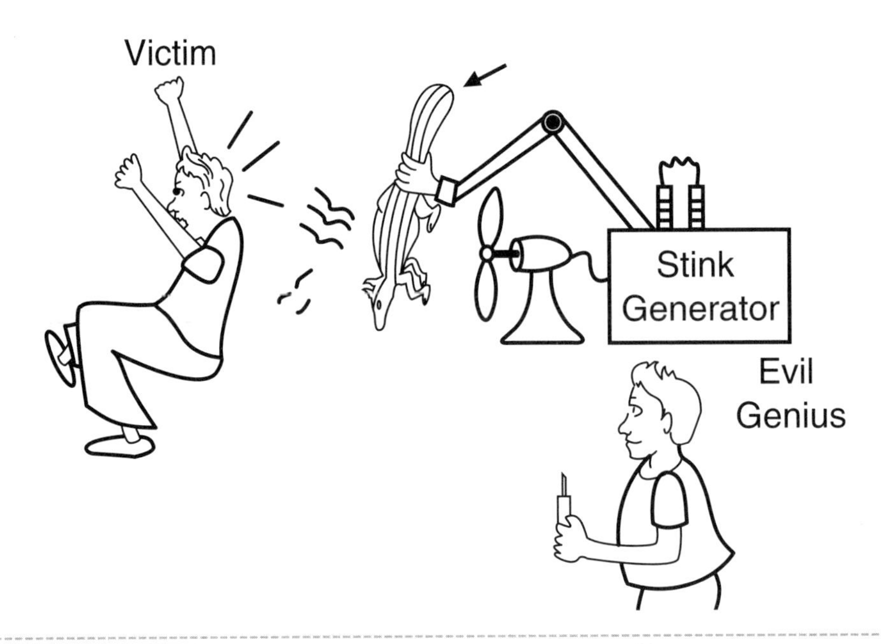


Figure 3.18.2 *Take care with the chosen smell for your project. It can affect you too.*

## Bionic Applications for the Project

Animals use smells to scare, attract, or even as a very efficient defense weapon. A device that can generate any smell can be used in many applications, such as the following:

- Produce pleasant smells in places where naturally bad smells are present (kitchens, restrooms, places where people smoke, etc.).
- Show how smells can be produced by a bionic device.
- Create fantastic smells by the combination of basic smells.
- Function as part of other projects that create effects in ambient (light, sounds, etc.).

## How the Circuit Works

The smells are produced by a small bottle of essence where a wick (like one from a candle) is heated by a resistor. When the resistor is powered, the heat vaporizes the essence, and smell is spread into the air, thanks to the presence of a small fan, as shown in Figure 3.18.3.

To prevent the resistor from being constantly heated, a timer circuit powers it at regular intervals. This circuit is formed by a 555 *integrated circuit* (IC) in the astable configuration with a frequency-adjusting potentiometer. Using this potentiometer, the evil genius can adjust the intervals between the smell productions according to the application.

The circuit is powered from the AC power line, and for small and medium areas power consumption is very low. The only concern the evil genius has is filling the bottle of essence when it becomes empty.

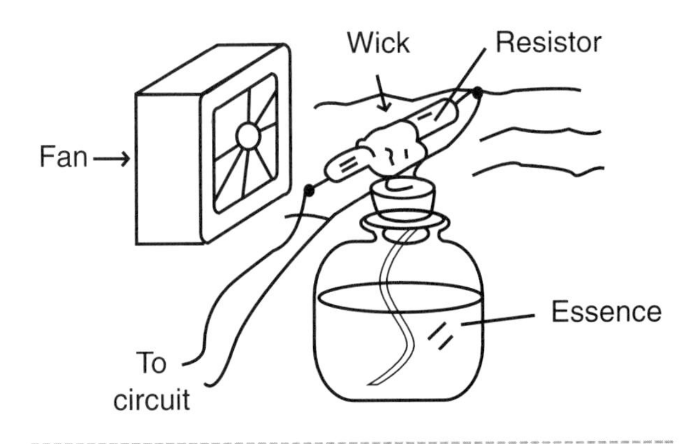


Figure 3.18.3 *Spreading the essence into the air*

## How to Build

The electronic part of the project is very easy to mount because no critical parts are used. Figure 3.18.4 shows the complete diagram of the bionic smell generator.

The circuit mounts on a *printed circuit board* (PCB), as shown by Figure 3.18.5.

R1 is the critical component of the project. It must produce heat that is hot enough to vaporize the essence in the wick.

For the 117 VAC power line, which must use a wire-wound resistor between 5 and 10 watts, values that range from 470 ohms to 2.2 kilo-ohms will be a good starting point for an experiment. Depending on the essence used, the vaporization can present different degrees of difficulty.

The suggested fan is a small type found in a computer's power supply, powered by a 12-volt source. This voltage is taken from the circuit that powers the timer.

The transformer has a 12-volt secondary with currents that range from 500 to 800 milliamps, enough to drive the relay and the fan.

The transducer resistor is powered directly from the AC power line, so the relay must have double contacts because the heater and fan are powered from different sources.

P1 adjusts the interval between the resistor heatings, and P2 adjusts the time in which the heater is on.

The values indicated in the parts list allow time intervals from some seconds to minutes. When mounting, be careful with the isolation because the heater (R1) is not isolated from the AC power line. We suggest that you install the device into a box, as shown in Figure 3.18.6.

The device must be placed in an area where the smells can easily be dispersed. Elevated places are ideal.

## The Smell Transducer

The transducer is formed by a small bottle where the wick is installed as shown in Figure 3.18.7. The wick involves the resistor so that the heat can vaporize the essence. Cotton wicks are ideal because they can easily absorb the essence.

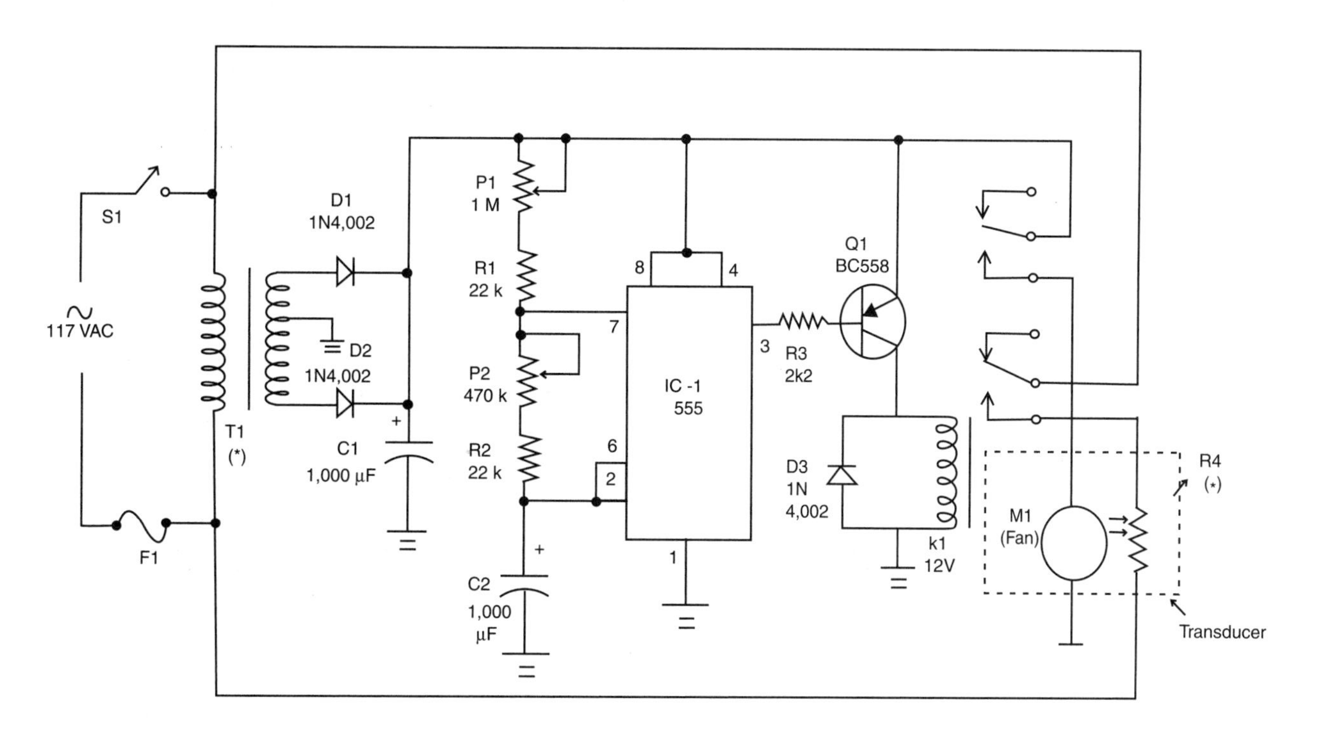


Figure 3.18.4 *Schematic diagram for the bionic smell generator*

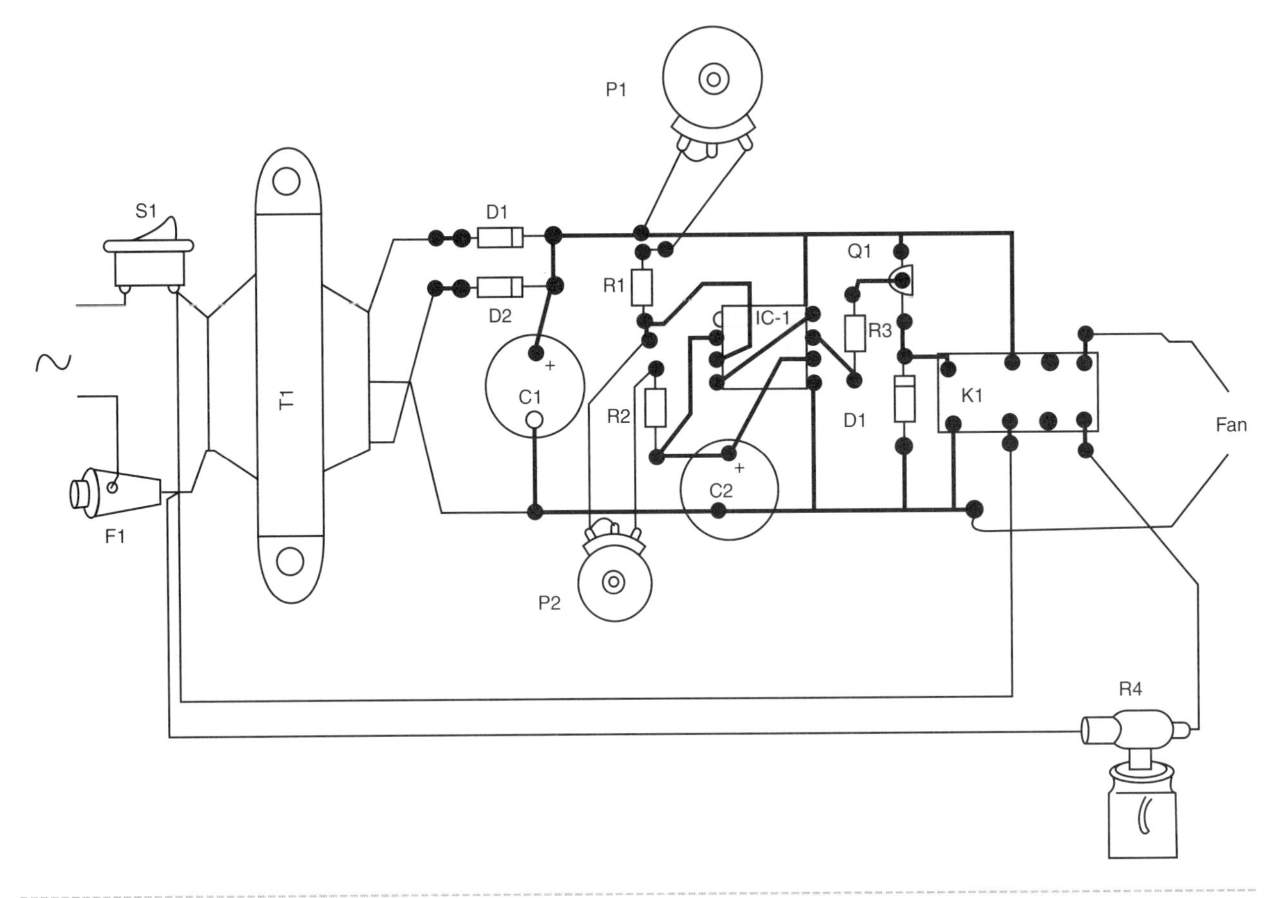


Figure 3.18.5 *The circuit is mounted on a PCB.*

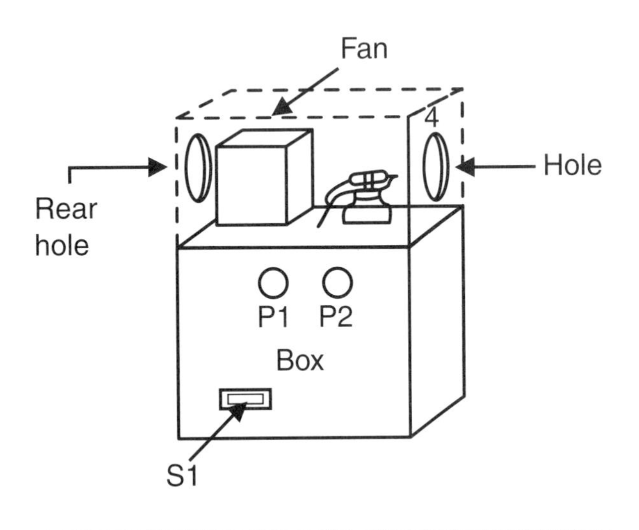


Figure 3.18.6 *The final mounting for the smell generator*

## The Essence Choice

Choosing the smell you want to create is also an important part of the experiment. Floral essences can be found in supermarkets, along with essences of smells of food. Depending on the application, you can use bottles that contain different smells.

Of course, the evil genius will probably be thinking about producing something that stinks. You can have some bottles of these essences (argh!), but take care, because if the experiment escapes your control, we are not responsible for the consequences.

## Testing and Using

Fill the transducer with the essence and turn on the circuit. Then adjust P1 and P2. The fan will turn on and off in regular intervals. Adjust the desired intervals to fill the area with the desired smell.

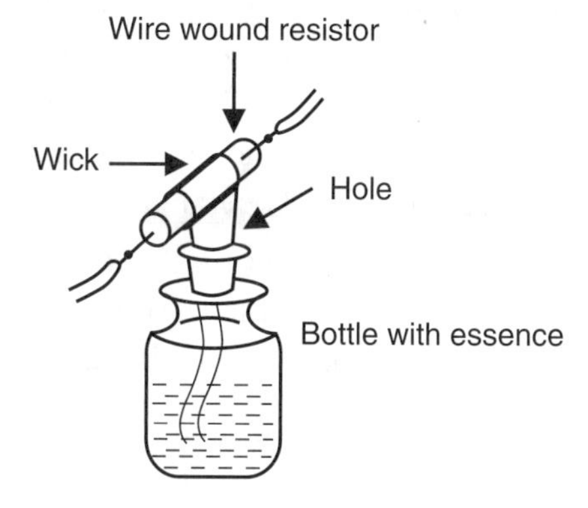


Figure 3.18.7 *The smell transducer*

If the resistor doesn't heat up enough to vaporize the essence, reduce its value. (Remember that reducing the resistance increases the current and therefore the temperature.) Conduct experiments according to the essence, and also be careful not to produce excess heat, which will burn the resistor. Temperatures up to 100°C are normal for wire-wound resistors, but they can consume too much essence.

After the tests, the evil genius can install the smell generator.

## Parts List

- IC-1: 555 integrated circuit timer
- Q1: BC558 general-purpose PNP transistor
- D1, D2, D3: 1N4002 silicon rectifier diodes
- R1, R2: 22 kΩ x 1/8 W resistor, red, red, orange
- R3: 2.2 kΩ x 1/8 W resistor, red, red, red
- R4: 470 Ω x 2.2 kΩ x 5 W wire-wound resistor (see text)
- P1: 1 MΩ potentiometer, lin or log
- P2: 470 kΩ potentiometer, lin or log
- C1, C2: 1,000 μF x 25 V electrolytic capacitors
- K1: 12 V x 50 mA DPDT relay
- T1: 12 + 12 V x 500 mA to 800 mA transformer
- S1: On/off switch
- M1: 12 V fan (see text)
- F1: 1 A fuse
- Other: PCB, box, fuse holder, material for the transducer (bottle, essence, wick, etc.), power cord, wires, solder, etc.

## Additional Circuits and Ideas

The device described in this project is only a basic idea. Starting from here, the evil genius can easily upgrade the project, adding many resources, and some additional ideas are provided in this section.

### Selecting Smells

A simple circuit that allows the manual selection of smells is shown in Figure 3.18.8. You must mount the resistor and fans with many bottles of essences, which are selected by switches.

Another idea is to create an automatic selector controlled by a computer, as shown by Figure 3.18.9. This is a version that would be ideal for theaters, where realism can be added to the experience of watching a movie.

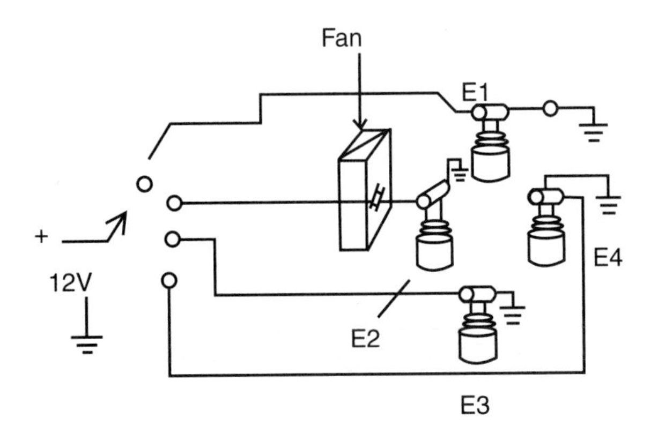

Figure 3.18.8 *Selecting smells*

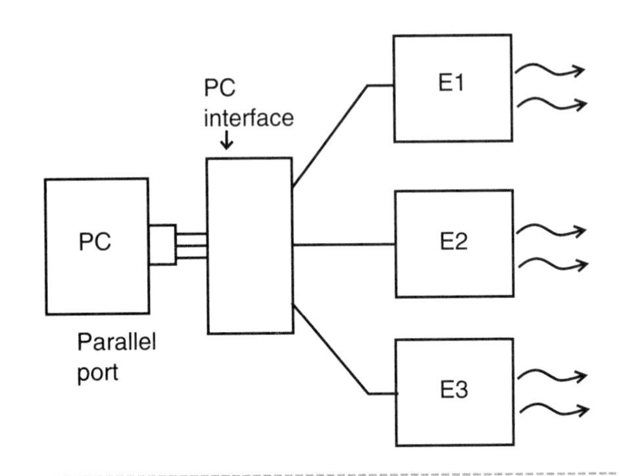

Figure 3.18.9 *Computer-controlled smell generator*

## Other Circuits

The circuit shown in Figure 3.18.10 is a simple dimmer that allows the evil genius to control the amount of smell produced by a sensor. This circuit can stand alone as a complete bionic generator or be added to the basic project for a timed operation.

The resistor in the transducer is the same as in the basic project, and you also must be careful with its installation because its parts are not isolated from the AC power line.

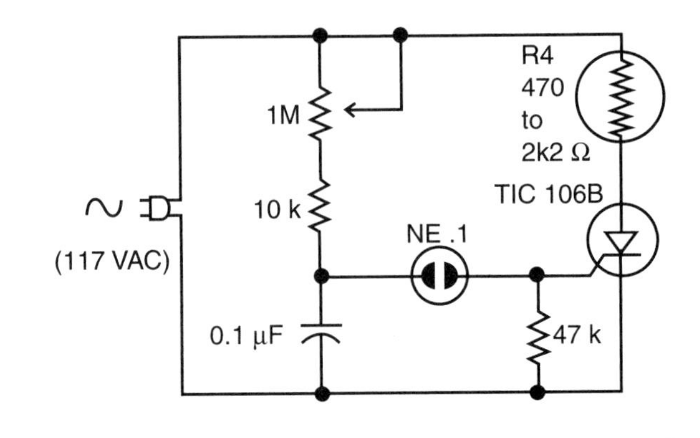

Figure 3.18.10 *A dimmer for the smell control*

# Project 19—Experimenting with Oscillators



Low- and high-frequency oscillators can be used in many experiments involving bionics. Using an oscillator, the evil genius can test how living beings react under certain stressful vibrations, stimulate plants to grow more quickly, or determine how flowers and fruit react to certain tests. Since the influence of many types of vibrations on living beings is not well known, this project provides a good opportunity to discover something new.

In previous projects, we worked with oscillators that generated sounds, ultrasounds, and electric and magnetic fields. Here we are going to give a circuit more capabilities that can be used in experiments and applications interacting with living beings.

## Experiments and Applications in Bionics

The curious evil genius can use the oscillator described here to see how many living beings react to medium- and high-frequency fields. He or she can also use it on insects or plants to see how they change their habits or growth.

## The Circuit

After having described other circuits in this book, this additional configuration is dedicated to the generation of high frequencies that range from 100 kHz to 20 MHz. The circuit is a Hartley oscillator, generating up to 5 watts of high-frequency signals that range from 100 kHz to 20 MHz, depending on the coil (L1).

In a Hartley oscillator, L1 and CV determine the frequency of the signals generated by the circuit. C2 and R1 form the feedback network that keeps the oscillator running.

The signal generated by the circuit is picked up from a second coil in the same form of L1. This coil acts as the secondary of a transformer where L1 is the primary.

The power supply comes from a battery or a supply with output voltages ranging from 6 to 12 volts. The maximum drained current is around 800 milliamps and depends on the tolerance of the components and the transistor gain.

## How to Build

Figure 3.19.1 shows the diagram for the medium/high-frequency oscillator. The circuit can be mounted using a small *printed circuit board* (PCB), but the evil genius without the resources to make a PCB can use other techniques, such as a solderless board or even a terminal strip.

Figure 3.19.2 shows a PCB for this project.

The coils (L1 and L2) depend on the frequency. Their characteristics are given in Table 3.19.1.

**Table 3.19.1**

Coil characteristics

| *Frequency* | *L1* | *L2* |
|---|---|---|
| 100 to 500 kHz | 75 + 75 turns | 50 turns |
| 500 to 1,500 kHz | 50 + 50 turns | 30 turns |
| 1.5 to 4 MHz | 25 + 25 turns | 15 turns |
| 4 to 10 MHz | 15 + 15 turns | 10 turns |
| 10 to 20 MHz | 10 + 10 turns | 8 turns |

The coils are made with 26 to 30 AWG wire on a ferrite rod 15 to 20 centimeters long and with a diameter of 0.8 to 1 centimeters. Also, L2 is placed on L1.

CV is a common variable capacitor, such as ones found in old AM radios. The maximum capacitance must range between 200 and 400 pF. The transistor is mounted on a heatsink and a power supply suitable for this oscillator is given in Figure 3.19.3.

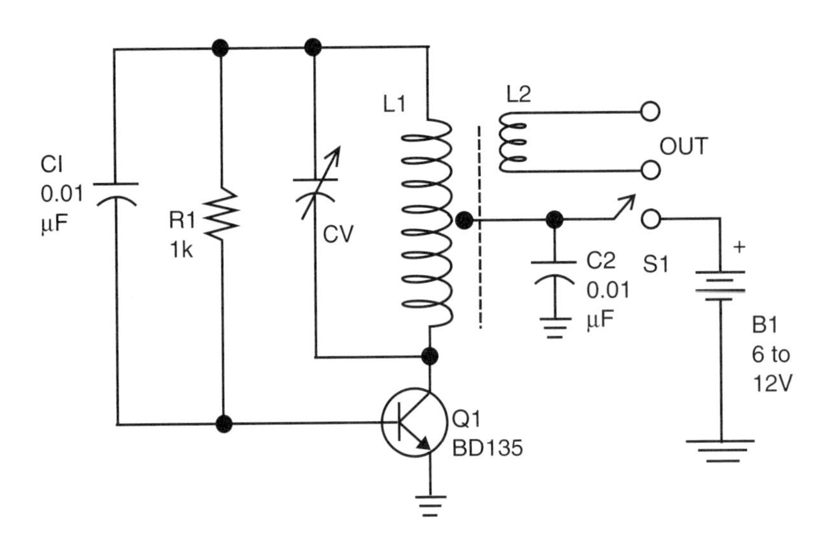


Figure 3.19.1 *The schematic diagram for the oscillator*

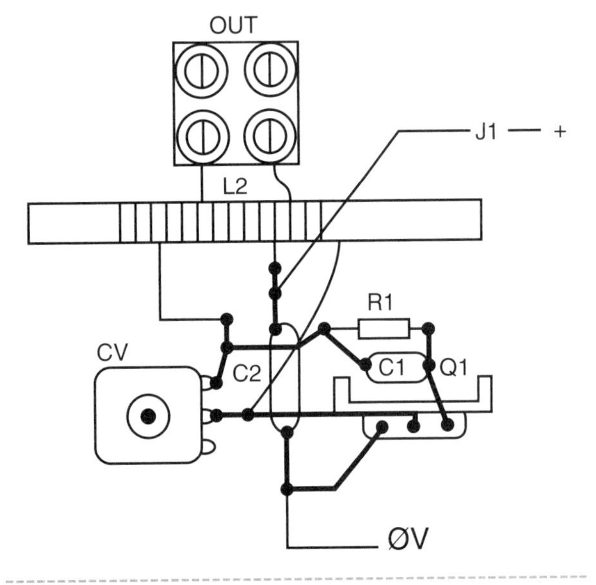


Figure 3.19.2 *PCB for the project*

## Testing and Using

Testing is very easy. Simply place a radio receiver that can tune the frequency of the oscillator near the circuit. Any short-wave receiver, tuning from the *long wave* (LW) to the *short wave* (SW) bands, can be used in the test according to the coil used.

Place the oscillator near the receiver and turn on S1. Tuning CV, the signal produced by the oscillator will be heard like a wind in the loudspeaker of the receiver. The range will depend on the sensitivity of the receiver but will not be more than some meters.

Since the circuit drains a high current, do not use AA cells. Use four C or D cells. For long-term experiments, use the power supply suggested in Figure 3.19.3.

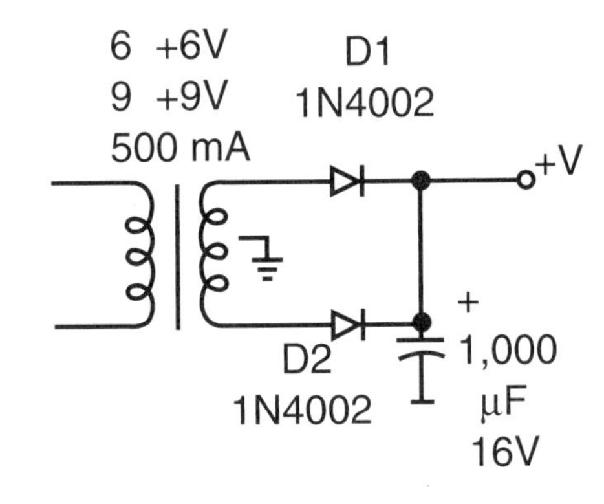


Figure 3.19.3 *Power supply for the oscillator*

## Parts List

Q1: BD135 medium-power NPN silicon transistor

L1/L2: Coils (see text)

R1: 1 kΩ x 1/2 W resistor, brown, black, red

C1: 0.01 μF ceramic capacitor

C2: 0.1 μF ceramic capacitor

CV: Variable capacitor (see text)

Other: Power supply or batteries, ferrite rod, wires, solder, etc.

# Experiments and Applications in Bionics

Some experiments using oscillators were suggested in previous projects, but here are a few more.

## Experiments with Plants and Insects

Figure 3.19.4 shows how to apply the signals generated by this oscillator to specimens via radiation.

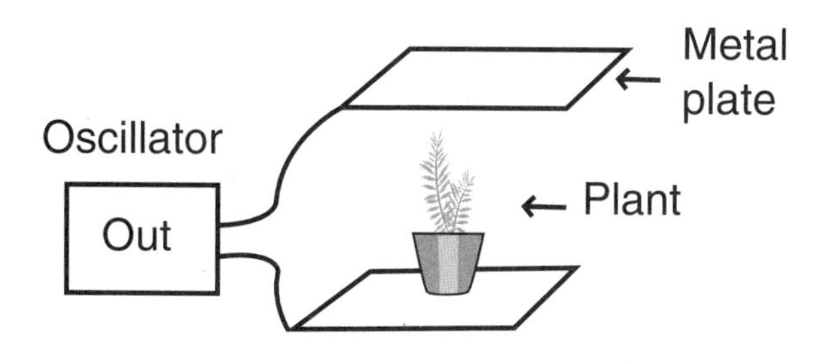


Figure 3.19.4 *Irradiating signals for experiments in bionics*

## Direct Application of the Signals

Direct application of the signals can be made by connecting the electrodes wired to the output of the circuit, as shown by Figure 3.19.5.

This circuit includes an amplitude control that regulates the amount of power sent to the specimens. The power in the circuit's output can reach a certain number of watts, which would be enough to kill small creatures and plants. Be sure to not apply the signal to humans.

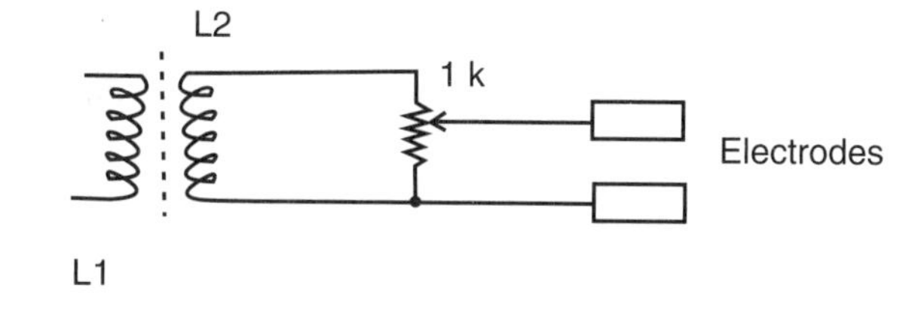


Figure 3.19.5 *Direct application of the signals*

# Project 20—Ionizer

The effect of negative and positive ions in the air is not well known. Some researchers say that negative ions can help in cases of panic, allergies, and other diseases associated with the respiratory system. In some cases, a small degree of negative ionization in the ambient air is recommended to add a sensation of comfort and even eliminate the allergy crisis.

According to the *Handy Science Reference Book* (Carnegie Library of Pittsburgh, 1994), "A negative ion generator is an electrostatic air cleaner that sprays a continuous fountain of negatively charged ions into the air. Some researchers claim that these ions cause a feeling of well being, increased stamina and relief from headaches, allergies and asthma."

It is not difficult to build an ion generator for experimental purposes and even make experiments with living beings. The reader can make experiments to see how the presence of ions in an area can affect plant growth or the behavior of insects or other small creatures.

Our basic project is a simple ion generator that, although powered from the AC power line, is very safe and can be used for medicinal purposes. The circuit will produce negative ions, but with a simple inversion of the diodes it can be changed to produce positive ones. The use of the device requires professional observation (a medical doctor, for instance, because it is for medical purposes).

The production of negative and positive ions is done by very high-voltage sources that range from 1,500 to 30,000 volts, but the current drain is very low, which does not represent the additional demand on your energy bill at the end of the month. The power demand of the circuit is low, less than 2 watts.

## Bionic Application of the Circuit

Producing negative (or positive) ions in an area is not a proper interaction between a living organism and an electronic circuit. The purpose of the project is to create conditions in order to investigate how electric charges can alter the behavior of living beings. The experiments can be made with

- Plants
- Insects
- Microorganisms

The device can also be used to investigate how ions can affect people with allergic diseases. In this case, with the supervision of a doctor, the device can be used at home or in the car.

## How It Works

The circuit is a voltage multiplier with 13 stages, which increase the peak of the AC power line voltage up to 1,800 volts. This voltage is enough to generate a constant flux of ions in the electrode. Of course, the reader is free to increase the number of stages, producing higher voltages.

When the circuit is powered on, each capacitor charges with the peak voltage of the AC power line, something near 150 volts. If the circuit is powered from the 220/240 VAC power line, the output voltage doubles. In this case, it is possible to use fewer stages to generate enough voltage for a good flux of ions. Since the capacitors are wired in a series, between semicycles, they are discharged through the output, appearing as the sum of their voltages in the electrode.

The resistors are important for limiting the current in case of accidental contact with the electrodes, because the circuit is powered directly from the AC power line.

The capacitors are not critical in this circuit. Values between 0.1 and 0.47 $\mu$F can be used, and they determine the amount of ions generated by the circuit.

The electrode is a needle or another metallic object with a sharp point. The reader can add a fan, such as one found in a computer, to spread the ions in the area.

## Building the Ionizer

Figure 3.20.1 shows the complete schematic diagram of the basic version.

The circuit can be mounted on a small *printed circuit board* (PCB), as shown in Figure 3.20.2. As explained before, the number of stages can be altered if the reader wants higher or lower voltages for the application in mind.

The circuit fits in a plastic box, and the electrode must be placed in a manner that won't cause any accidents. The box should protect the ionizer and contain a hole to let the ions escape if the ionizer is placed somewhere out of reach. Figure 3.20.3 shows how the electrode can be protected against accidental contact.

It is also possible to add more electrodes to the same circuit. A fan can be used to disperse the ions.

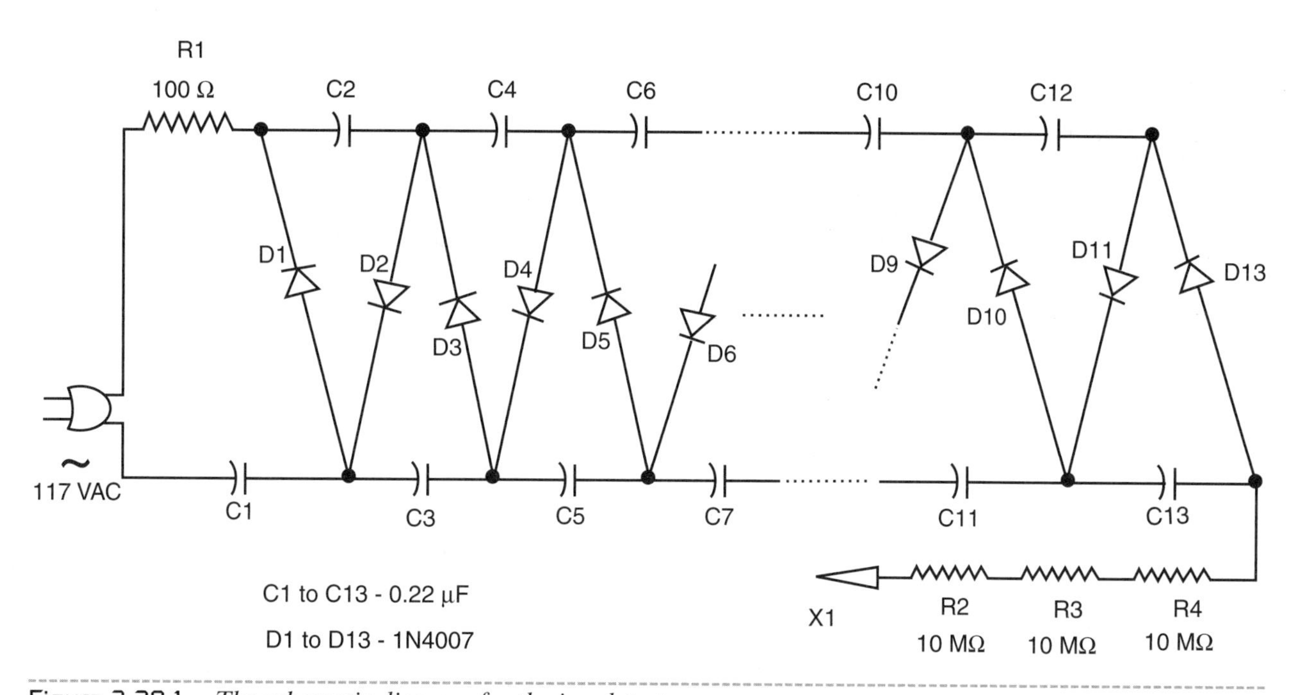


Figure 3.20.1 *The schematic diagram for the ion detector*

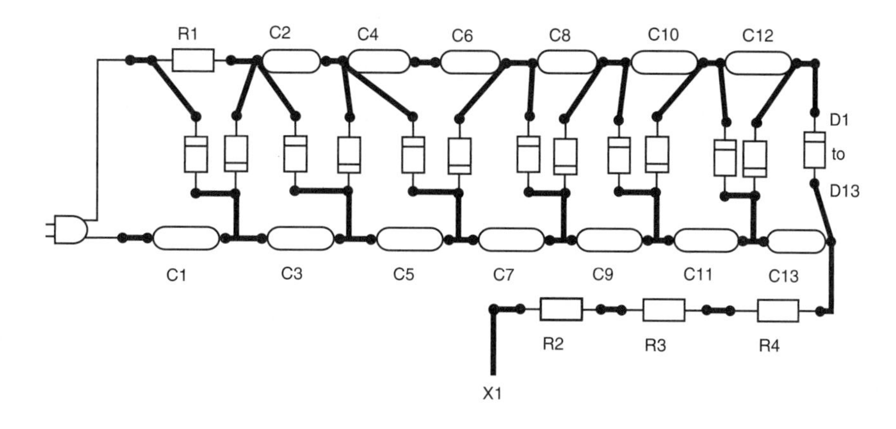


Figure 3.20.2 *The circuit mounted on a PCB*

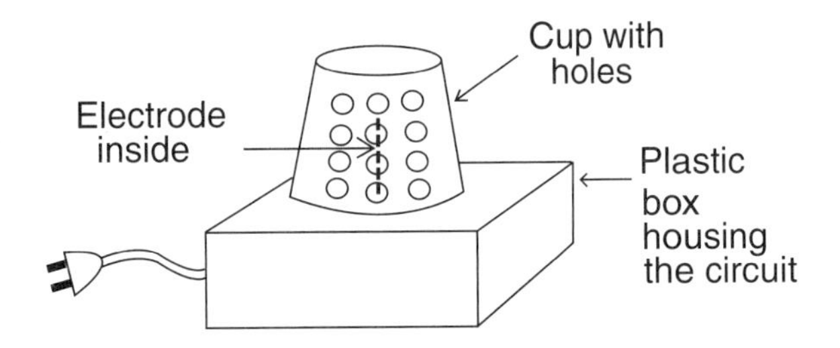


Figure 3.20.3 *Protecting the electrode X1*

## Testing and Using

First, plug the circuit to the AC power line. Then place a small neon lamp near the electrode and, from the influence of the high-voltage field, it should glow, as shown in Figure 3.20.4.

At the same time the ozone gas is produced, the presence of the ions will also be indicated by a characteristic smell being emitted.

### Parts List

- D1 to D13: 1N4007 silicon rectifier diodes
- C1 to C13: 0.22 $\mu$F x 400 V polyester capacitors
- R1: 100 $\Omega$ x 1 W resistor
- R2, R3, R4: 10 M$\Omega$ x 1/8 W resistors, brown, black, blue
- X1: Electrode (see text)
- Other: PCB, power cord, wires, plastic box, solder, etc.

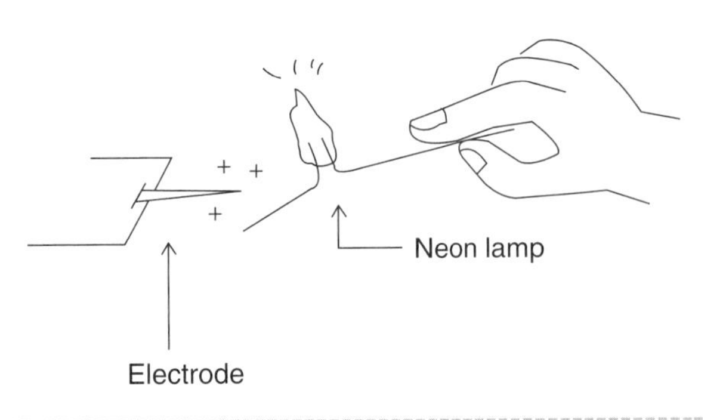


Figure 3.20.4 *Testing with a small neon lamp*

## Additional Circuits and Ideas

Some other configurations can be used to generate a high voltage and therefore a flux of ions. Remember that all the configurations given here are negative-ion generators. Inverting the diodes is the procedure for generating positive ions.

It is also important to say that the amount of ions produced by the circuit cannot be controlled, unless you are working on a medical experiment with specialized supervision. All the projects here should only be used for experimental purposes.

### High-Power Version Using an SCR

The *silicon-controlled rectifier* (SCR) circuit is ideal for experiments in a lab or when producing ions at home or in other places.

The circuit in Figure 3.20.5 is an ionizer powered from the AC power line. The values between the brackets in the parts list are for the 220/240 VAC power line.

The circuit consists of a relaxation oscillator driving a flyback transformer, such as one found in an old TV set or video monitor. Find one with part of the ferrite core exposed to make the primary winding, as shown by Figure 3.20.6.

The primary is formed by 7 to 15 turns of common covered wire 22 or 24 AWG. The circuit can be mounted on a terminal strip or a PCB and is housed in a plastic box. C3 is a glass capacitor mounted as shown in Figure 3.20.7. Replace the 1N4004 with a 1N4007 and the TIC106B with a TIC106D when plugging the device into a 220/240 AC power line.

The size of the glass piece is not critical. Aluminum foil or copper foil can be used as armatures for this capacitor.

C1 determines the power of the circuit, and values between 1 and 16 $\mu$F can be used. A high-voltage capacitor can be salvaged from old TV sets, but before using it, test it because this kind of component

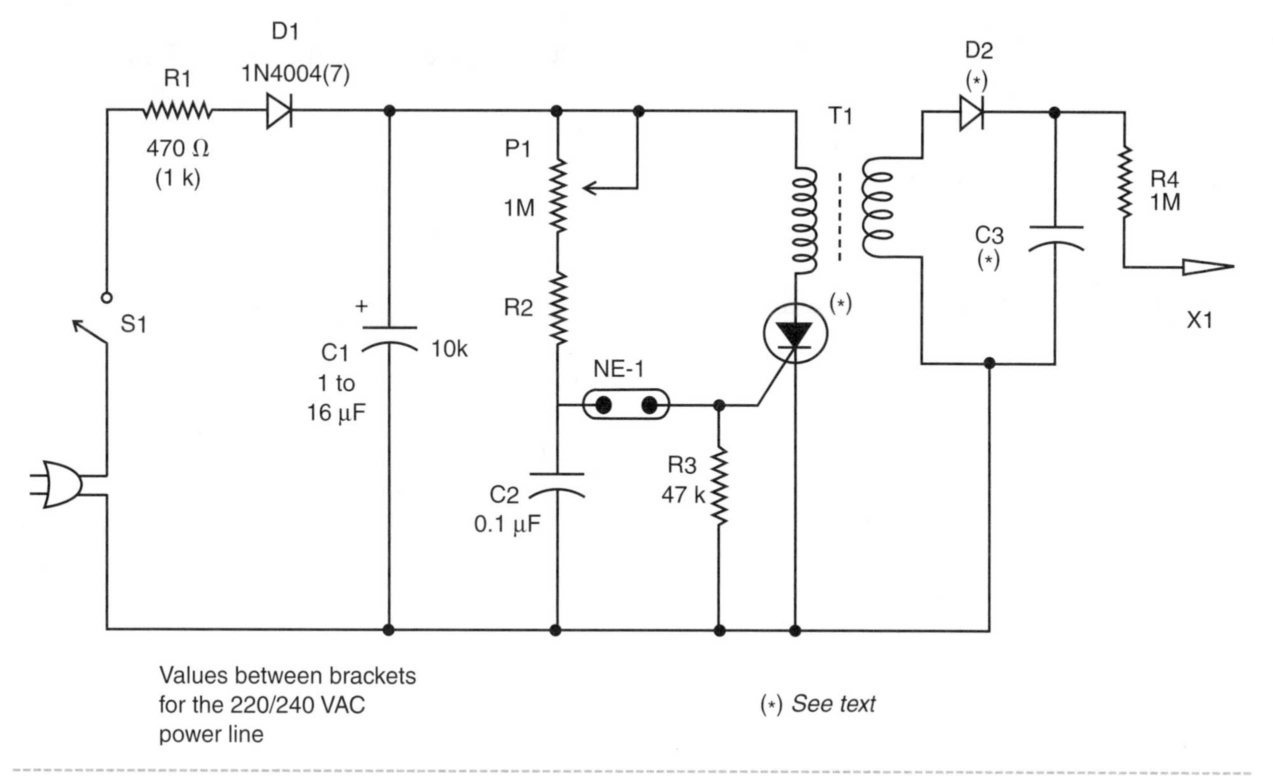


Figure 3.20.5 *Circuit using SCR*

will suffer from loss capacitance if it hasn't been used for a long time interval.

The SCR doesn't need a heatsink, because it conducts for short intervals of time without dissipating large amounts of heat. The only component that heats a little is R1, so you should mount it leaving 1 or 2 millimeters of separation from the PCB.

D2 is any high-voltage rectifier, which can also be found in old TV sets. It is usually connected to the flyback. Also, when using this ionizer, take care to not touch the electrode.

Figure 3.20.6 *The primary coil for T1*

## Parts List

- SCR: TIC106B (D) SCR
- D1: 1N4004 (7) silicon rectifier diode
- D2: High-voltage rectifier, 15 kV or more
- NE-1: NE-2H or equivalent neon lamp
- R1: 470 Ω (1 kΩ) x 10 W wire-wound resistor
- R2: 10 kΩ x 1/8 W resistor, brown, black, orange
- R3: 47 kΩ x 1/8 W resistor, yellow, violet, orange
- R4: 1 MΩ x 1/8 W resistor, brown, black, green
- P1: 1 MΩ lin or log potentiometer
- C1: 1 to 16 µF x 200 V (400 V) electrolytic capacitor
- C2: 0.1 µF x 100 V or more ceramic or polyester capacitor

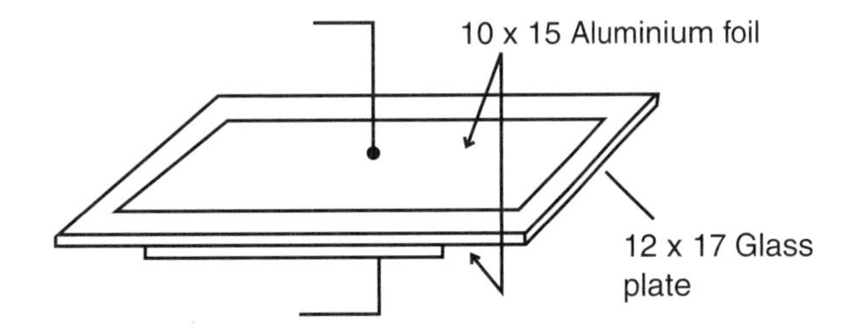


Figure 3.20.7 *The glass capacitor*

```
C3: Glass capacitor (see
  text)
S1: On/off switch
T1: Flyback transformer
  (see text)
X1: Electrode (as in the
  basic project)
Other: Power cord, PCB
  or terminal strip,
  plastic box, knob for
  P1, wires, solder,
  etc.
```

## Low-Voltage Circuit for Home and Car

The ion generator stage used in this project can operate with voltage sources as low as 12 volts. This means that it can be powered from car batteries or a power supply producing low DC voltages at home.

In the car, negative ions can be important for reducing the effects of the positive charges accumulated in the metallic structure, causing discomfort mainly to those with allergies. The complete circuit is shown in Figure 3.20.8.

Figure 3.20.9 shows a PCB suitable for this project.

If powered from a battery, the power supply sector is not used.

T2 is any flyback obtained from an old nonfunctioning TV set or another piece of equipment using a TRIAC circuit. You must be sure to choose a flyback with an exposed ferrite core where the primary is wound. This primary coil is formed by 7 to 15 turns of any wire between 22 and 18 AWG. You can also use a car ignition coil of any type.

C3 is a glass capacitor, as in the previous project.

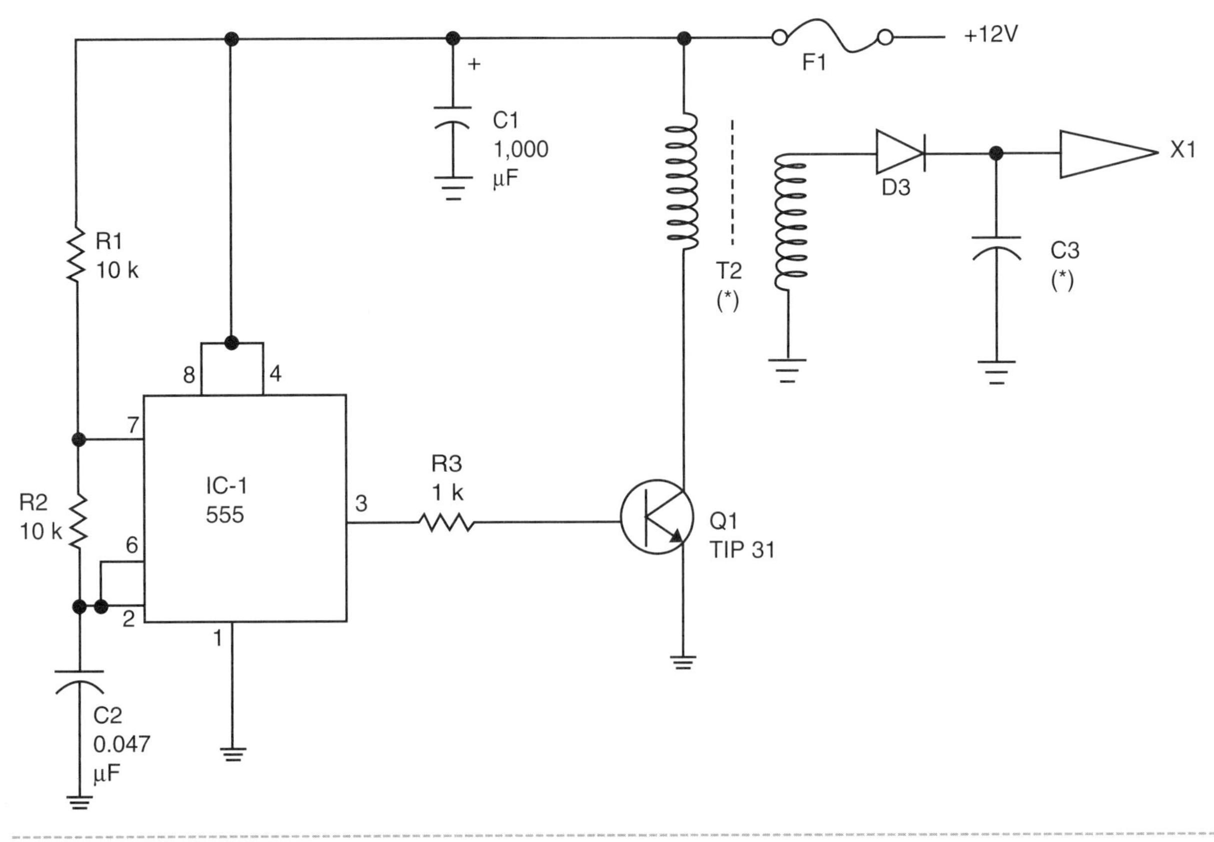


Figure 3.20.8 *Low-voltage circuit with power supply*

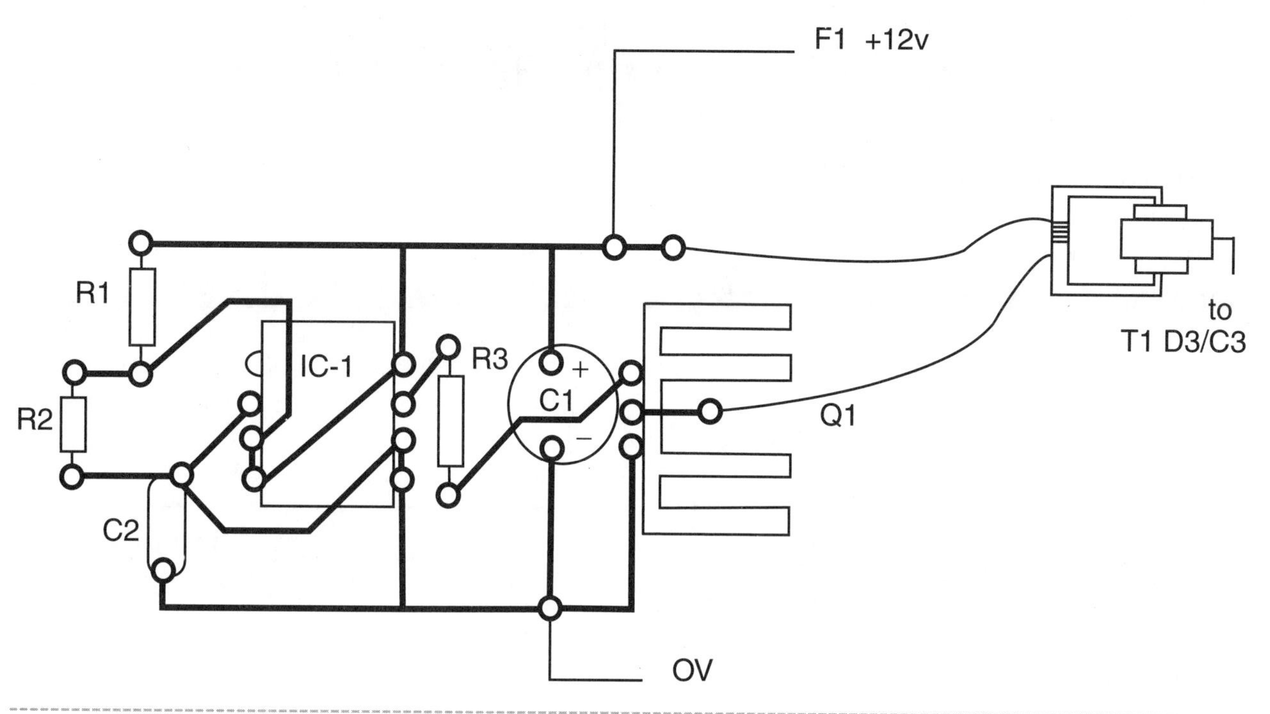


Figure 3.20.9 *The PCB for this project*

## Parts List

The parts list is divided into two sections, because the second block is necessary only when powering the circuit from the AC power line.

### Oscillator and Power Stage

IC-1: 555 integrated circuit timer

D3: High-voltage rectifier (20 kV)

Q1: TIP31 silicon NPN power transistor

R1, R2: 10 kΩ x 1/8 W resistors, brown, black, orange

R3: 1 kΩ x 1/8 W resistor, brown, black, red

C1: 1,000 μF x 25 V electrolytic capacitor

C2: 0.047 μF polyester or ceramic capacitor

C3: Glass capacitor (see text)

X1: Electrode (as in the basic project)

F1: 1A fuse

T2: Flyback transformer or ignition coil (see text)

Other: PCB, fuse holder, heatsink for Q1, wires, solder, plastic box, etc.

### Power Supply for the AC power line

D1, D2, D3, D4: 1N4002 or equivalent silicon rectifier diode

T1: Transformer, the primary according to the power supply line and 12 V x 1A secondary

S1: On/off switch

F1: 1A fuse

Other: Power cord, fuse holder, wires, solder, etc.

# Project 21—Tactile Hearing Aid

A tactile hearing aid can be used to facilitate communication between the profoundly deaf and the hearing. The basic idea is to convert sounds into mechanic vibrations or an electric stimulus that can be sensed by the skin or other sensory organs.

A common microphone picks up the sounds, amplifies them, and applies them to a transducer. Of course, other sound sources such as recorders, computers, and so on, can be used.

Two kinds of transducers can be used. The first is a vibrator that converts the sounds into mechanical vibrations that can be sensed by the fingers or other parts of the body. Research reveals that an interesting part of the body on which to place the electrodes, due to the sensitivity to vibrations, is the belly.

The other type of sensor is a high-voltage converter that produces an electric stimulus sent to electrodes that can be placed on any part of the body. Figure 3.21.1 shows these possibilities.

The simplest way to make a tactile hearing aid is to use a microphone, an amplifier, and a transducer, which is just what we are going to describe here.

Of course, many improvements can be made to facilitate communication between the hearing world and the deaf. Research is being made using signal processors and analog-to-digital converters to transform a sound signal to a form that can be easily recognized by the tactile senses. This includes filtering, the separation of the sound spectrum into bands, and even making changes in the wave shapes. This is an interesting bionics field to be investigated.

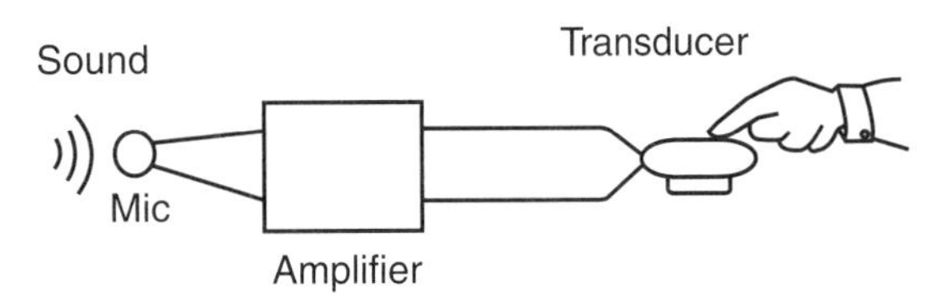


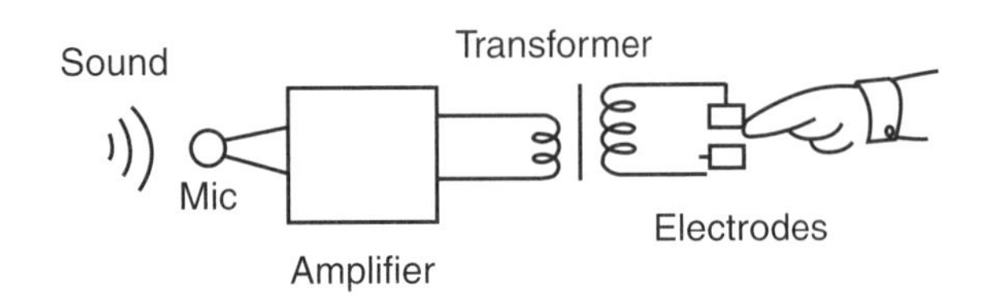


Figure 3.21.1 *Hearing through electric and mechanic vibrations*

The evil genius, even if not deaf, can undertake this bionic tactile hearing project to achieve other purposes, as we will suggest in the next section.

## The Projects in Bionics

The circuit described here is only a portion of a complete application. The evil genius can change the basic idea and work with other senses or even convert sounds to other kinds of signals, such as light or electric shock. Some ideas of applications are given in the following lines.

- The evil genius can use this circuit to communicate with friends using an invisible language of vibrations.
- Different sensations can be added to music reproduction so that you can hear through the belly (and other parts if you want).
- Help deaf people sense sounds and learn to speak (Many deaf people are also mute. They have not learned to speak because they cannot hear the sounds).

Figure 3.21.2 shows an interesting possibility for sound interfacing using vibrations, which is not recommended for practical purposes.

It is also possible to use the circuit in experiments with other living beings such as plants, mammals, and insects.

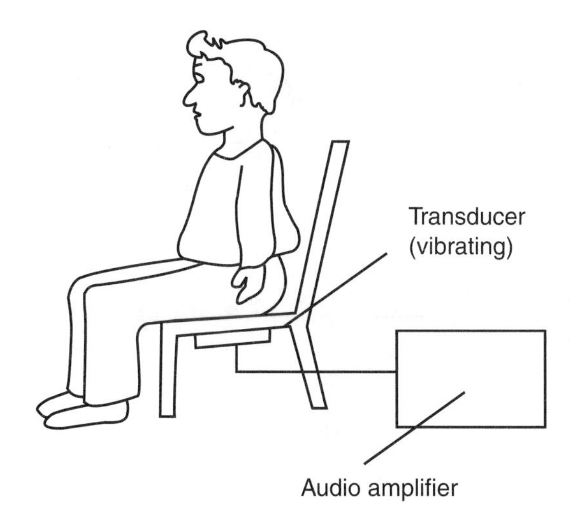


Figure 3.21.2 *Interfacing people with sound sources using mechanical vibrations*

## How the Circuit Works

The electronic part of the project is very simple. All we need is an electret microphone, an amplifier, and a transducer.

The amplifier can be the same one used in our first project and in the bionic ear, the unit using the LM386 *integrated circuit* (IC). It is an easy-to-find part and will have enough gain for the application we have in mind.

The transducer can be any electret microphone. This type of microphone has a *field effect transistor* (FET) inside, giving the device a high sensitivity. Finally, the transducer is the only part that the evil genius will have to mount with care.

To convert the sounds into mechanical vibrations, even a small loudspeaker can be used. It is enough to touch the diaphragm to sense the vibrations, as shown by Figure 3.21.3, but details of another transducer will be given later.

The other transducer is a transformer that generates high voltage from the audio signals in the output of the amplifier. Applied to electrodes, the high-voltage signal can be sensed as a light shock in the finger or other places. An amplitude adjuster is also added to control the level of the stimulus, according to the place where the electrodes are used.

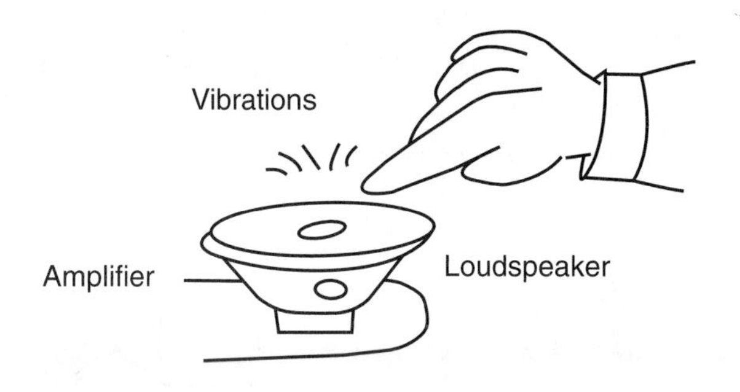


Figure 3.21.3 *A loudspeaker used as transducer*

## How to Build

Figure 3.21.4 shows the schematic diagram for the tactile hearing aid.

The circuit can use the same *printed circuit board* (PCB) described in the electric fish and bionic ear projects. Figure 3.21.5 shows how to connect the microphone and the transducer.

The power supply consists of four AA cells installed in a holder. When mounting, observe the position of the polarized components, such as the electrolytic capacitors, IC, and power supply. The electret microphone is also a polarized component.

If the wire that connects this component to the circuit is long, a shielded cable must be used. The screen must be connected to the 0-volt line or ground line of the circuit.

## The Transducers

The simplest transducer is a small plastic loudspeaker placed on a base, and you must touch the cone to sense the vibrations. Another transducer is shown in Figure 3.21.6.

The coil is formed by 200 to 500 turns of 28 to 32 AWG wire on a small screw. The diaphragm is a plate made of iron, steel, or another magnetic metal.

The distance between the plate and the magnet is kept as low as possible to transfer the vibrations, which can be sensed when the fingers are placed on the sensor.

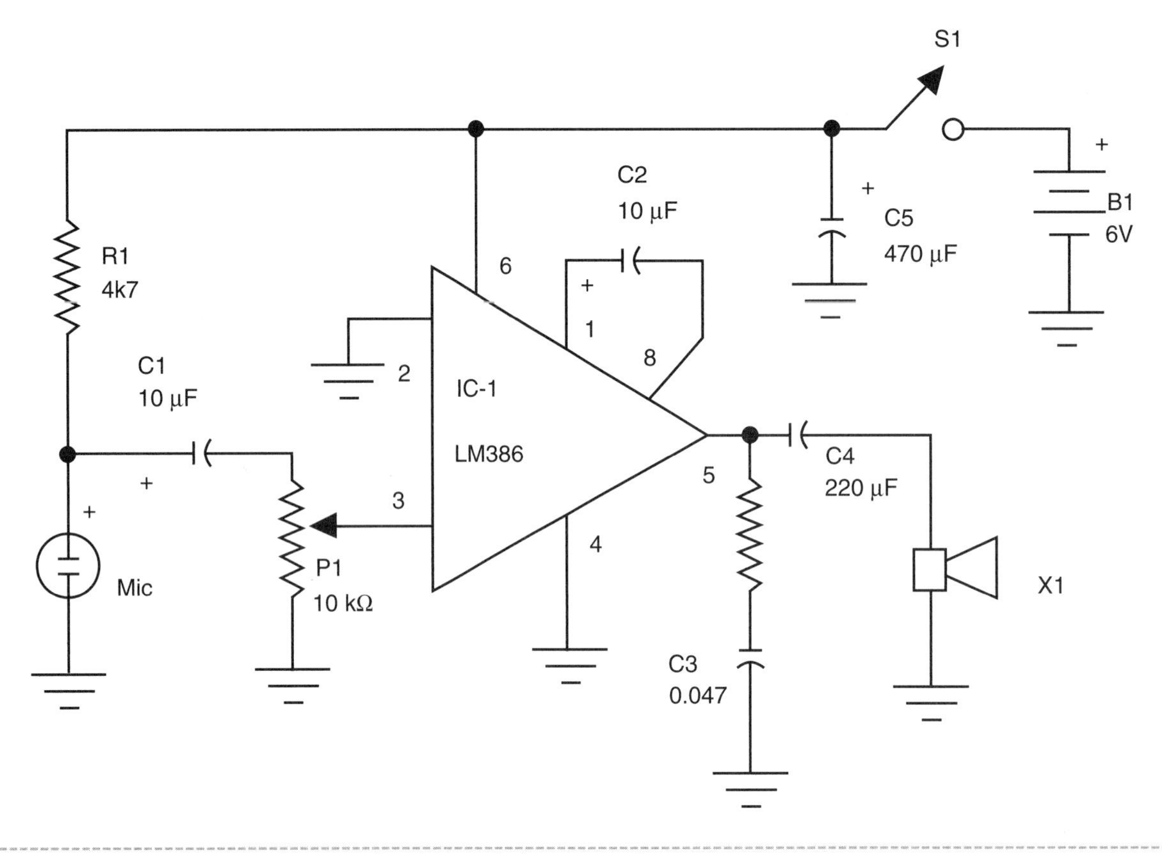


Figure 3.21.4 *Schematic diagram for the tactile hearing aid*

This sensor and the loudspeaker can be used to transfer the vibrations to other parts of the body, as suggested at the beginning of this project.

The electric transducer is a transformer with a 117-volt or other high-voltage primary and a 6-volt secondary with currents between 150 and 300 milliamps.

As shown in Figure 3.21.7, the subject must place his or her fingers on the electrodes and adjust P2 to the ideal electric stimulus from the sounds picked up by the microphone.

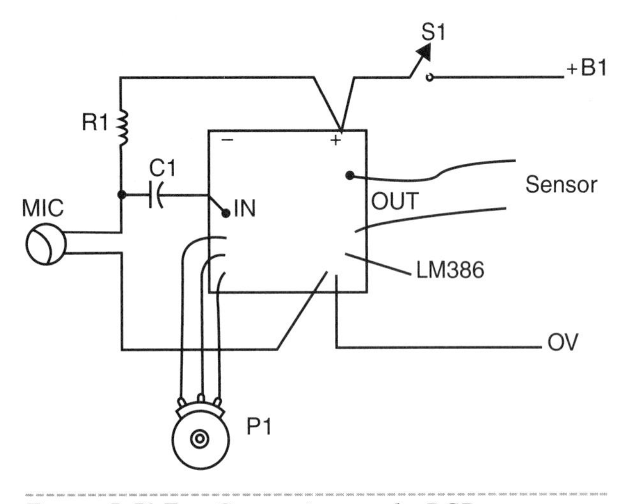


Figure 3.21.5 *Connections to the PCB*

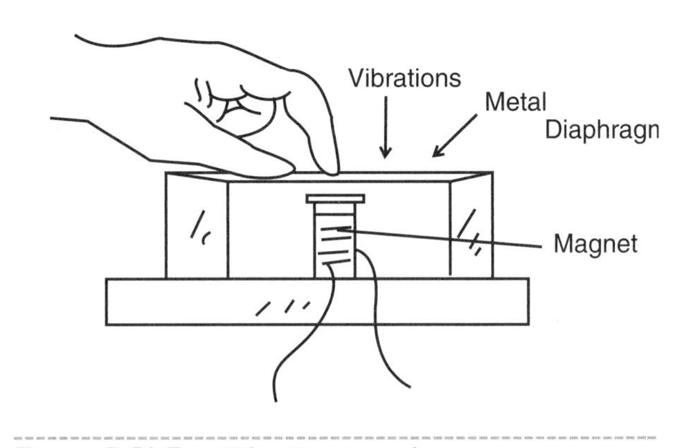


Figure 3.21.6 *Vibration transducer*

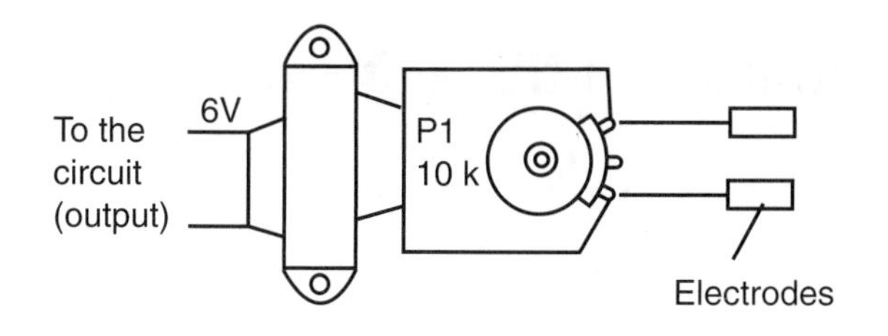


Figure 3.21.7 *The electric transducer*

## Testing and Using

Place the cells in the holder and turn on the power supply, closing S1. Then open P1 and P2. Place your fingers on the sensor and, when speaking near the microphone, the vibrations will be sensed.

## Parts List

IC-1: LM386 (any suffix) integrated circuit

P1, P2: 10 kΩ log potentiometer

R1: 4.7 kΩ x 1/8 W resistor, yellow, violet, red

R2: 10 Ω x 1/8 W resistor, brown, black, black

C1, C2: 10 μF x 12 V electrolytic capacitor

C3: 0.047 μF ceramic or polyester capacitor

C4: 220 μF x 12 V electrolytic capacitor

C5: 470 μF x 12 V electrolytic capacitor

X1: Transducer (see text)

S1: On/off switch

B1: Four 6 V AA cells

Other: PCB, material for the transducer, cell holder, plastic box, wires, solder, etc.

## Additional Circuits and Ideas

In principle, any common audio amplifier can be used to transfer the vibrations sensed by a microphone to a transducer. Two ideas are given in this section.

## Circuit Using a Common Audio Amplifier

Figure 3.21.8 shows how to use a common audio amplifier to convert sounds into vibrations or electric signals. Audio amplifiers with power outputs from 1 to 50 watts can be used.

When using a high-voltage transformer, the evil genius must be careful to not use all the power provided by the high-power amplifier (more than 10 watts). This can cause severe shocks or burn the transformer. Slowly open the volume control to reach the ideal point for the stimulus of the subject.

Figure 3.21.8 *Using a common audio amplifier*

## Other Circuits Using the TDA7052

Figure 3.21.9 shows how to use another IC as an audio amplifier in the experiments described here.

This amplifier can produce more than 100 megawatts to a small loudspeaker or even a transformer, providing vibrations or electric stimulus for a subject.

## Upgrading the Project

Deaf people, as well as the blind, can develop an acute tactile sense because in most cases this sense replaces the one that is missing. Experiments can reveal the ideal stimulus for transferring the correct sensation of sounds to a deaf person.

This experiment can include the use of filters or tone controls, or even frequency converters. The curious evil genius can discover many interesting things when developing a communication system using tactile vibrations.

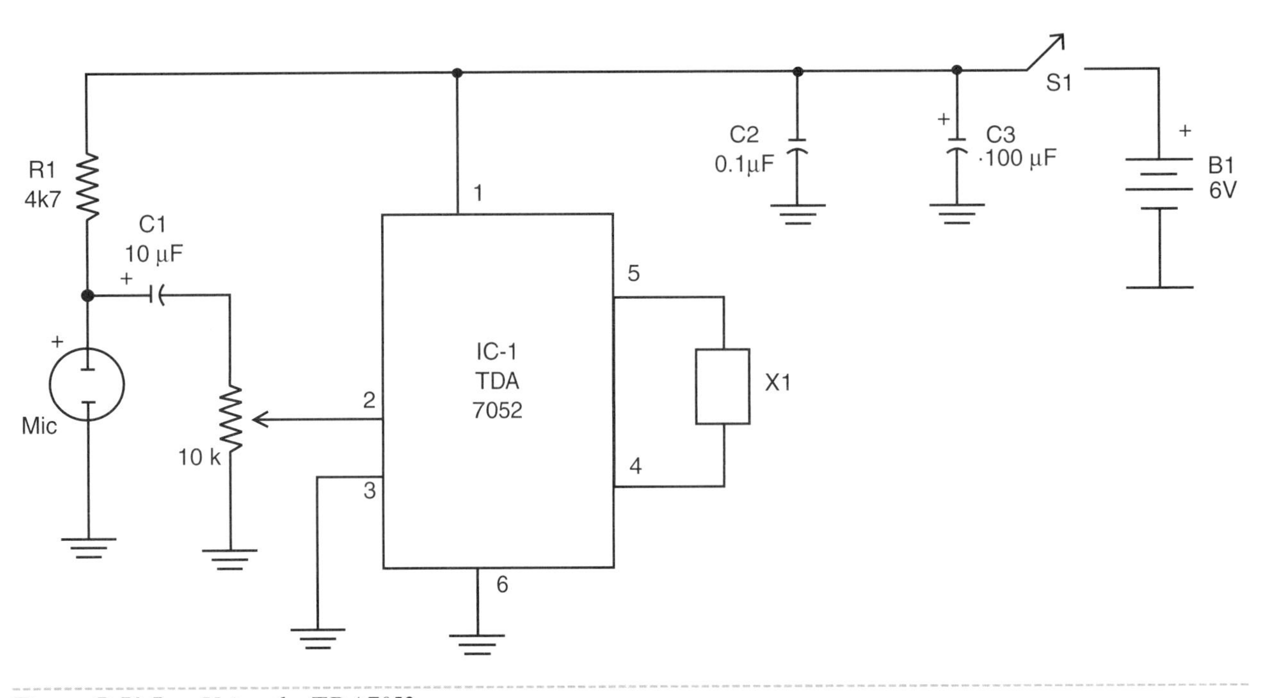


Figure 3.21.9 *Using the TDA7052*

# Project 22—Using the Multimeter in Bionic Experiments

Low-cost analog or digital multimeters are sensitive enough to detect changes in the skin resistance of living beings, especially humans, and even changes in a medium, or environment, where a living creature is placed.

This chapter describes a series of applications and experiments where a common multimeter can be used to detect the biologic activity of living beings. The reader can use this multimeter to monitor changes in the environment, using the creature as a living sensor. Other applications include studies on the behavior of many creatures, and even stress or other states of mind in the lives of humans.

The important point to arise in this chapter is that even a very low cost multimeter can be used in many interesting applications without any need for hard-to-find or expensive additional parts.

## The Project

Many biologic activities include electric changes. Resistance changes, the generation of electric potential, or even the production of electric signals or current, such as in the case of electric fish, are examples of these changes. Some of those changes are also strong enough to be detected by even a common multimeter, such as the ones found in supermarkets for making general-purpose tests in your home or car (see Figure 3.22.1).

Digital types are more expensive but can also be used in the same experiments. Figure 3.22.2 shows a digital multimeter that can be used with the same purposes as the analog multimeter.

Using any of these multimeters, we can detect small changes of resistance in living systems or the generation of low voltages. Low voltages starting from a fraction of a volt and resistances with a large range of values can be measured using these multimeters.

The sensitivity of low-cost multimeters can also be increased with a few external components, such as transistors that act as amplifiers. The other possibility, explored here, is adding a sensor to the multimeter to interface both of them with the biologic system being studied. *Light-dependent resistors* (LDRs), transistors, and pressure sensors can be connected to the multimeter, acting as transducers. They can convert other physical magnitudes such as light, color, temperature, or pressure into electricity measured by the multimeters.

Figure 3.22.1 *Low-cost analog multimeter suitable for the experiments described here*

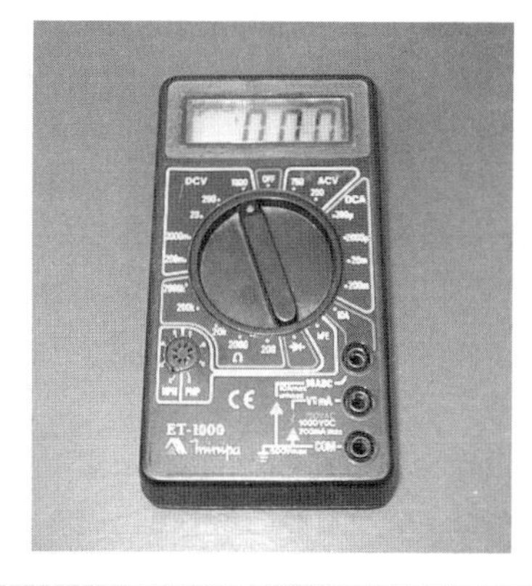

Figure 3.22.2 *A low-cost general-purpose digital multimeter*

It is also possible to add some electronic circuits for converting magnitudes and increasing the sensitivity in order to detect other biological phenomenon. Operational amplifiers and even transistors can perform these tasks, as described in the next section.

## Bionic Experiments

Using the multimeter and some accessories as the electronic part of the project, the biologic part can be any living being or biologic system:

- Bioactivity of microorganisms
- Sensing the presence of life
- Heartbeat observation
- Breath detector
- Temperature sensor

## The Multimeter and Its Characteristics

Common analog multimeters can detect voltages of up to 0.1 volts (100 millivolts) and resistances with a large range of values. Multimeters with sensitivities of 5,000 or 10,000 ohms per volt are also available but are more expensive.

The sensitivity in ohms per volt expresses the amount of resistance introduced by the multimeter when plugged into a specimen. For instance, when using a multimeter with a sensitivity of 1,000 ohms per volt in the scale of 1 to 1.5 volts, this represents a resistance of 1,500 ohms in parallel with the measured circuit, as shown in Figure 3.22.3.

This resistance drains current from the circuit, causing a voltage drop or change in the real voltage to be measured. The less current that is drained, the better the multimeter is because the less of a change it causes in the voltage to be measured.

Since biologic currents are very low, representing sources with high internal resistances, the presence of

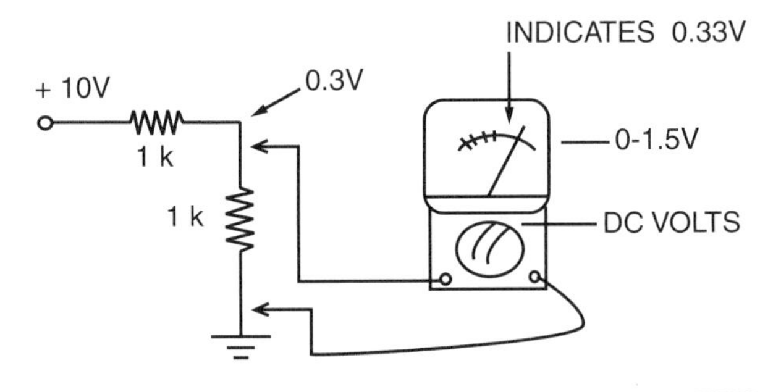


Figure 3.22.3 *The influence of the multimeter*

a common multimeter normally will change the real voltages being measured. This is because, as high as the sensitivity of the multimeter is, it is better for biologic experiments. However, these limitations don't prevent the use of low-cost multimeters in many different experiments. This chapter will explain how to increase the sensitivity and how to use it in experiments or applications where low sensitivity is not a limitation.

Another point to observe is that, besides the voltage scales of the multimeter, we can also measure resistances. The changes in the resistance of a specimen can reveal many things about it. To measure resistances, the multimeter applies a voltage and measures the current flow across it, as Figure 3.22.4 shows.

In this case, the sensitivity of the multimeter is highest, and the current flowing across the specimen is lowest. Low currents are highly desirable because they don't cause changes in the biological subject.

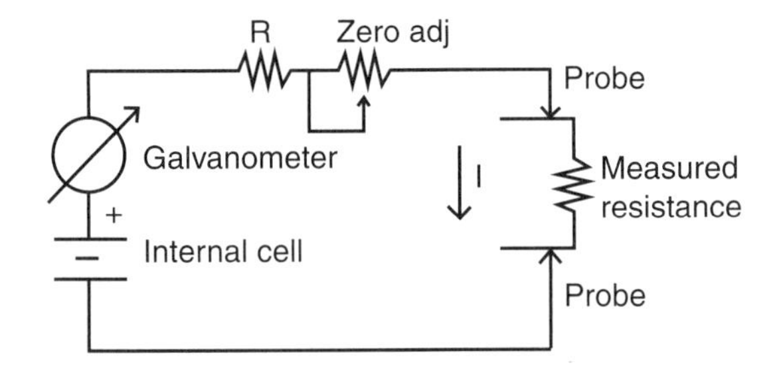


Figure 3.22.4 *Measuring resistances*

What happens is that when a current flows in a living being, within the cells of a specimen, it causes galvanic effects or chemical reactions that change the specimen. The result is that, if the current is applied for long time intervals, the specimen can die.

Also, digital multimeters are much more sensitive than analog multimeters. Digital multimeters have a very high input impedance (22 MΩ), meaning that they introduce a very low change or none in the circuit being analyzed.

Analog multimeters also are unable to detect fast changes in currents or voltages due to the fact that the sample rate used to acquire data in the analog form is not fast enough. This means that some experiments described in this chapter can't be performed with analog multimeters.

Generally speaking, digital meters have high impedance and introduce low changes in the measured voltage; analog multimeters are low-impedance devices, and introduce changes in the voltage to be measured.

## The Projects

Most of the following projects use low-cost multimeters. In some cases (as indicated in the project), more sensitive multimeters are recommended, as well as digital multimeters.

The accuracy of the project or experiment depends on several factors besides the multimeter, such as the biological specimen, the skills of the reader, and local and unpredictable conditions. Although the author is careful to point out when harmful materials and substances are used, be sure to prevent contact with one's skin and eyes when using these materials.

## Bioactivity of Microorganisms

Changes in the transparency of a culture of microorganisms can be detected easily using a multimeter and an LDR, as shown by Figure 3.22.5.

This principle was used in one of the first Mars expeditions. Food for microorganisms was placed in a liquid. If life exists on Mars, the microorganisms would eat and reproduce in the liquid, changing its transparency. A sensor and a light source were used to measure these changes and send the information back to Earth.

The multimeter for this experiment will be adjusted to a resistance scale according to the level of light used. The evil genius must experiment to find the best light scale to be used.

Other experiments involving changes in transparency and color can be made using the same configuration. Light filters and polarizers can also be used to perform experiments involving organic substances associated with biologic activity.

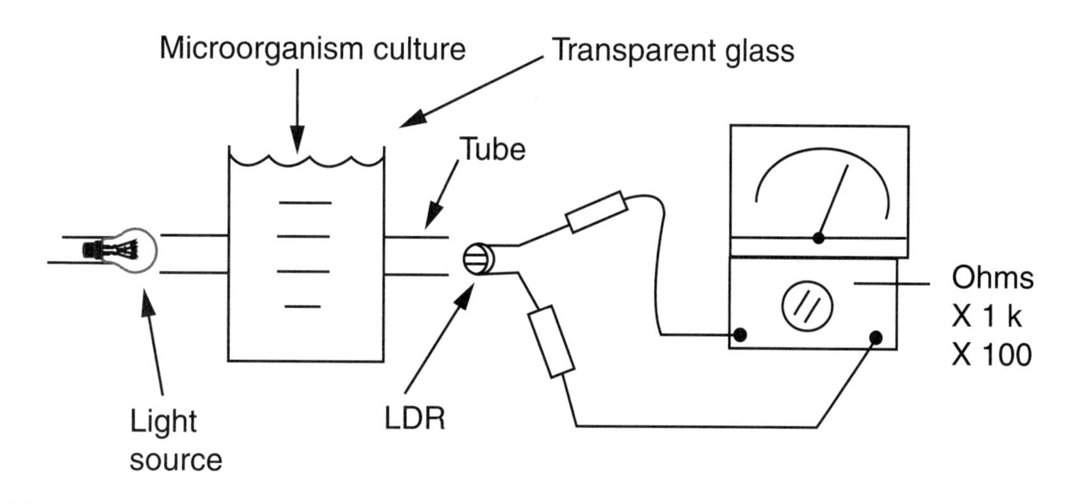


Figure 3.22.5 *Using the multimeter as biosensor*

## Heartbeat Observation

The LDR and a light source placed as shown in Figure 3.22.6 can monitor a heartbeat.

The heart acts as a pump, forcing the blood to flow across the circulatory system. Its action is made by a process that produces waves of blood, propagating through the vessels and changing the transparency of the parts of the vessles where the pressure is high.

So by using a light source and a light sensor, as shown in the figure, small changes in the transparency due to the blood waves can be detected. Small oscillations picked up by a multimeter will indicate which waves are associated with the heartbeat. The project is critical because no external light can interfere with the sensor and the ideal scale must be experimented with according to the light source.

## Breath Detector

When the air enters our lungs, it is cold, but when we expel this air it is warm. We can detect a person's breath by the difference of temperature between the air that enters and the air expelled.

The sensor is a common diode that has an inverse resistance that depends on the temperature. Since the current is too weak to be measured by the multimeter, the sensor made by a diode must have a transistor for amplification purposes.

The complete circuit for a hot air detector is shown in Figure 3.22.7. The scale is the highest for the multimeter (ohms × 1 k or ohms × 10 k).

Testing and using it is very easy; just breathe on the sensor. The multimeter will indicate a fast fall in the resistance. After waiting a few seconds, the resistance will rise again because the sensor is now returning to the original temperature.

As a temperature sensor, the device can be used to detect bioactivity. Fermentation, the decomposition of organic matter, and other phenomenon associated

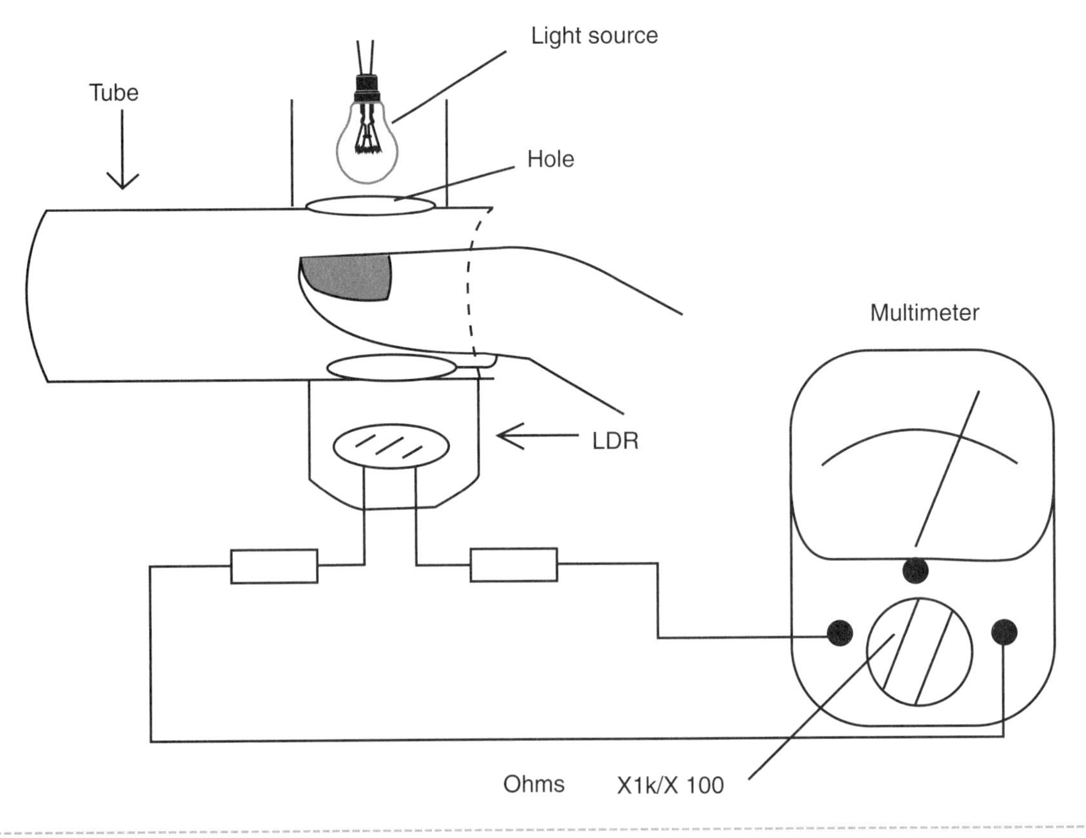


Figure 3.22.6 *Monitoring a heartbeat*

with biologic activities normally are accompanied by changes in temperature. The multimeter with a temperature sensor, such as the one described here, can be used to monitor such changes.

An important fact to be considered when using the multimeter is that the sensor can be placed far from the indicator. The multimeter can be wired to the sensor by cables up to 100 meters long.

However, digital multimeters are not suitable for some of the experiments described here because they do not use a power supply with enough current to power the sensors. Their impedance is very high and in some cases the changes in the measured magnitudes are not detected as fast as we need for the application.

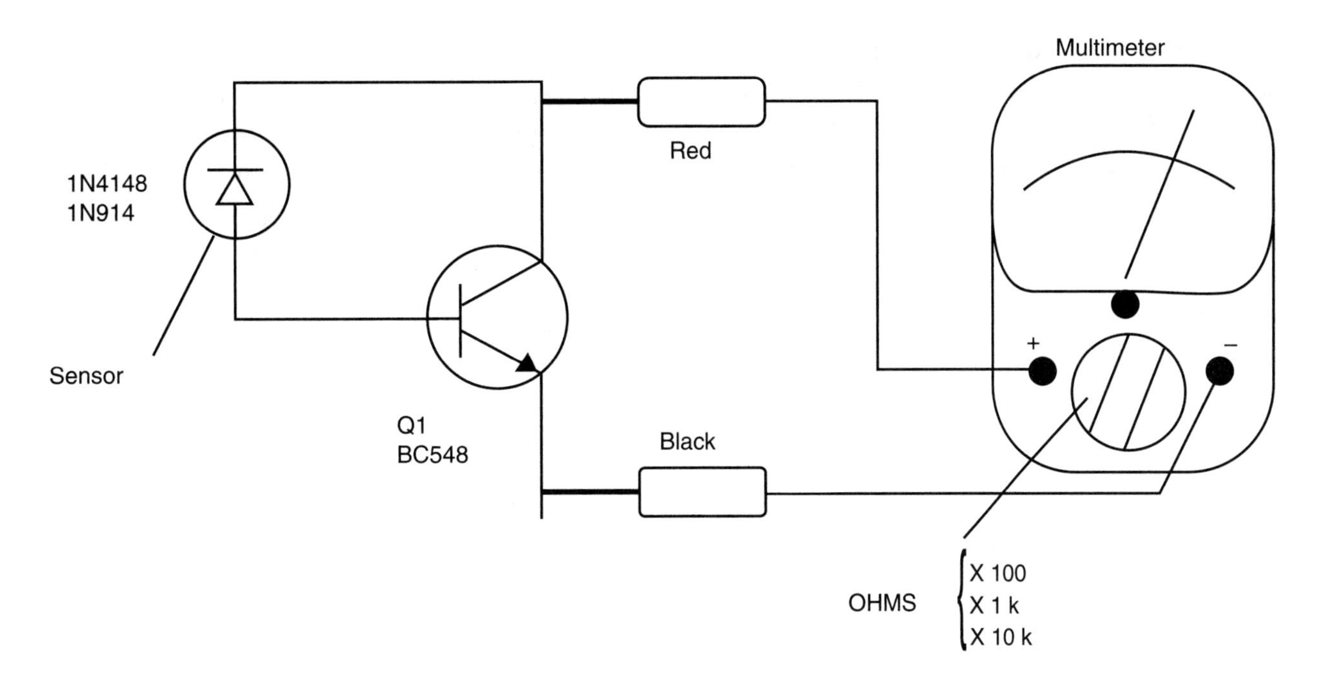


Figure 3.22.7 *The breath detector using a diode as a sensor*

# Project 23—Bionic Vision

The bionic man we mentioned earlier can see in the dark and even through walls. Devices to see in the dark are common, but no X-ray system has been created yet to allow humans to see through walls.

What we are proposing here is an interesting project that allows blind people to sense their environment by using a device that converts light into sound. Like an elementary eye (a very simple, one-pixel version of an eye), the project creates complex sounds because it uses four sensors, the same way insects have eyes that are multifaceted. Here we will be using light sensors, each one acting as a single image point or pixel sensor.

The number of points of image seen by an insect depends on the number of facets in its eye, as shown in Figure 3.23.1. The number of facets in common insects varies between 50 and 5,000, but its eyesight would be considered to suffer from poor definition compared to humans or even a common *charge-couple device* (CCD) photographic camera that has some millions of pixels of definition.

The idea of bionic vision, replacing natural vision, is to reproduce the image sensor (the eye), connecting it to the optical nerve or sending the information to the brain some other way. In this way, a damaged eye can be replaced by an electronic image sensor connected directly to the optical nerve.

Our project works with an interesting idea that can be explored starting from its simplicity. The project consists of a four-pixel sensor that converts the amount of light picked up by each facet into a sound, as shown by Figure 3.23.2.

Of course, the evil genius is free to increase the number of facets and even create a high-definition project using CCD image sensors or optical resources, such as image-scanning systems.

The tones produced by each facet are combined to produce an output signal in a small loudspeaker or earphone. The eye of the device is placed in the person's hand so that it can explore the environment, generating a sound image. With some experience, it will be possible to know what each change in a tone represents. The person will be able to recognize simple objects such as an open door or window or the presence of moving things.

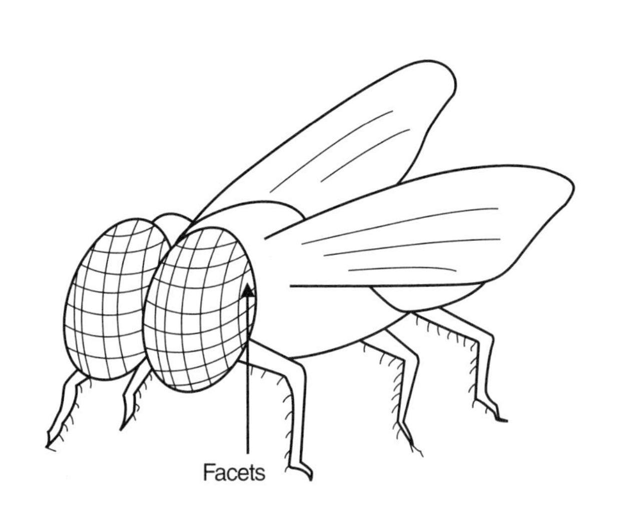


Figure 3.23.1 *Facets of an insect eye*

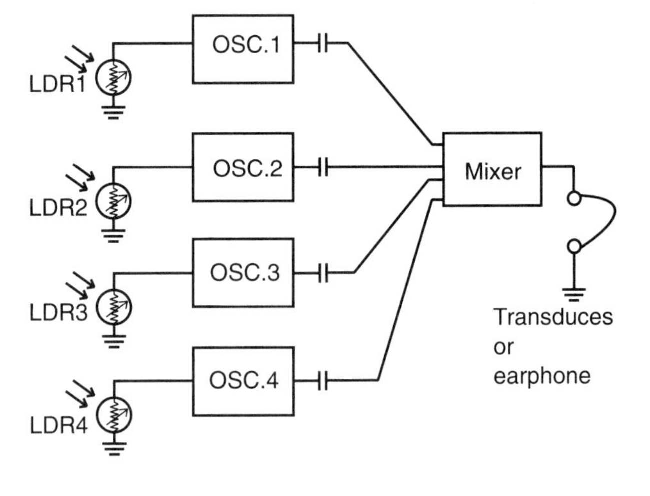


Figure 3.23.2 *The four-faced bionic eye*

## Bionic Applications

The project is experimental and will appeal to the curious. The evil genius can use it to create a game where a blindfolded friend must find his or her way to a group of people using only the signals given by the bionic eye, as suggested by Figure 3.23.3.

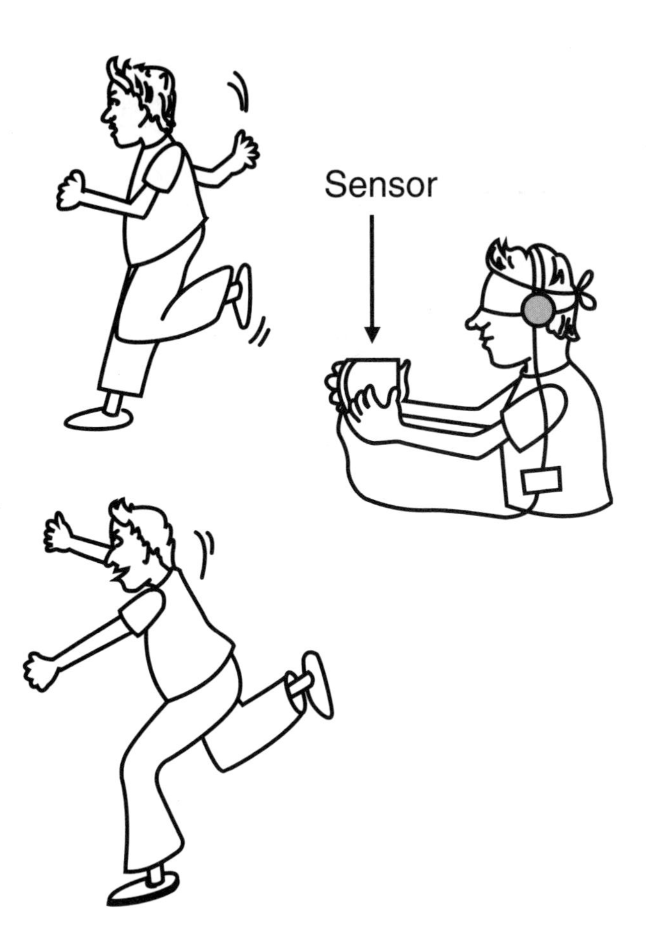


Figure 3.23.3 *Playing with the bionic eye*

Other applications include the following:

- Experiments with artificial vision
- Create a simple vision system interfaced with the computer
- Design an alarm that can recognize forms

## How the Circuit Works

The sensors consist of *light-dependent resistors* (LDR) that control the frequency of the oscillators. A 4093 *integrated circuit* (IC) formed by four NAND gates is used for the four independent oscillators, which are controlled by the four LDRs.

The basic frequency of each oscillator is determined by the correspondent capacitor. The signals in the output of each oscillator are mixed and applied to a high-impedance transducer.

Since the oscillator drains low currents and the transducer doesn't need much power to be activated, power consumption for the circuit is very low. The four AA cells that power the circuit can be used during many weeks or even months without the need for replacement.

The sensors (LDRs) are mounted together, forming a multifaceted eye. A lens and small cardboard tubes are used to add directivity and sensitivity to the eye.

## How to Build

Figure 3.23.4 shows the schematic diagram of the bionic eye.

The basic circuit can be mounted on a *printed circuit board* (PCB) or a solderless board. Figure 3.23.5 shows a suggestion for a PCB for this mounting.

The LDRs are common, round types with diameters that range from 1 to 2.5 centimeters. Smaller types are better because they can be easily mounted inside thin cardboard tubes. When mounting, take care with the positions of the polarized components, such as the IC and power supply.

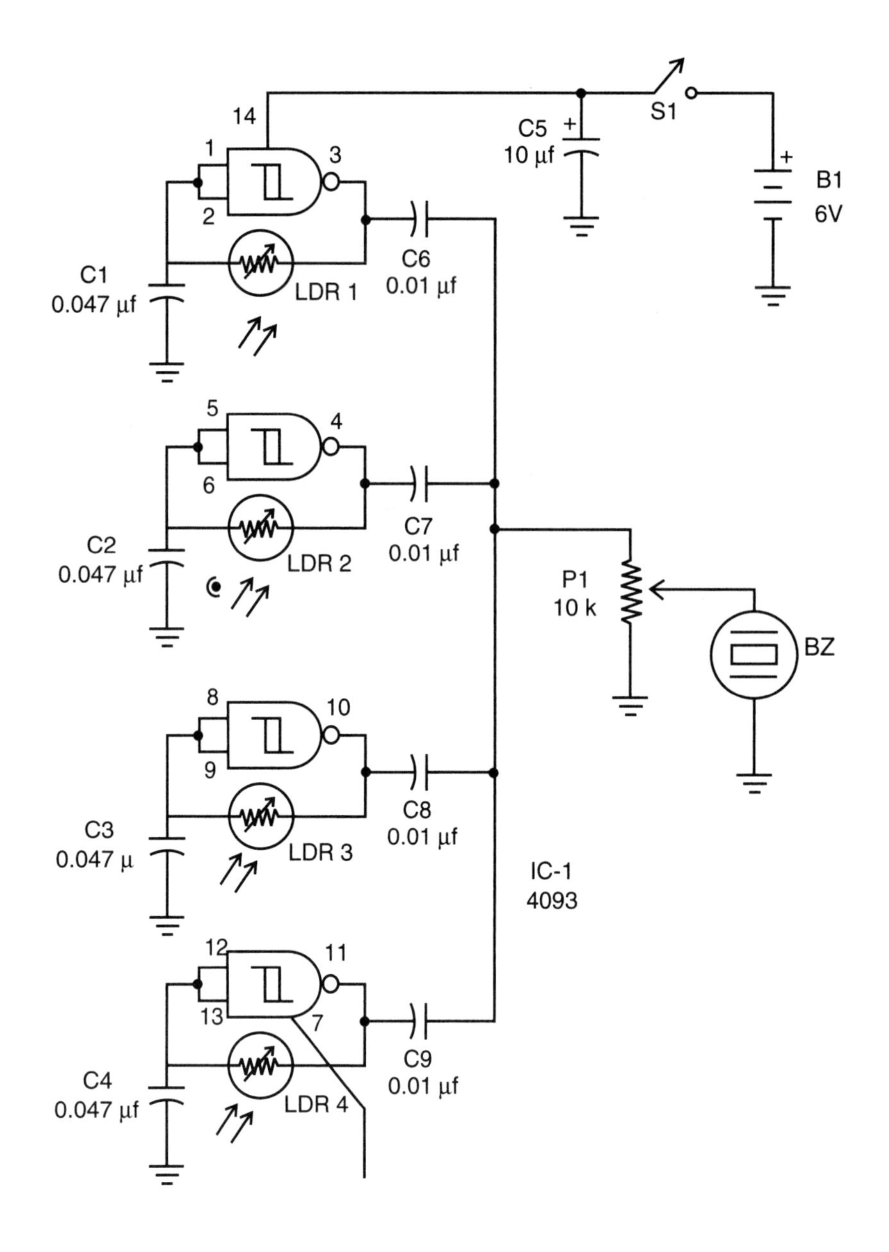


Figure 3.23.4 *Bionic eye schematic diagram*

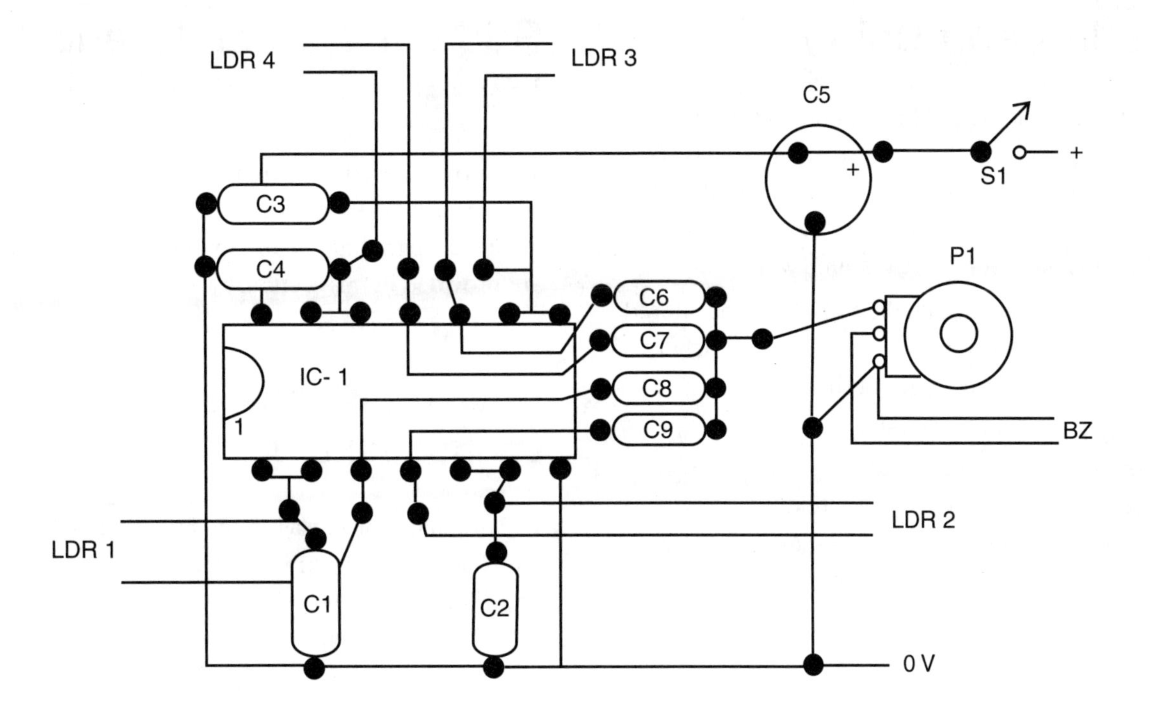


Figure 3.23.5 *The PCB for the bionic eye*

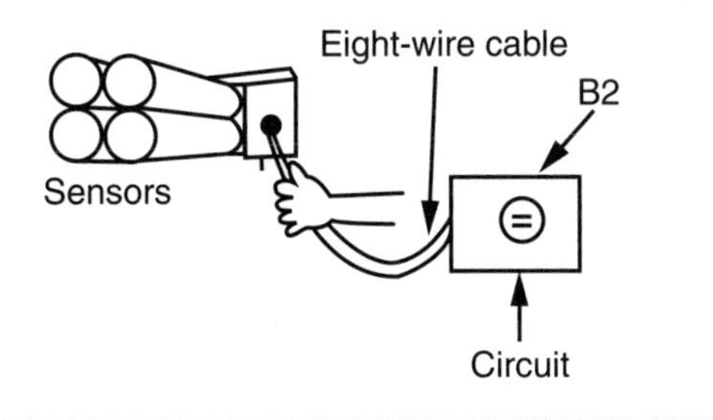


Figure 3.23.6 *The final mounting*

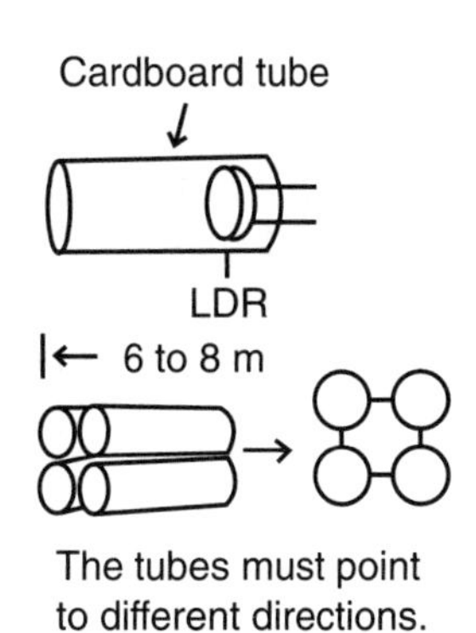


Figure 3.23.7 *Details of the sensor*

The electronic part of the circuit fits inside a plastic box, and the sensor is connected to the circuit by a short cable, as shown in Figure 3.23.6.

The sensor consists of four cardboard tubes mounted as shown in Figure 3.23.7.

To add sensitivity and directivity, a small plastic lens can be placed in front of each sensor. The LDR must be placed in the correct position near the focus of each lens. The transducer is a small loudspeaker or, if the reader prefers, an earphone.

## Testing and Using

Place the cells in the cell holder and turn on the power supply, closing S1. Then open the volume control. Sounds will be produced according to the light picked up by the sensor. Moving the sensor to explore the environment, the sound will change.

If the evil genius wants, the sounds can be altered, replacing the capacitors in each oscillator. Higher values will lower the frequency of the sounds.

Using is very simple: Explore the surrounding area and try to associate the sounds with the image sensed by the circuit. You could try blindfolding yourself with a mask and explore the environment, identifying open doors, windows, or people by the sounds produced by the device.

## Parts List

- IC-1: 4093 CMOS integrated circuit
- BZ: Piezoelectric transducer
- S1: On/off switch
- B1: A 6 V source of four AA cells
- LDR1 to LDR2: Any LDR (see text)
- P1: 10 kΩ potentiometer, lin or log
- C1 to C4: 0.047 μF ceramic or polyester capacitor
- C5: 10 μF x 12 V electrolytic capacitor
- C6 to C9: 0.01 μF ceramic or polyester capacitor
- Other: PCB or solderless board, plastic box, material for the transducer, cell holder, wires, solder, etc.

## Additional Circuits and Ideas

The project described here is a very simple version of a bionic eye. As suggested, many improvements can be made, upgrading the original project. Some ideas are provided in this section.

## Driving a Phone or Loudspeaker

The original project was created to drive high-impedance piezoelectric transducers. To drive low-impedance loads, the evil genius needs a transistor stage like the one shown in Figure 3.23.8.

With a low-impedance load such as earphones or a loudspeaker, power consumption increases, reducing the battery life.

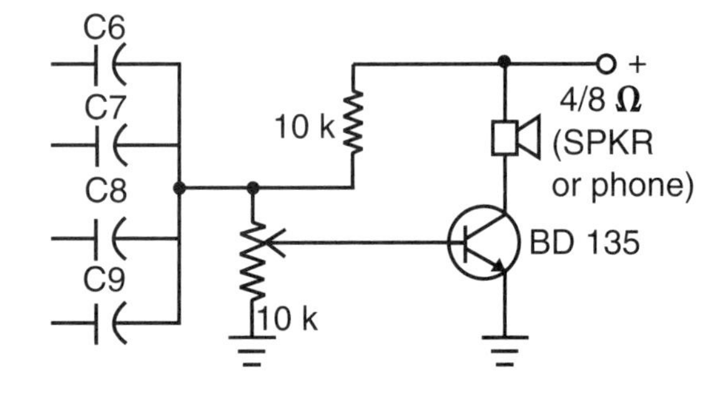


Figure 3.23.8 *Driving a low-impedance load*

## Other Light-Dependent Oscillators

Figure 3.23.9 shows another light-dependent oscillator using the 555 IC that is suitable for a bionic eye.

The main disadvantage when using this circuit is that you will need one chip for each oscillator, increasing the size and cost of the project.

Consumption can be reduced if the *complementary metal oxide semiconductor* (CMOS) version of the 555 (7555) is used. The current drain of each oscillator in this case will fall to less than 1 milliamp.

## A Two-Sensor Oscillator

Another idea for a bionic eye project is to implement two sensors controlling one oscillator, as shown in Figure 3.23.10.

The interesting characteristic of this circuit is that when light increases on sensor 2, the frequency of the oscillator also increases. Otherwise, when light increases on sensor 1, the frequency falls to a point where the oscillations are cut. This means that the circuit can be adjusted to sense a determined light intensity range.

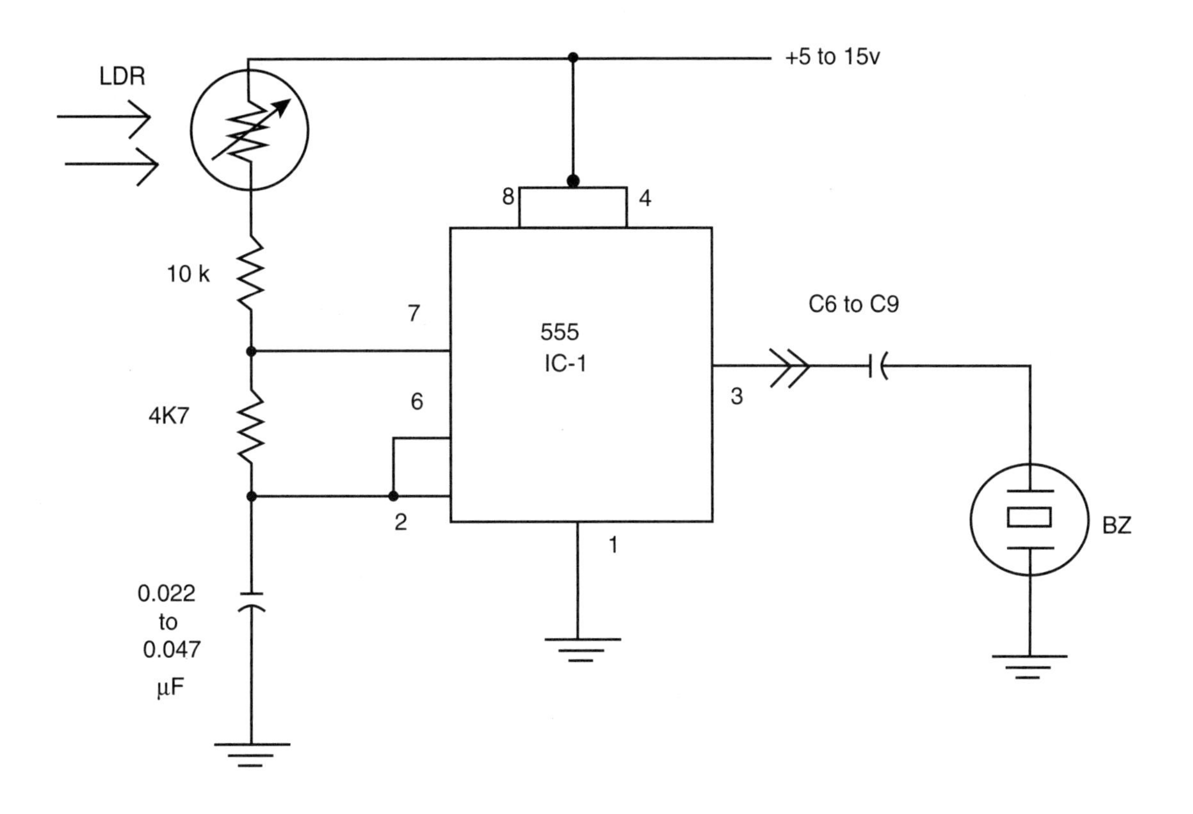


Figure 3.23.9 *A light-dependent oscillator using the 555 IC*

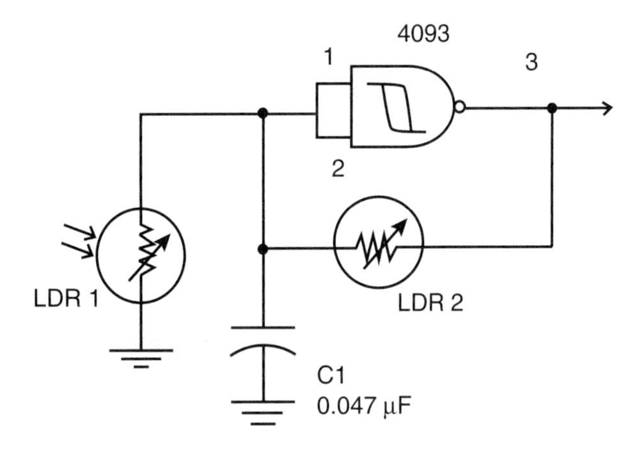


Figure 3.23.10 *Two LDRs are used to control one oscillator.*

## Seeing in the Dark

Phototransistors and many LDRs can see *infrared* (IR), or invisible, light. This means that, with the use of appropriate filters at the front of the sensor, it is possible to use the device in the dark. Figure 3.23.11 shows how to control the oscillators using phototransistors instead of LDRs.

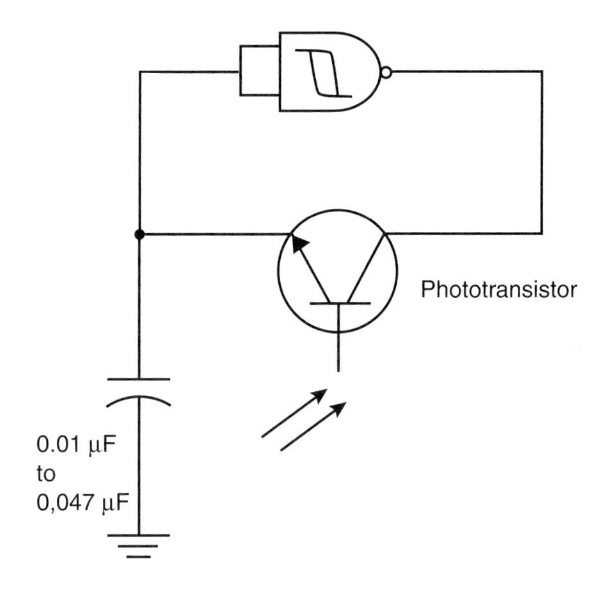


Figure 3.23.11 *Using phototransistors and IR filters to see in the dark with the bionic eye*

## Using the Computer

Powering the circuit with a 5-volt supply, the information picked up by the sensor can be translated into a digital form, as shown in Figure 3.23.12.

Using the *input/output* (I/O) port (parallel port) configured to receive data, the evil genius can develop a program to create an image, starting with the information scanned from the bionic eye.

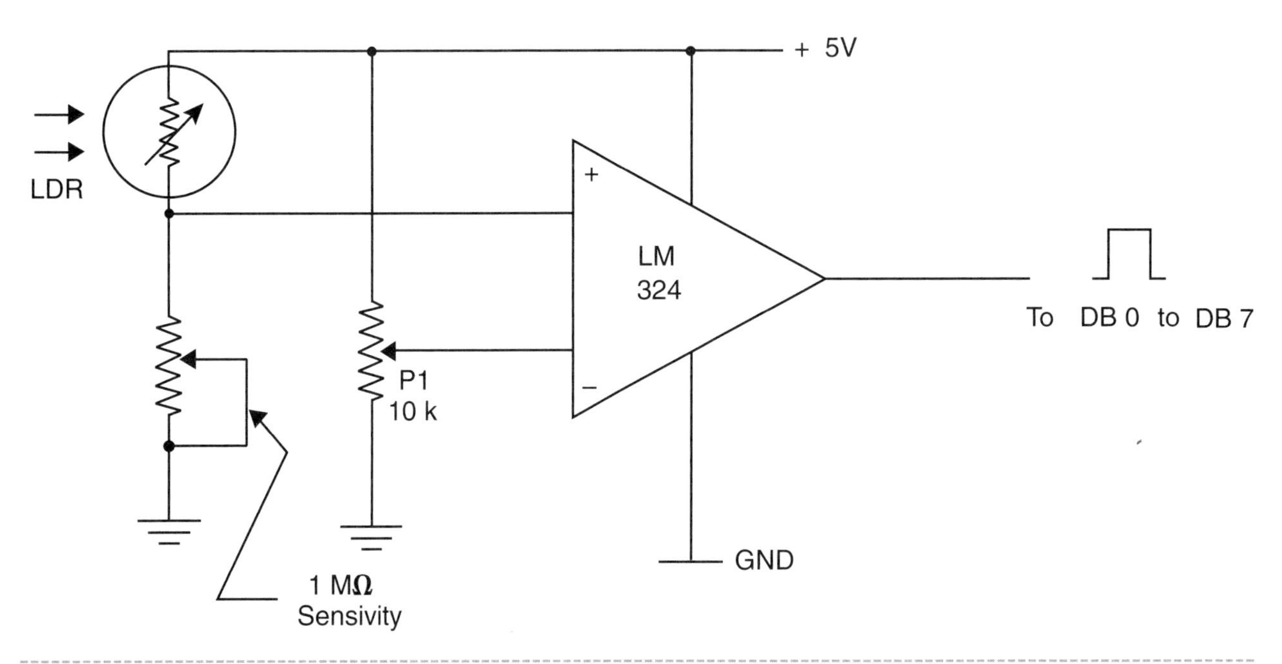


Figure 3.23.12 *Converting images in bits*

# Project 24—Ecological Monitor

Scientists researching the possibilities of life on other planets define *ecosphere* as a region where temperature and other conditions can allow some kind of life to develop. On Earth, we know that life can only exist within a narrow range of temperatures. If too hot or too cold, living beings cannot survive.

Nature has equipped all creatures with sensitive temperature monitors that tell them when they reach an uncomfortable state. The sensors can also trigger defense mechanisms such as the ones that cause your hair to stand up or that cause perspiration.

What we will describe in this chapter is a simple project that monitors the environment's temperature, like the sensors in our skin, and that triggers an alarm if it rises to an uncomfortable value or falls below another fixed value. The circuit can be installed any place where sensitive creatures live, such as greenhouses, aquariums, or areas where insects and microorganisms are common. In other words, the device monitors the ecosystem, triggering an alarm if the temperature is not appropriate for its existence.

By changing the sensor, the circuit can also be used to monitor the amount of light in an environment, triggering an alarm if the place becomes too dark or too bright. The alarm is a beeper driving a piezoelectric transducer or a loudspeaker.

## The Bionics

The device imitates the sensors found in many living creatures and can be used in the following applications:

- Control the temperature of a greenhouse or other monitored area.
- Sound an alarm in case of fire.
- Alert someone if heaters or refrigeration systems fail due to energy cuts or other causes.

## How the Circuit Works

Voltage comparators are very high gain operation amplifiers, having two inputs, as shown in Figure 3.24.1.

Wiring a voltage divider at the inverting input (−IN) determines a reference voltage (REF) at this point. If the input voltage is less than the reference voltage, the output remains low (0V). But if the input voltage is higher than the reference voltage, the output goes high (Vcc).

We can connect the device so that it operates in the opposite manner. If the reference voltage is applied to the noninverting input (+IN), the output is high (Vcc) when the input is less than the reference voltage, and the output is low when the input voltage is higher than the reference. This performance is show in Figure 3.24.2.

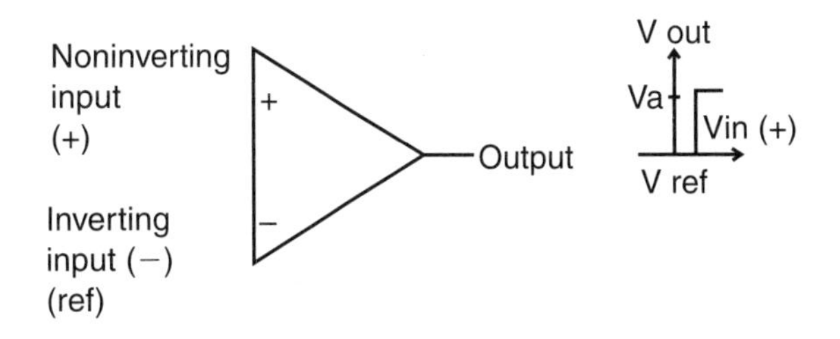


Figure 3.24.1 *A voltage comparator*

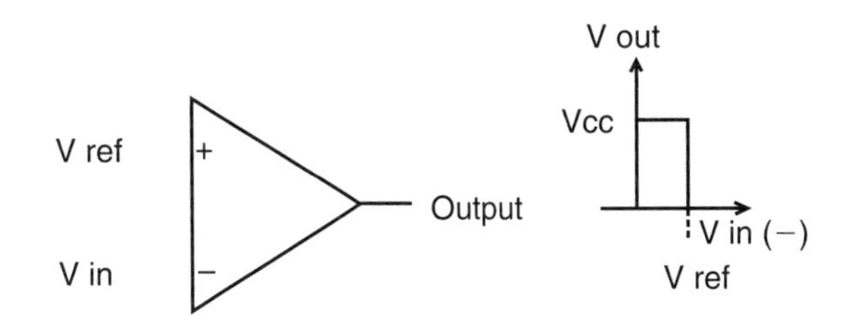


Figure 3.24.2 *Operating in the inverse mode*

Two comparators operating in different modes can be combined so that they become what is known as a window comparator. This circuit is shown in Figure 3.24.3.

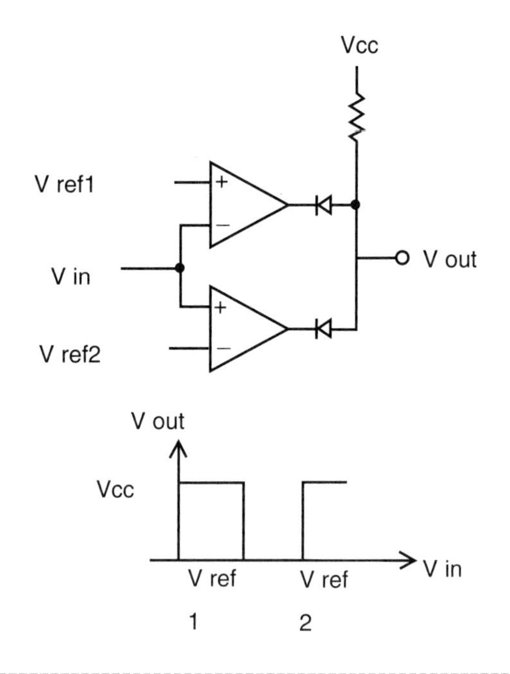


Figure 3.24.3 *The window comparator*

As we can see, two reference voltages are applied to the comparators, V1 and V2. If the input voltage is lower than V1 or higher than V2, the output is low. The output is high only if the voltage is in the range comprised by V1 and V2.

If a temperature transducer provides the input, we can convert voltage into temperatures. The output of the circuit will be at the high level if the temperatures are between the values adjusted by the references. The output of the circuit passes to an inverter so that a two-tone oscillator is triggered when the temperatures are out of the adjusted range.

The evil genius can make some improvements to this circuit, such as adding a relay to power a heater or a cooler according to the application.

## How to Mount

Figure 3.24.4 shows the complete circuit of the ecological monitor.

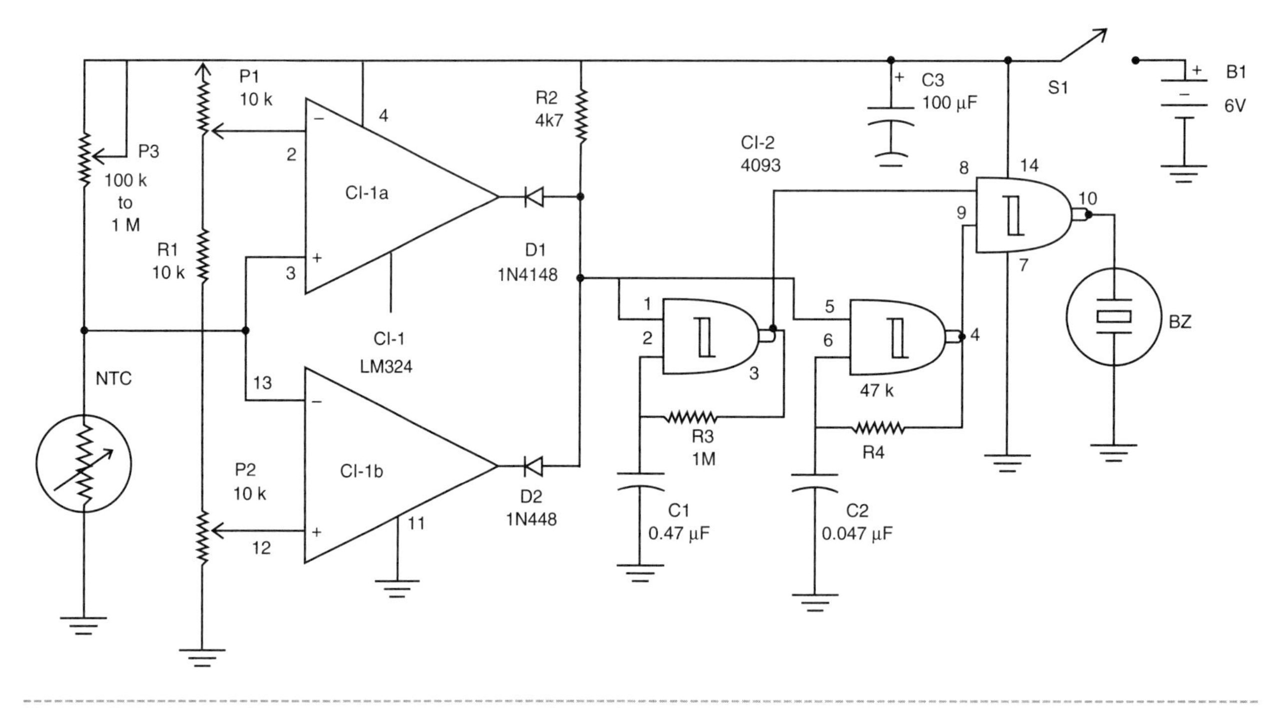


Figure 3.24.4 *Schematic diagram for the ecological monitor*

The circuit is mounted on a *printed circuit board* (PCB) or a solderless board. A PCB pattern is shown in Figure 3.24.5.

The sensor is a common *negative coefficient resistor* (NTC), a device that changes the resistance as a function of temperature. Essentially it is a temperature sensor.

P3 must have values that are twice the NTC resistance or more at a normal temperature. For instance, if the NTC is a 10 k type, use a 22 to 47 k potentiometer. Types with resistances at normal temperatures (ambient or 20° C) between 10 and 100 kilo-ohms can be used.

The operational amplifier can be replaced by an equivalent device. Even common types such as the 741 can be used. The only concern is choosing an operational amplifier that operates with low-voltage sources (6 volts).

Most of these devices have a linear characteristic (that is, the resistance changes in direct ratio to the temperature), making easier adjustments based on a common thermometer.

The sensor can be placed far from the circuit, and if exposed to water or environmental conditions, it is recommended that you protect it using glass, epoxy, or another substances. Of course, when protected, the device will take longer to respond to temperature changes.

When mounting, the position of polarized components such as *integrated circuits* (ICs) and the power supply must be observed. The circuit has a very low current consumption and this means that the cells will have an extended life. If the evil genius wants, a power supply can be used to provide energy to this project.

## Testing and Using

Close S1 to power the circuit on. Adjust P1 and P2 to stop the alarm if it is activated. P3 adjusts the sensitivity of the circuit, so start from the point where P3 is in the maximum resistance.

First, place the sensor in a cold place (using some ice, for instance) and adjust P1 to trigger the alarm. Find a point where, breathing on the sensor, the alarm stops.

Now place the sensor near a soldering iron, as shown in Figure 3.24.6. Do not touch the sensor to the iron. Adjust P2 to trigger the alarm.

This is a simple, basic adjustment. A more accurate adjustment can be made, using a thermometer as a reference. Once adjusted, the sensor can be installed where the temperature must be monitored.

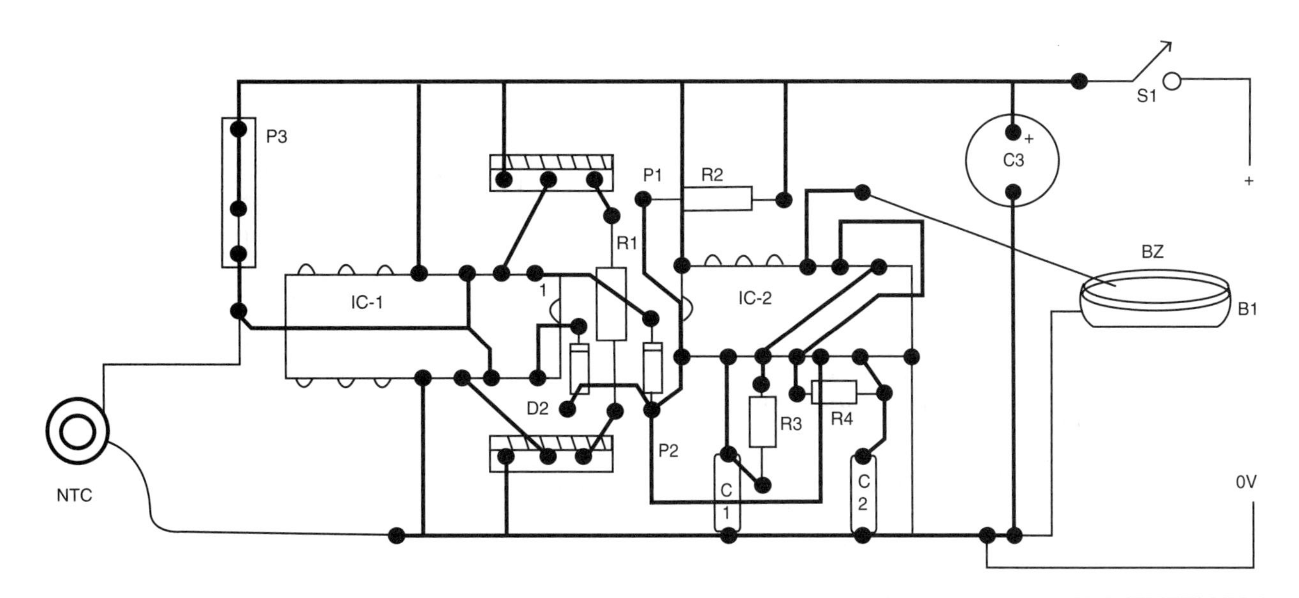


Figure 3.24.5 *PCB used for the mounting*

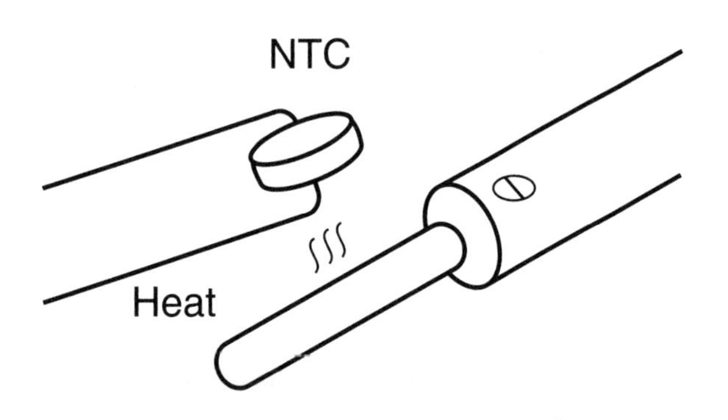


Figure 3.24.6 *Adjusting the circuit*

## Parts List

IC-1: LM324 integrated circuit operational amplifier

IC-2: 4093 CMOS integrated circuit

D1, D2: 1N4148 general-purpose silicon diode

P1, P2: 10 kΩ trimmer potentiometer

P3: 10 kΩ to 1 MΩ trimmer potentiometer (see text)

R1: 10 kΩ x 1/8 W resistor, brown, black, orange

R2: 4.7 kΩ x 1/8 W resistor, yellow, violet, red

R3: 1 MΩ x 1/8 W resistor, brown, black, green

R4: 47 kΩ x 1/8 W resistor, yellow, violet, orange

C1: 0.47 μF ceramic or polyester capacitor

C2: 0.047 μF ceramic or polyester capacitor

C3: 100 μF x 12 V electrolytic capacitor

NTC: 10 kΩ to 470 kΩ NTC

BZ: Piezoelectric transducer

S1: On/off switch

B1: A 6 V source or four AA cells

Other: PCB or solderless board, wires, cell holder, solder, etc.

# Additional Circuits and Ideas

The basic circuit can be altered to perform other functions, and some suggestions are given in this section.

## Driving a Relay

Figure 3.24.7 shows how to drive a relay when the temperature falls below an adjusted value or rises above the value.

Six- or 12-volt relays can be used. The only concern in this case is to use the power supply voltage according to the relay. Types with a 50-milliamps coil are the best.

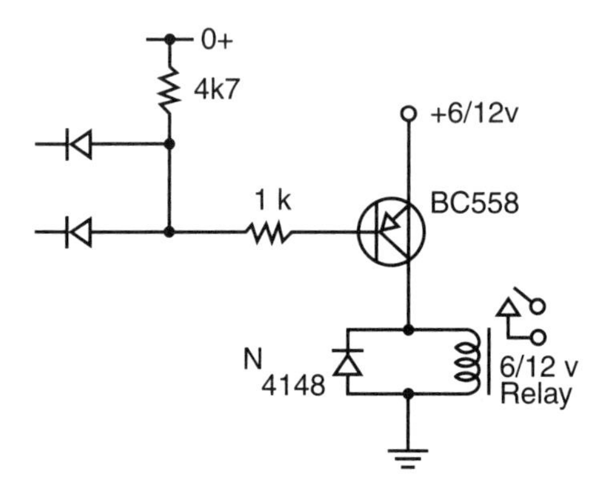


Figure 3.24.7 *Driving a relay*

## Using a Diode as Sensor

Common diodes can also be used as sensors, as shown in Figure 3.24.8.

When using diodes, a transistor must be added to increase the current in the circuit to a value that can drive the operational amplifier. Any silicon diode can be used as a sensor. Adjustments are made the same way as in the basic version.

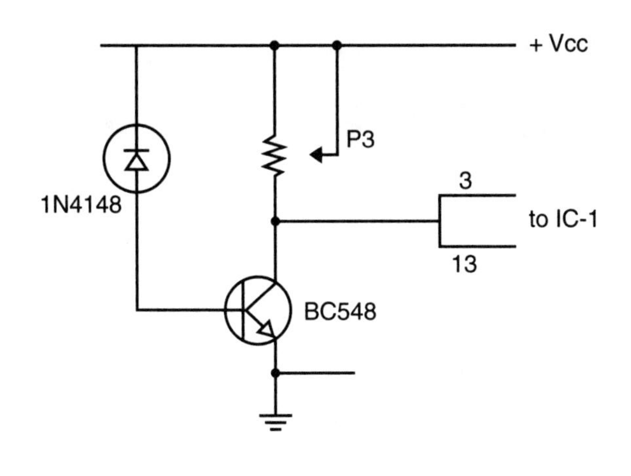


Figure 3.24.8 *Using a diode as sensor*

## Monitoring Light

Figure 3.24.9 shows how to replace the NTC with a *light-dependent resistor* (LDR) to monitor the light of a particular environment.

Any common LDR can be used. P3 adjusts the sensitivity of the circuit according to the light range to be monitored. Other resistive sensors such as pressure sensors or position sensors (potentiometers) can be used in this project.

Figure 3.24.10 shows how to use a potentiometer as a position sensor, monitoring the level of water in a reservoir.

The evil genius can also find many other applications for this circuit in monitoring bioprocesses.

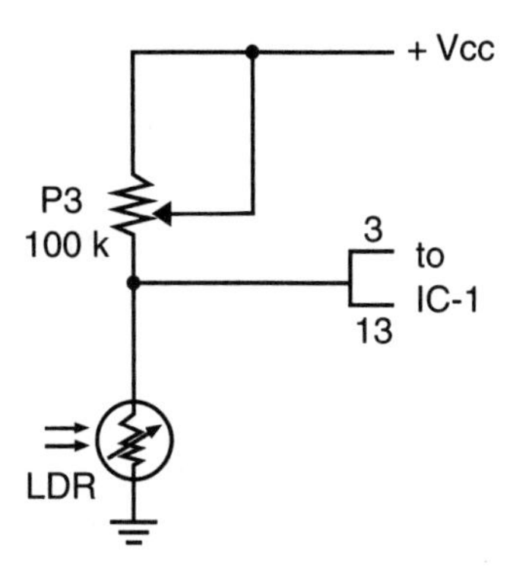


Figure 3.24.9 *Monitoring light*

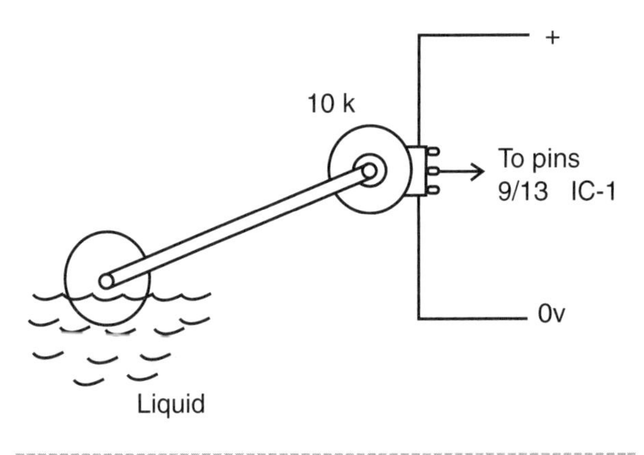


Figure 3.24.10 *Monitoring position*

# Project 25—Bat Ear

Bats can hear ultrasounds reaching frequencies as high as 100,000 Hz. Dogs and dolphins can also hear sounds with frequencies above 18,000 Hz, the human limit of hearing. Of course, the bionic man can hear ultrasounds as well.

The project described here is a device that converts ultrasounds into sounds, allowing the evil genius to have a "bat ear." Many sounds in one's environment that cannot be heard by humans will be revealed with this device. The circuit is only experimental, so it is open to many improvements.

The circuit divides the frequency of the ultrasounds but does not maintain their original waveshape. This means that what the evil genius will hear is not exactly the original sound, but something near it.

The sensitivity of the device depends on the transducer used to pick up the sounds. For experimental purposes, even a piezoelectric tweeter can be used for this task, reaching frequencies of up to 25,000 Hz.

The circuit is powered from common cells, which would allow portable use. The evil genius can use it to explore nature, as suggested by Figure 3.25.1. The sounds of bats and other animals can be heard using this circuit.

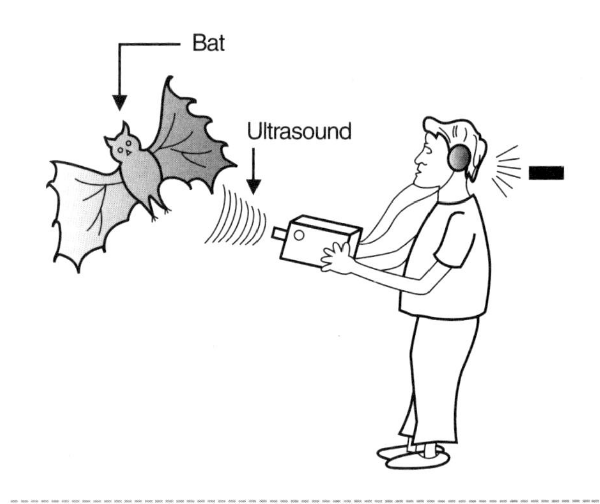


Figure 3.25.1 *Exploring nature with the bat ear*

## Applications in Bionics

Many applications exist for a circuit that converts ultrasounds into sounds. The evil genius can explore nature and sounds produced by mechanisms in his or her house or in an industrial environment. Some applications follow:

- Explore the sounds of nature.
- Hear curious sounds produced by mechanisms.

## How the Circuit Works

Figure 3.25.2 shows a block diagram representing the stages of the bat ear. The input stage is formed by a high-pass filter. The function of this filter is to block the low frequencies, letting only the high frequencies corresponding to the ultrasounds pass through.

A small piezoelectric transducer is connected to this stage. Common tweeters are not ultrasonic transducers, but they can perform well with sounds up to 25 kHz, which is enough for our purposes. If the reader wants, he or she can look for piezoelectric ultrasonic transducers that provide higher frequencies.

The output of the filter is applied to the input of an audio amplifier. Again, the choice was the LM386, which can operate with signals as high as 300 kHz.

With a 10 $\mu$F capacitor in the feedback circuit, the gain will be 200 with frequencies of up to 30 kHz. The output of this amplifier is also an ultrasound signal, not heard by our ears.

The frequency conversion is made, passing the signal across a 4093 *integrated circuit* (IC) to convert it into square waves, and then a 4020 performs the division by 16.

The result is shown in Figure 3.25.3, a square wave that has $^1/_{16}$ of the original frequency picked up by the microphone. The output signal obtained in this stage is applied to a transducer for reproduction.

In the simplest way, the transducer is a high-impedance piezoelectric type, but the evil genius can add a transistor stage, such as the one shown in Figure 3.25.4, to drive a low-impedance earphone or a small loudspeaker. Also, the circuit is powered from AA cells.

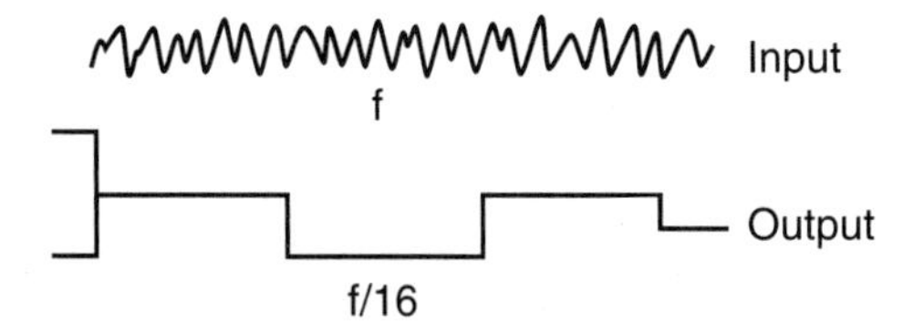


**Figure 3.25.3** *Waveshapes in the circuit*

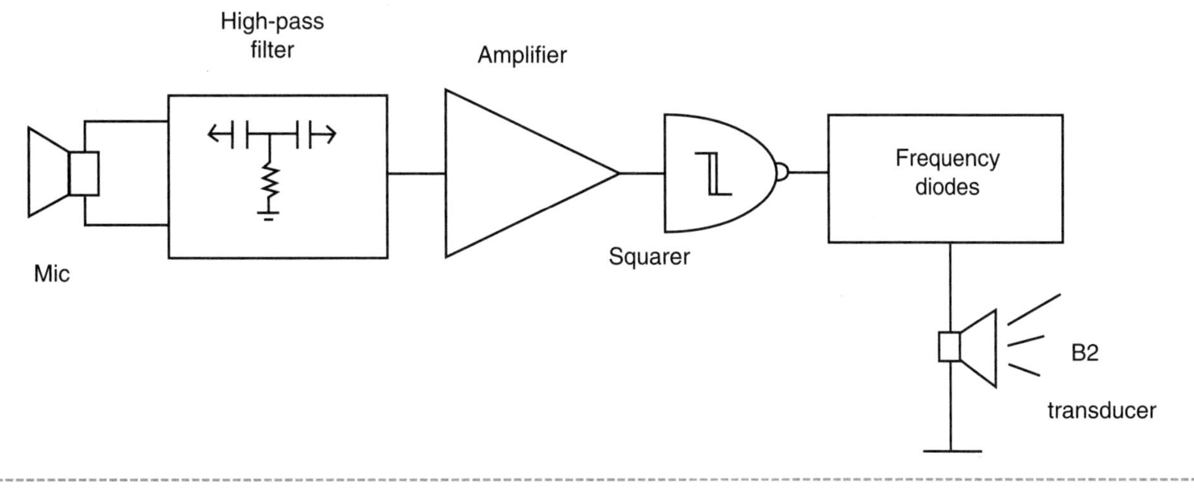


**Figure 3.25.2** *Block diagram for the bat ear*

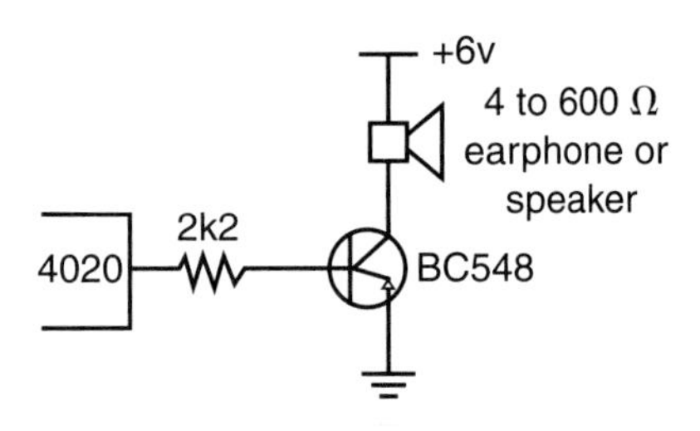


Figure 3.25.4 *Driving a low-impedance load*

## How to Build

Figure 3.25.5 shows the schematic diagram of the bat ear.

The circuit can be mounted on a *printed circuit board* (PCB), and the board suggested for this project is shown in Figure 3.25.6.

When mounting, the position of the polarized component must be observed. Any inversion of the ICs, the electrolytic capacitors, and the power supply can damage the components and prevent the circuit from operating.

For the transducer, any piezoelectric tweeter can be used. If a common low-impedance tweeter is used, a transformer is necessary (T1). Any output transformer found in an old-time transistor radio is suitable. The high-impedance coil is plugged into the circuit and the low-impedance coil to the tweeter.

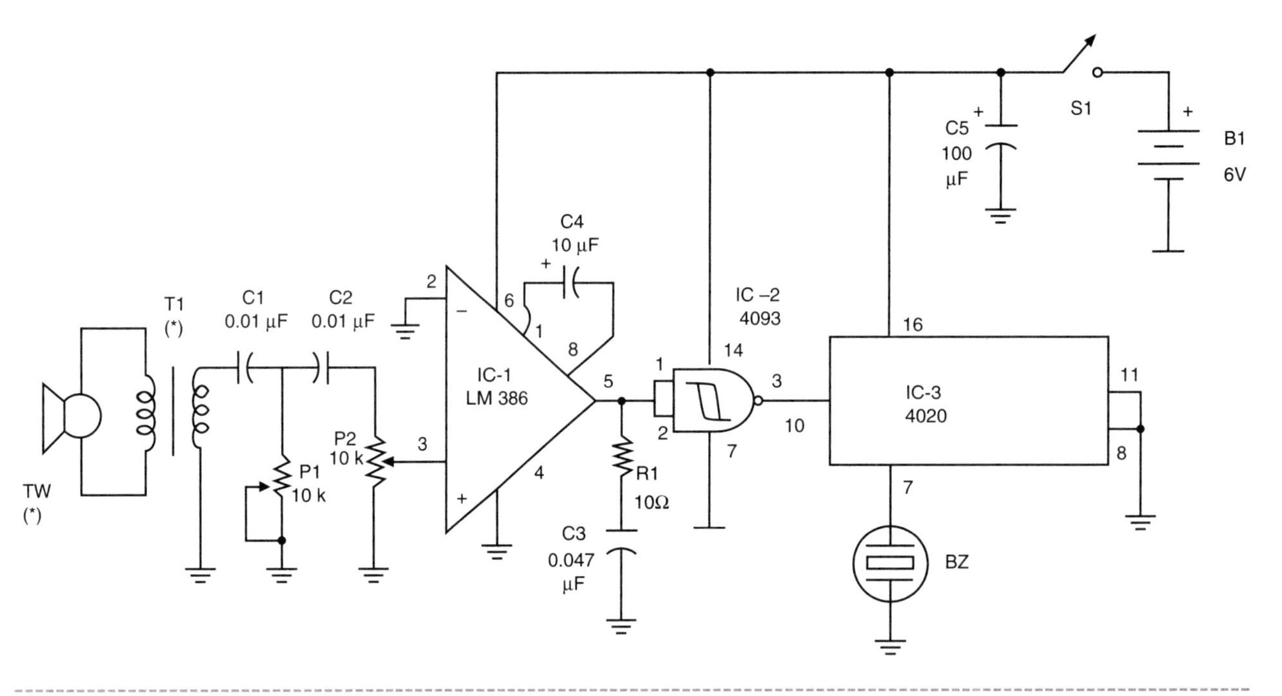


Figure 3.25.5 *Schematics for the bat ear*

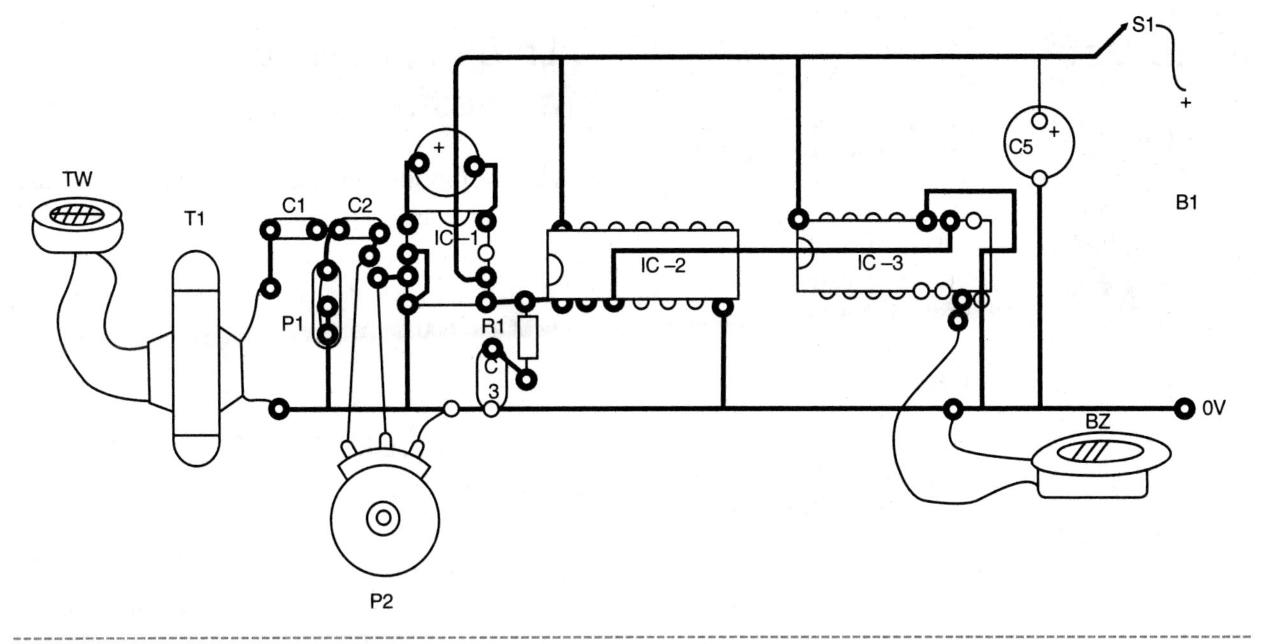


Figure 3.25.6 *PCB suggested for this project*

Another solution is to take off the small transformer found inside the tweeter and plug the transducer directly into C1 and GND without the need of a transformer. According to the characteristics of the transformer and the tweeter, the evil genius may have to make experiments with C1 and C2, altering the value and finding the best adjustment for P1.

## Testing and Using

Place the cells in the cell holder and close S1. Put P1 in the position of maximum resistance and open the gain control (P2).

Beating on the tweeter with one's fingers, the sound from BZ will be reproduced in the form of a lower-frequency signal. Sounds from one direction can be concentrated in the sensor (tweeter) using a parabolic reflector, as described in the bionic ear project.

## Parts List

IC-1: LM386 integrated circuit audio amplifier

IC-2: 4093 CMOS integrated circuit

IC-3: 4020 CMOS integrated circuit

T1: Transformer (see text)

BZ: Piezoelectric transducer or earphone

TW: Tweeter (ultrasonic microphone; see text)

P1: 10 kΩ trimmer potentiometer

P2: 10 kΩ log potentiometer

R1: 10 Ω x 1/8 W resistor, brown, black, black

C1, C2: 0.01 μF ceramic or polyester capacitors

C3: 0.047 μF ceramic or polyester capacitor

C4: 10 μF x 12 V electrolytic capacitor

C5: 100 μF x 12 V electrolytic capacitor

S1: On/off switch

B1: A 6 V source or four AA cells

Other: PCB or solderless board, plastic box, knob for P2, wires, solder, etc.

## Additional Projects and Ideas

Many improvements can be added to the original version of the bat ear, and some of them are suggested in this section.

## A Low-Impedance Preamplifier

Figure 3.25.7 shows a preamplifier that allows the use of a low-impedance microphone as the tweeter without the need for a transformer. This circuit will work well with transducers used as microphones in the range of 4 to 600 ohms.

## Driving a Low-Impedance Phone

Driving a low-impedance source as a small loudspeaker or a magnetic earphone needs a special drive stage. This stage is shown in Figure 3.25.8.

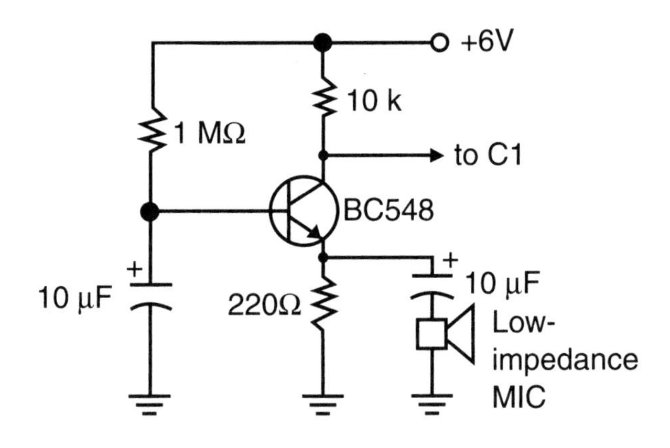


Figure 3.25.7 *Preamplifier for low-impedance sources*

## Upgrading the Circuit

Other upgrades for the circuit include a low-pass filter to convert the square pulses in the output to a smooth signal, which is better for hearing the sounds. Figure 3.25.9 shows how to add this filter. The best values for C1 must be tested in the range 0.01 to 0.47 $\mu$F.

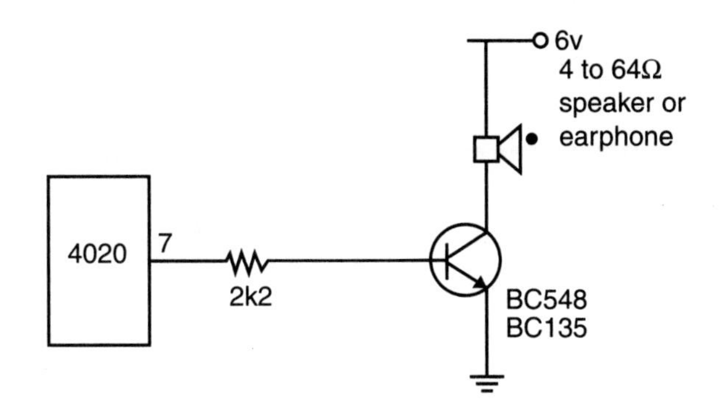


Figure 3.25.8 *Driving low-impedance sources*

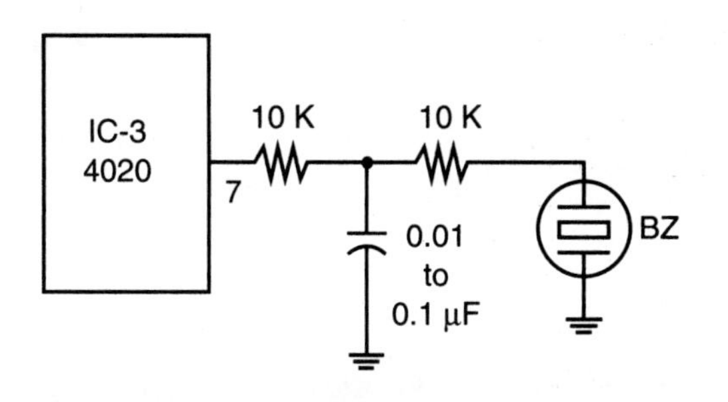


Figure 3.25.9 *Adding an output filter*

# Additional Information



The evil genius who wants to mount bionic projects must have some resources for finding components. Many dealers can be located on the Internet and send your parts by mail. You can pay using your credit card and receive your shipment in a few days, depending on where you are.

When buying parts for the projects described in this book, the evil genius must take care to not buy the wrong components. Make sure you're aware of the following points before purchasing some parts:

- Resistors have values given by a code color. Take a good look at all the resistors before using them in your project. It is not easy to remove a resistor from a *printed circuit board* (PCB) after you discover it is incorrect.
- Capacitors also have special codes to be identified, so be careful when conferring the values. In particular, take care with ceramic capacitors, which are specified by a three-digit code. 4k7 is not the same as 4K7!
- The voltages indicated in a parts list for the electrolytic capacitors are the minimum recommended for the application, but you can use larger voltages for the same capacitance. For instance, if you cannot find a 100 $\mu$F × 12 V for a project, you can use a 100 $\mu$F × 16 V, or even a 100 $\mu$F × 25 V, replacing the original.
- Transistors have a suffix indicating its gain or voltage. Be careful not to use one with characteristics below the minimum required for the project.
- Dissipation for a resistor follows the same rule: If you don't find a 1 kΩ x 1/8 W resistor for a project, you can use a 1 kΩ x 1/4 or even a 1/2 W resistor.
- If you intend to use parts obtained from old, nonfunctioning equipment, test them beforehand. Electrolytic capacitors' use will diminish with time. These capacitors will not have a capacitance or be able to function anymore. Although transistors, diodes, and other semiconductors are not affected by time, it would be better to test them before using.

# Resources



## Internet Resources

The Internet is a rich source of information about bionics and electronic projects that can be applied to bionics. This section provides some bionics-related Web sites:

- Many links to sites containing information about bionics are given at www.aleph.se/Trans/Individual/Body/bion_page.html.
- The University of Berlin's site provides basic information about bionics at www.bionik.tu-berlin.de/institut/xstart.htm.
- Many links to useful information on bionics are located at http://mercury.sfsu.edu/~swilson/emerging/artre332.bionics.html.
- Also, a site on bionics and the Six Million Dollar Man can be found at www.chiprowe.com/articles/bionics/bionic3.html.

## Bibliography


Barbarello, James. *Handbook for Parallel Port Design*. Prompt Publications, 1999.

Braga, Newton C. *CMOS Projects and Experiments*. Boston: Newnes, 1999.

——. *CMOS Sourcebook*. Boston: Prompt Publications, 2001.

——. *Curso Básico de Electronica.* Sao Paulo, Brazil: Editora Saber, 1980.

——-. *Eletronica Basica Para Mecatronica*. Sao Paulo, Brazil: Editora Saber, 2005.

——-. *Electronics Projects from the Next Dimension*. Boston, MA: Newnes, 2001.

——-. *Fun Projects for the Experimenter*. Indianapolis, IN: Prompt Publications, 1998.

——-. *Mechatronics for the Evil Genius*. New York: McGraw-Hill, 2005.

——-. *Robotics, Mechatronics and Artificial Intelligence*. Boston: Newnes, 2001.

Chklovsky, I. *Univers Vie et Raison*. Moscow: Editions de la Paix.

Iovine, John. *Robots, Androids and Animatrons*. New York: McGraw-Hill, 1997.

McComb, Gordon. *The Robot Builder's Bonanza*. New York: Tab Books 1987.

Mironov, I. *La Bionique*. Moscow, Soviet Union: MIR Publications, 1970 (in french).

Many articles published between 1980 and 2005 in Brazilian, European, and American magazines such as *Eletronica Total*, *Popular Electronics*, *Revista Saber Eletronica*, *Eletronique Practique*, *Mecatronica Facil*.

# Index

## D



## E



## F



## G



## H

## I



## J



## L



## M



## N



## O

## P



## R



## S

## T



## U



## V

## W